Dénes König

Theory of Finite and Infinite Graphs

Translated by Richard McCoart
With Commentary by W.T. Tutte

With 112 Illustrations

Birkhäuser
Boston · Basel · Berlin

Translator
Richard McCoart
Department of Mathematics
Loyola College
Baltimore, Maryland 21210-2699
USA

Author of commentary
W.T. Tutte
Department of Mathematics
University of Waterloo
Waterloo, Ontario
Canada N2L 3G1

Originally published as "Theorie der endlichen und unendlichen Graphen"
Akademische Verlagsgesellschaft Leipzig 1936. German edition © 1986
by B.G. Teubner GmbH.

Library of Congress Cataloging-in-Publication Data
König, D. (Dénes), 1884–1944.
    [Theorie der endlichen und unendlichen Graphen. English]
    Theory of finite and infinite graphs / Dénes König ; translated by
Richard McCoart ; with commentary by W.T. Tutte.
      p.   cm.
    Translation of: Theorie der endlichen und unendlichen Graphen.
    Includes bibliographical references.
    ISBN 0-8176-3389-8 (alk. paper)
    1. Graph theory. 2. König, D., (Dénes), 1884–1944. I. Title.
QA166.K6613 1989
511'.5—dc20   89-39380

Printed on acid-free paper.

ISBN 0-8176-3389-8
ISBN 3-7643-3389-8

Camera-ready copy provided by the translator.
Printed and bound by Edwards Brothers, Inc., Ann Arbor, Michigan.
Printed in the U.S.A.

9 8 7 6 5 4 3 2 1

Contents

Commentary

1. Introduction.

To most graph theorists there are two outstanding landmarks in the history of their subject. One is Euler's solution of the Königsberg Bridges Problem, dated 1736, and the other is the appearance of Dénes König's textbook in 1936. "From Königsberg to König's book" sings the poetess, "So runs the graphic tale ... " [10].

There were earlier books that took note of graph theory. Veblen's *Analysis Situs,* published in 1931, is about general combinatorial topology. But its first two chapters, on "Linear graphs" and "Two-Dimensional Complexes", are almost exclusively concerned with the territory still explored by graph theorists. Rouse Ball's *Mathematical Recreations and Essays* told, usually without proofs, of the major graph-theoretical advances of the nineteenth century, of the Five Colour Theorem, of Petersen's Theorem on 1-factors, and of Cayley's enumerations of trees. It was Rouse Ball's book that kindled my own graph-theoretical enthusiasm. The graph-theoretical papers of Hassler Whitney, published in 1931-1933, would have made an excellent textbook in English had they been collected and published as such. But the honour of presenting Graph Theory to the mathematical world as a subject in its own right, with its own textbook, belongs to Dénes König.

Low was the prestige of Graph Theory in the Dirty Thirties. It is still remembered, with resentment now shading into amusement, how one mathematician scorned it as "The slums of Topology". It was the so-called science of trivial and amusing problems for children, problems about drawing a geometrical figure in a single sweep of the pencil, problems about threading mazes, and problems about colouring maps and cubes in cute and crazy ways. It was too hastily assumed that the mathematics of amusing problems must be trivial, and that if noticed at all it need not be rigorously established. Students tempted by Graph Theory would be advised by their supervisors to turn to something respectable or even useful, like differential equations. I am reminded that my own most recent research in Graph Theory has involved differential equations. Mathematics is One, after all.

There were other fads of that sort. Interest in the infant science of Astronautics was popularly regarded as a sign of insanity. Nuclear Physics was disparaged as being unlikely to have any significant impact on practical affairs. So Dénes König did not just introduce Graph Theory. He strove to lift it out of a Slough of Despond and to set it upon a height. I would say that he, and those his book inspired, ultimately succeeded in this task, even though some undue scorn of Graph Theory persisted into the fifties.

2. König's First Chapter.

We find in König's first Chapter a proper axiomatization of the subject, an axiomatization that has required little change in the succeeding fifty years. Alas, he did not succeed in standardizing the terminology. This was partly because he wrote in German, a language that was about to decline as a medium of mathematical communication. So today one school of graph theorists speaks of "vertices" and "edges", while another prefers to describe "points" and "lines". And those coming to the subject from general Topology may still speak of "0-cells" and "1-cells". Nor are the "nodes" and "branches" of an earlier time completely forgotten. Wherever a student goes in Graph Theory he is likely to find competing terminologies, among which he must choose according to his own preference and convenience.

3. König's Second Chapter.

König next discusses Eulerian trails and Hamiltonian circuits. The Eulerian problem is the following. When can we trace a path in a graph that traverses each edge once and once only, and when can we arrange for such a path to return to its starting point? The problem is old; we recognize it in Euler's puzzle of the Seven Bridges, and we hail his work on the problem as the origin of Graph Theory. Readers who wish to know what Euler actually proved, and who shrink from the original Latin, would do well to consult the historical textbook *Graph Theory, 1736-1936* by Biggs, Lloyd and Wilson, [5] and to read Robin Wilson's article in the commemorative issue of the Journal of Graph Theory [52].

The solution of our problem is easily stated. We can construct the required path if and only if the graph is connected and has at most two odd vertices. If there are two odd vertices then the path goes from one of them to the other. In the alternative case there is no odd vertex at all, and the path terminates at its starting point.

König noted that the Euler Problem for infinite graphs "of finite degree" was still unsolved. But a solution was given by P. Erdös, G. Grünwald, and E. Vázsonyi in 1938 [12]. Here we may notice some more examples of the chaotic state of graph-theoretical terminology. By the "degree" of a vertex König means the number of incident edges, loops if allowed being counted twice. Many graph theorists still use the term "degree" for this number. But others, including the present writer, prefer to use "valence" or "valency" for the sake of the chemical analogy. By an infinite graph "of finite degree" König means a graph in which the valency of each vertex is finite. To some readers the expression suggests that the valencies of the vertices have a finite upper bound, but this is not intended. Some of us try to avoid the suggestion by speaking instead of a "locally finite" infinite graph.

In the third section of the chapter König writes about Hamiltonian circuits. A path in a graph is called Hamiltonian if it visits each vertex once and once only, except that it may finally return to its starting point to complete a closed circuit. The Eulerian and Hamiltonian problems are superficially similar. In the first we desire to traverse each edge just once; in the other each vertex. Yet the second problem is far more difficult than the first. There is as yet no simple and useable necessary and sufficient condition for the existence of a Hamiltonian circuit in a given graph.

König says he will mention only the two most studied parts of the theory of these circuits. One is the old problem of the Knight's tour of the chessboard. There we need a Hamiltonian circuit in the graph whose vertices are the squares of the board and whose edges are the possible Knight's moves. In König's Figure 12 he shows a solution due to Euler.

The second part is more general. It is the study of the Hamiltonian properties of the "cubic" graphs, the graphs in which each vertex has degree or valency 3. König shows a Hamiltonian cubic graph in Figure 9 and a non-Hamiltonian one in Figure 10. He goes on to point out that the graph of the regular dodecahedron is Hamiltonian. Indeed it was Hamilton's interest in the circuits of this figure that gave "Hamiltonian Circuits" their name.

Historical research [4, 5] suggests that it would be fairer to call them "Kirkman's Circuits", since T. P. Kirkman had already studied them as part of his extensive work on the convex polyhedra. Moreover in view of Euler's work on Knights' tours they might easily have been called "Euler paths of the Second Kind".

In the latter half of the nineteenth century the Hamiltonian circuits of cubic graphs were studied by P. G. Tait. He was, I think, primarily interested in the Four Colour Problem, and by his time that puzzle had been reduced to the case of cubic planar graphs. When such a graph is known to have a Hamiltonian circuit the Four Colour Problem for the corresponding map becomes trivial. The faces or "countries" inside the circuit can be coloured alternately red and blue, and those outside alternately green and yellow. Then no edge will separate two faces of the same colour. This observation suggested to Tait that the Four Colour Theorem might be proved by way of a theory of Hamiltonian circuits of cubic graphs.

Accordingly Tait put forward his famous conjecture about "true polyhedra". This asserts that if a cubic graph can be represented as the grid of edges and vertices of a convex polyhedron, then it must be Hamiltonian. If this Conjecture could be proved then the Four Colour Theorem would follow as a simple corollary. But König comments that the conjecture is still not proved or disproved at his time of writing.

We should comment here on the notion of "3-connection" for cubic graphs. A cubic graph can be called "1-connected" if it is connected. Such a graph may have what König calls a "bridge". This is an edge whose removal disconnects its graph. The graph falls apart into two pieces, each connected and each containing one end of the bridge. The 1-connected cubic graphs with no bridges are called "2-connected" or "non-separable". A graph with a bridge has, obviously, no Hamiltonian circuit.

I prefer another term in place of König's "bridge", say "isthmus" or "1-bond".

It may be that a 2-connected cubic graph G has a pair of edges whose deletion destroys its connection. It falls apart into two connected pieces each containing one end of each member of the deleted pair. Let us call such a pair a "2-bond". A connected cubic graph having no 1-bond or 2-bond is said to be "3-connected".

Now the 3-connected planar cubic graphs correspond to Tait's "true polyhedra". This is not obvious: it is part of the very difficult Steinitz Theorem relating graph theory to 3-dimensional geometry [15, 34]. However to a graph theorist Tait's Conjecture is simply that every 3-connected planar cubic graph is Hamiltonian.

The non-Hamiltonian planar cubic graphs shown in König's Figures 10 and 14 are 2-connected but not 3-connected.

A counterexample to Tait's Conjecture was published [35] in 1946. It appeared together with a proof of Smith's Theorem that in any cubic graph, planar or otherwise, the number of Hamiltonian circuits passing through any given edge is even. (The reader is reminded that zero is an even number). Smith's Theorem has the curious corollary that if a cubic graph has one Hamiltonian circuit it must have at least three.

In succeeding years there were attempts to improve upon the counterexample. They involved the notion of "cyclic n-connection". A 3-connected cubic graph may contain a triad of edges whose deletion separates it into two connected pieces each containing a circuit. A 3-connected cubic graph having no such triad is "cyclically 4-connected". Such a graph may still have a quartet of edges separating two circuits. But if it does not then it is cyclically 5-connected, and so on. This writer described a cyclically 4-connected counterexample in 1960 [40], H. Walther described a cyclically 5-connected one in 1965 [47]. Both were rather complicated. In a sense there was no need to go further. Euler's Polyhedron Formula ensures that no planar cubic graph can be cyclically 6-connected, unless we count some trivial graphs too simple to have separable circuits.

These large counterexamples were put into the shade by the very simple theory of E. Ya. Grinberg. He showed that it was still possible to get results in Hamiltonian theory that were both simple and interesting. Studying his proof one wonders why Kirkman and Tait never happened upon it [14].

Oddly enough the Grinberg theory, though stated only for cubic planar graphs, applies equally well to planar graphs in general. The main result goes like this.

Consider a planar graph G in which there is a Hamiltonian circuit H. In the map determined by G we can classify the faces by the numbers of their incident edges, and by whether they lie inside or outside H. Let us write f'_i for the number of i-sided faces inside H, and f''_i for the number of such faces outside H. We need three more symbols, h for the number of edges of H, d' for the number of edges of G crossing the region inside H, and d'' for the number crossing the outside. The following two equations are easily established.

$$\text{(1)} \qquad \sum_i f'_i = d' + 1.$$

$$\text{(2)} \qquad \sum_i i f'_i = h + 2d'.$$

From these we eliminate d' to obtain

$$\text{(3)} \qquad \sum_i (i - 2) f'_i = h - 2.$$

Similarly

$$\text{(4)} \qquad \sum_i (i - 2) f''_i = h - 2.$$

From (3) and (4) we have

$$\text{(5)} \qquad \sum_i (i - 2)(f'_i - f''_i) = 0.$$

Some sets of numbers f'_i and f''_i that are combinatorially possible turn out to be incompatible with (5). Suppose for example that all faces of the map, with a single exception, have either 5 or 8 edges, and that the exception is a nonagon. Without loss of generality we can suppose that $f'_9 = 1$ and $f''_9 = 0$. Then, by equation (5) we must have

$$(9 - 2)(1 - 0) \equiv 0 \ (\mathrm{mod} 3),$$

which is false. But Grinberg exhibits a cyclically 5 -connected planar cubic map whose faces are 18 pentagons, 3 octagons and 1 nonagon. By the above reasoning it can have no Hamiltonian circuit. This map, with another of Grinberg's examples, appears in the twelfth edition of *Mathematical Recreations and Essays* [2].

Since 1936 some interesting theorems have been discovered having to do with Hamiltonian circuits in graphs that are not necessarily cubic. Some of them are to the effect that if a graph has sufficiently many edges, reasonably evenly distributed, then it must be Hamiltonian. The fact that every complete graph of three or more vertices has a Hamiltonian circuit can be taken as an extreme and trivial example. A more impressive theorem was published by G. A. Dirac in 1952 [11]. It is concerned with what I call "strict" graphs, that is graphs without loops or multiple edges. Some writers call these simply "graphs" and deny that name to anything with a loop or a 2-circuit. Dirac's Theorem asserts that if, in a strict graph of n vertices, the valency of each vertex is at least $n/2$, then the graph is Hamiltonian.

A later theorem of L. Pósa deserves mention [27]. Its date is 1963. Pósa considers a strict graph of n vertices wherein, for each integer $k < (n-1)/2$, the number of vertices of valency $\leq k$ is less than k. He requires also that if n is odd then the number of vertices of valency $\leq (n-1)/2$ shall be at most $(n-1)/2$. Such a graph, his theorem asserts, must be Hamiltonian. Proofs of these theorems can be found in the monograph *Über Kreise in Graphen*, by H. Walther and H. -J. Voss [48].

Another theorem on Hamiltonian circuits, published in 1956, asserts that every 4-connected planar graph is Hamiltonian [39]. In this context the graph is understood to be loopless. Its "4-connection" means that at least 4 vertices are required to separate any two non-adjacent ones.

In his fourth section König invites us to consider analogous problems for directed graphs. In a directed graph, or "digraph", each edge has a definite direction. This is shown in a diagram by an arrow, going from one end of the edge, called its "tail", to the other end, called its "head". For a loop head and tail coincide, but there are still two ways of directing it. The Euler problem for directed graphs is that of finding an Euler trail that follows the directions of the edges. Similarly with the Hamiltonian problem.

König observes that a reentrant Eulerian path in a digraph D is possible only if D has the following property: at each vertex there are just as many incoming edges as outgoing ones. I regard a directed loop as both incoming and outgoing. I shall describe a digraph with this property as "balanced". König shows that a balanced digraph has an Eulerian trail if and only if it is connected. Much work has been done on balanced digraphs since 1936. It has produced a result called the BEST Theorem, a formula for the number of Euler cycles in a given connected balanced digraph. The BEST Theorem involves directed trees as well as directed paths, and I defer discussion of it until I come to König's chapters on trees.

The device of replacing each edge of an undirected graph by two oppositely directed edges turns it into an "equivalent balanced digraph". Sometimes it is convenient to infer a theorem about graphs from a known and analogous theorem about digraphs. Often this is done by applying the theorem to the equivalent balanced digraph.

König next discusses what we now call a "tournament". This is a strict graph of n vertices and $\frac{1}{2}n(n-1)$ directed edges, each pair of vertices being joined by just one directed edge. He then states a theorem of L. Rédei, that each tournament has a directed Hamiltonian path, not necessarily reentrant. Much work has also been done on tournaments since König's time. There is a book about them by J. W. Moon [22]. A convenient summary of papers on tournaments and other graph-theoretical matters is provided by "Reviews in Graph Theory", culled from the combinatorial section of "Mathematical Reviews", and edited by W. G. Brown [7].

4. König's Third Chapter.

König now brings us to the Labyrinth Problem. Someone finds himself in a labyrinth. He wants to make his way through it, either to the exit or to some central point. For our purposes the labyrinth becomes a finite graph. We have to construct a path from a given initial vertex A so that it will certainly lead us to another distinguished vertex B. We are supposed to know nothing about the structure of the graph, save only that it is connected. When we arrive at some vertex we must decide which edge to traverse next in the light only of information available at that vertex. It is supposed however that we have some way of marking our passage through the graph or labyrinth. So at each vertex we know which of its incident edges we have already traversed and, in some versions of the problem, in which direction we have traversed them. A solution of the problem consists of a set of rules for choosing the next edge, together with a proof that the resulting path visits every vertex of the graph, and therefore gets sometime to B.

König describes three solutions, corresponding to different assumptions about how we mark our path. They are associated with the names of Wiener, Trémaux and Tarry. Wiener's solution is of historical interest as being the first. But in this Commentary I will concentrate initially on the method of Trémaux.

In that method we are allowed to mark each passage in the labyrinth as we traverse it. Should we happen again upon that passage we will know that we have been there before, and how many times we have already traversed it.

We start the path from A by proceeding along any edge that is available, making an arbitrary choice among such edges if there are more than one. At any later step let us suppose that we have come along an edge E to its end v. We lay down a system of rules to govern our choice of the next step.

First we decide, by checking for marked edges, whether we have been at v before. If not we proceed along any unmarked edge we please, if such an unused edge exists. If there are no such edges, that is if v is a vertex of valency 1, then perforce we turn back along the edge E by which we came.

Suppose however that we have been at v before. Then if this has been our first traversal of E we turn back along that edge.

If we have traversed E at least once before our procedure is different. If possible we leave v by some edge as yet unused. If there is no such edge we leave v, if possible, by some edge that has been traversed only once before. If even this is not possible we terminate the path at v.

We note some simple properties of the resulting path P. It traverses no edge more than twice, and it terminates at the initial vertex. The first of these properties ensures that P is only of finite length. The second property can be proved by the following parity argument. Whenever we arrive at a vertex v other than A we have used up an odd number of half-edges there. For at each previous visit we arrived by one half-edge and left by another. Since the number of half-edges at v is even we still have left at least one way out. I expect to need this sort of argument again, but hereinafter I will condense it into some such phrase as "by parity".

I digress to note that Trémaux' method is a graph-theoretical example of an algorithm, a set of rules guaranteed to solve a given problem if only we will follow them faithfully and mechanically. Algorithms are at least as old as Euclid, and their name commemorates an Arabian gentleman of the ninth century. But they have acquired a new importance in this age of electronic computers. If we can find an algorithm for solving a problem then we can program a computer to do the rest of the work for us. So there has grown up a rich theory of good and bad algorithms in the borderland between Graph Theory and Computer Science. The interested reader is referred again to "Reviews in Graph Theory" [7], and in particular to the many algorithmic papers of J. Edmonds.

I return to Trémaux' algorithm and the proofs that justify it. Let me say that I do not much like the proof given by König. Regarded as a discoverer's pioneering effort it is excellent. But when it comes to enshrinement in a textbook one hopes for something easier to follow. Something shorter and simpler, or at least something giving us a clearer understanding of the mathematical necessity of the theorem. Such easier solutions are now possible. One such is found in the proof of the BEST Theorem, which has a labyrinthine problem at its heart. I think a treatment of Trémaux' method, involving only the graph theory of König's first three chapters, should go along the following lines.

We suppose the path P completed, and now discuss its properties. We classify the edges appearing in the path as unicursal or bicursal, according to whether they are traversed once or twice. Edges not traversed at all can be "acursal".

We next prove that there are in fact no unicursal edges. For suppose the contrary, and let E be the last unicursal edge in the path, traversed to a vertex v. This must be the first visit of P to v, for otherwise the rules would require us to go back along E and that edge would not end up as unicursal. So P then leaves v along some previously unused edge F. There may also be later visits of P to v. But by parity there is some edge E' incident with v, but other than E, that is traversed only once by P. But then E' ends up as a unicursal edge occurring later in P than E, which is impossible.

Next we show that there are no acursal edges. For suppose at least one such edge to exist. By connection there is a vertex v incident with at least one acursal edge and at least one bicursal. Now P did not terminate at v since an unused "acursal" edge was still available there. But when P left v for the last time it did so along an edge that had been traversed once before. This was against the rules because of the availability of the unused edge.

We have now proved that P traverses every edge twice, and therefore visits every vertex. Our solution is thus complete. However it can also be shown that P traverses no edge twice in the same direction, and König gives a proof of this theorem. We can deal with it as follows.

Let us call an edge of the graph G one-way or two-way according to whether P traverses it twice in the same direction or once in each direction. A one-way edge can naturally be regarded as directed, its direction being that in which P traverses it. If a one-way edge is traversed for the first time as the nth step of P we define its "epoch" in P as the number n. Now P enters a given vertex v just as many times as it leaves v. Hence the number of one-way edges directed to v is equal to the number of those directed from v. Hence if any one-way edge exists it is possible to trace out a directed circuit C of one-way edges. In this circuit there must be an edge E which goes to a vertex v and is there followed by an edge E' of an earlier epoch. We here detect a violation of the rules: when P first traversed E to v it should have turned back at once through E, since v had been visited at least once before.

We conclude that a path P constructed according to Trémaux' rules must traverse each edge of G twice, once in each direction.

Tarry's solution of the Labyrinth Problem is another algorithm. In this version we are supposed to mark each edge as we traverse it, showing not only the fact of traversal but also the direction. We are also permitted and required to make a special mark on one edge coming to a vertex v to show that it was along this edge that we reached v for the first time. That edge is the "entry-edge" of v. The initial vertex A has no entry-edge.

We start the path P from A as before. Thereafter whenever we traverse an edge E to a vertex v we continue the path along any edge we please, subject to the following two conditions:

(i) *No edge may be traversed twice in the same direction.*

(ii) *P is not to leave v along the entry-edge if any other exit satisfying* (i) *is available.*

If there is no way out that satisfies these conditions, then the path must terminate at v.

As with Trémaux' solution it is clear that P traverses no edge more than twice, and that it terminates at A.

To justify Tarry's algorithm we treat our hypothetical completed path P as follows. Let U be the set of vertices of G such that P leaves, necessarily for the last time, along the entry edge, together with the initial vertex A. Let V be the set of all other vertices of G.

We proceed to show that V is null. But suppose that it is not. Then we can choose an edge E with one end u in U and the other end v in V in the following way. If P visits V then E is the edge by which it does so for the first time. E is then the entry-edge of v. If P does not visit V, then E is an arbitrarily chosen edge with one end in U and one in V. At least one such edge exists, by the connection of G.

We note that P never traverses E to u. Hence there is some edge incident with u that is not traversed by P from u, since P leaves u just as often as it enters. So by Tarry's rules the entry-edge of u is not traversed by P from u, if such an entry-edge exists. We deduce from the definition of U that u must be the initial vertex A. But then P has been terminated too soon, since an exit from u still exists. This completes the proof that V is null. Accordingly P visits every vertex of G and Tarry's solution is validated.

We can, if we please, prove a little more, namely that P traverses each edge just once in each direction. First we observe that P can terminate at A only when it has left A by every incident edge, and therefore returned by every incident edge. But P leaves every other vertex u by its entry-edge, and it can do this only after leaving, and returning, by every other incident edge.

5. König's Fourth Chapter.

An "acyclic" graph is a graph with no circuit. König defines a tree as a finite connected acyclic graph. He considers also infinite connected acyclic graphs. Most graph theorists would call these "infinite trees", but König does not do that. For him a "tree" is necessarily finite.

König shows that in a tree there is one and only one path from a given vertex to another. He remarks that this property can be used as an alternative definition of a tree.

There is a third definition or characterization of a tree that König does not mention. A tree is a connected graph in which every edge is a 1-bond, that is a "bridge" in König's terminology.

König points out that in any connected finite graph G there is a subgraph that is a tree, and that includes all the vertices. It is now usual to describe a subgraph of G that includes all the vertices as "spanning". If such a subgraph is a tree it is a "spanning tree".

The number $T(G)$ of spanning trees of a finite graph G has become prominent in graph theory since König's time. König does not discuss it, even though it appears in a generalized form in Kirchhoff's theory of electrical networks. The tree-number $T(G)$ can be calculated by a simple formula; it is the value of a certain determinant associated with G.

To construct this determinant we first enumerate the n vertices of G as $v_1, v_2, \ldots, v_n$. For any vertex v_i we define c_{ii} as the number of edges joining v_i to other vertices. For any two distinct vertices v_i and v_j we define c_{ij} as minus the number of edges joining v_i to v_j. Thus $c_{ij} = c_{ji}$. The $n \times n$ square matrix $K(G) = \{c_{ij}\}$ is called the "Kirchhoff matrix" of G.

It can be shown routinely that the elements of $K(G)$ sum to zero in each row and column, and that therefore all the first cofactors of $K(G)$ are equal. It can be shown further that the common value of these first cofactors is the tree-number $T(G)$ of G. That result is called the Matrix-Tree Theorem for undirected graphs. To calculate $T(G)$ by this method we usually strike out one row and the corresponding column from $K(G)$ and evaluate the determinant of what is left.

In the general theory of Kirchhoff's Laws c_{ii} is the sum of the conductances of the edges joining v_i to other vertices, and c_{ij} is the sum of the conductances, taken with a minus sign, of the edges joining v_i and v_j. The common value of the first cofactors of $K(G)$

is now a polynomial in the conductances. It has one term for each spanning tree, and that term is the product of the conductances of the edges of the tree. What we discussed in the two preceding paragraphs was the special case in which each conductance is unity.

All this, barring the modern notation, was known to Kirchhoff. But in 1936 it still needed to be drawn to graph-theorists' attention. This was done in 1940, by a paper of Brooks, Smith, Stone and Tutte. [6].

We note in passing that the tree-number of any disconnected finite graph is zero.

Tree-numbers satisfy an interesting recursion formula. Consider any edge A that is not a loop in the graph G. We can get another graph G'_A from G by deleting the edge A. We retain the two end-vertices of A as vertices of G'_A, even if one of them becomes isolated, that is it ceases to have incident edges. Another graph G''_A can be derived from G by contracting A, with its two end-vertices, into a single new vertex. It is not difficult to show that

$$(6) \qquad T(G) = T(G'_A) + T(G''_A).$$

This is because the spanning trees of G containing A contract into those of G''_A, while the others are the spanning trees of G'_A. Often this formula is used to calculate $T(G)$ from the known tree-numbers of simpler graphs. Indeed, in principle we can calculate the tree-number of any finite graph using only the recursion formula and the following initial propositions.

(i)   *The tree-number of the vertex-graph is* 1.

(ii)   *The tree-number of a disconnected graph is* 0.

(iii) *The tree-number is not altered by the deletion of a loop.*

Even the Matrix-Tree Theorem, giving $T(G)$ as a determinant, can be proved by combining this recursion formula with an analogous one for determinants.

The "vertex-graph", I should explain, is a graph with one vertex and no edges.

There are other functions of graphs, interesting for various reasons, that satisfy the above recursion or some essentially equivalent variation such as

$$(7) \qquad F(G) = F(G'_A) - F(G''_A).$$

This is the form that was pointed out for "chromatic polynomials" by
R. M. Forster in 1931 in a footnote to Whitney's paper *The coloring
of graphs.* The theory of those polynomials is based almost entirely
on the recursion formula [50].

Having observed that connected graphs have spanning trees
König goes on to define an important graph-theoretical invariant
that he calls the "connectivity". For a connected finite graph G this
is the (constant) number of edges left over when a spanning tree is
constructed. For a disconnected finite graph it is the sum of the
connectivities of the components.

Nowadays it is usual to call this invariant the "cyclomatic number"; most graph theorists prefer to reserve "connectivity" for something else. In topological terminology the cyclomatic number is
$p_1(G)$, the Betti number of dimension 1. There is also a Betti number $p_0(G)$ of dimension zero: it is simply the number of components
of G.

The cyclomatic number of G is the minimum number of edges
whose deletion destroys every circuit. It is the maximum number
of circuits such that each has an edge not belonging to any of the
others. It satisfies the identity

$$(8) \qquad p_1(G) = \alpha_1(G) - \alpha_0(G) + p_0(G),$$

where $\alpha_1(G)$ and $\alpha_0(G)$ are the numbers of edges and vertices of G
respectively. Indeed it can be defined quite conveniently by that
equation. Incidentally, $p_1(G)$ is always non-negative and it is zero
only for forests, that is for graphs whose components are trees.

A set of circuits of G such that each has an edge not belonging to
any of the others is called, if maximal, a fundamental set of circuits.
Such sets are very important in connection with the algebraic theory
of Chapter IX. As König shows, for a connected finite graph G there
is a fundamental set of circuits of G associated with every spanning
tree. Each circuit of the fundamental set corresponds to one of the
edges of G not in the tree. The circuit is made by combining that
edge with the arc that joins its end-vertices in the tree.

There is a theory of directed trees, a theory that has become
more important since König's time. Given a tree T let us mark one
vertex v_0 as the root. Let us now change T into a directed graph by
directing each edge towards the root. More formally, each edge A is

a 1-bond and when it is deleted the tree splits into two connected pieces H_A and K_A where we may suppose H_A to contain v_0. We are to direct A so that its head is in H_A and its tail is in K_A. The resulting digraph I shall call a "directed tree converging to v_0". If we had directed each edge the other way we would have had a "directed tree diverging from v_0".

For a given digraph D we may wish to find the number of spanning directed trees that converge to a given vertex r. There is a Matrix-Tree Theorem for digraphs that enables us to do this. As before we number the vertices from 1 to n and we define a "Kirchhoff matrix" $K(D) = \{c_{ij}\}$. This time c_{ii} is the number of edges of D directed from v_i to other vertices. And if $i \neq j$ then c_{ij} is minus the number of edges directed from v_i to v_j. We note that now c_{ij} is not necessarily equal to c_{ji}. The elements of $K(D)$ sum to zero in the rows but not necessarily in the columns. And it is found that the first cofactors of $K(D)$ are not necessarily all equal. However, the Matrix-Tree Theorem asserts that if we strike out the row and column of v_i and evaluate the determinant of the resulting matrix, then we shall get the number $T_i(D)$ of spanning directed trees of D converging to v_i. In general this number can vary from vertex to vertex. Moreover it can be zero at some vertices and non-zero at others.

The directed Matrix-Tree Theorem was uncovered by Brooks, Smith, Stone and Tutte during the researches leading to their paper of 1940, [6]. It is referred to, albeit obscurely, in Paragraph 10.3 of their paper. However the first publication of a proof seems to have been in 1948 [37].

Tree-numbers are important in connection with the BEST Theorem. This theorem applies to finite balanced digraphs, and it gives a formula for the number of directed Euler cycles. The acronym BEST comes from the initial letters of four authors. Van Aardennes-Ehrenfest and de Bruijn, who published the general proof in 1951 [1], contribute E and B respectively. S and T stand for Smith and Tutte, who dealt with a special case in 1941 [33].

Suppose we have an Euler path P in D, following the directions of the edges, and beginning and ending at a vertex r. For each vertex v other than r let us mark its last exit, the edge by which P leaves v for the last time. We shall then find that these edges of last exit make up a spanning tree converging to r.

To prove this, consider any edge E_1 of last exit, directed let us say from u_1 to u_2. If u_2 is not r there will be another edge E_2 of last exit directed from u_2 to a vertex u_3, and so on. It is clear that P traverses E_2 later than E_1, and so on along all this sequence of successive edges of last exit. Hence the sequence cannot bend back upon itself to form a directed circuit. It must continue to r, and there terminate. We infer that the edges of last exit define a directed spanning subgraph H of D in which there is no directed circuit. H has no other circuit either, since no two of its edges can have the same tail. It is a directed tree, and it converges to r.

The BEST Theorem requires also a converse of this result. Suppose we are given a balanced digraph D, and a directed spanning tree T_r of D converging to a vertex r. We ask if and how we can construct a directed Euler path P in D from r to r which has the edges of T_r as its own edges of last exit. This is a labyrinth problem, for a labyrinth of one-way streets, and its solution is reminiscent of Tarry's solution of an undirected labyrinth. We start from r, following any edge we please that has r as its tail. Thereafter whenever we traverse an edge to a vertex v we choose an edge that has its tail at v and is still unused. There is just one restriction on our choice: we are not to leave by the edge of last exit, that is the outgoing edge belonging to T_r, unless no other way out is available.

It is clear that a path P so constructed must terminate at r. For whenever it arrives at another vertex there is always a possible way out "by parity". Moreover when it terminates at r it must have used all the outgoing edges, and therefore all the incoming ones as well.

Suppose there is a vertex v of D such that P does not traverse all the edges incident with v. Then v is not r. Hence there is an edge E of T_r directed from v. By the rule of construction P does not traverse E. Let the head of E be v_1. Then P does not traverse all the edges incident with v_1. Hence v_1 is not r. We can now repeat the argument with v_1 replacing v. But we could have chosen v to be as close as possible to r in the tree T_r, and then the existence of v_1 would have been contradictory.

We conclude that any path P in D constructed according to our rule must be Eulerian. Clearly it must have the edges of T_r as its edges of last exit.

For each vertex of v let $d(w)$ be the number of edges directed from it. Since D is balanced $d(w)$ is also the number of edges directed to w.

When we started P we had $d(r)$ choices for the first edge. When we left r for the second time we had $d(r) - 1$ choices. The third time we had $d(r) - 2$, and so on. All in all we had $d(r)!$ choices for our sequence of outgoing edges at r.

At any other vertex w we had just one choice for the $d(w)^{th}$ departure. Apart from that we had $d(w) - 1$ choices for the first departure, $d(w) - 2$ for the second, and so on. So we had $(d(w) - 1)!$ choices in all for the sequence of outgoing edges at w.

Let us now enumerate the vertices of D as $v_1, v_2, \ldots, v_n$, with $v_n = r$. Then the number of Eulerian trails from r to r corresponding to the tree T_r is

$$d(v_n)! \prod_{i=1}^{n-1} (d(v_i) - 1)!$$

To get the total number of Eulerian trails from r to r in D we have only to multiply this expression by the number of spanning directed trees of D converging to r. Using our earlier notation we write this tree-number as $T_n(D)$.

We may prefer to talk of "Euler cycles" of D for which only the cyclic sequence of directed edges is important, rather than of Euler trails with a distinct first edge. To convert the number of Euler trails from r to r into the number of Euler cycles we have only to divide by $d(v_n)$. Letting $N(D)$ denote the number of Euler cycles in D we infer that

(9) $$N(D) = T_n(D) \prod_{i=1}^{n} (d(v_i) - 1)!$$

This is the BEST Theorem.

Let us note some interesting graph-theoretical consequences. Obviously $N(D)$ is not dependent on an initial choice of r. Hence $T_n(D)$ must be the same for all choices of $r = v_n$. Thus the number of spanning directed trees of D converging to a given vertex must be the same for every vertex. Moreover if D' is the balanced digraph obtained from D by reversing the direction of every edge, then $N(D') = N(D)$. For we obtain the Euler cycles of D' from those of D by taking their (reversed) edges in the opposite cyclic order. Hence $T_i(D') = T_i(D)$ for each vertex v_i. Accordingly each directed and balanced graph D has a unique tree-number $T(D)$ such that, for each vertex v, the number of directed spanning trees converging to v and the number of spanning trees diverging from v are each equal to $T(D)$.

These results can be proved algebraically in terms of the Kirchhoff Matrix $K(D)$ of D. The essential observations are that for a balanced digraph the elements of $K(D)$ sum to zero in columns as well as rows, so that all the first cofactors become equal, and that $K(D')$ is the transpose of $K(D)$.

The analogy between the proof of the BEST Theorem and Tarry's treatment of the labyrinth problem is very close. We can make this evident by replacing Tarry's labyrinth-graph by its equivalent digraph, which is balanced. Tarry's algorithm gives a directed Euler path from A to A in that digraph. Instead of specifying an edge E as the "entry-edge" of a vertex v we may equally well distinguish the corresponding directed edge leaving v as the edge of last exit. The algorithm builds the Eulerian trail, together with its tree of last exits, as it goes along.

6. König's Fifth Chapter.

In this Chapter König points out that every (finite) tree has a uniquely determined central vertex or edge. He writes of paths in a tree T. These are what I call "simple paths", paths in which no edge or vertex is repeated. The subgraph consisting of the edges and vertices of a simple path, with at least one edge, is called an "arc".

König considers the paths of maximum length in T. He shows that each one has its ends at monovalent vertices ("end-vertices") of T. Let the maximum length be s. König shows that if s is an even number $2n$ then all the paths of length $s = 2n$ have the same middle vertex, and he calls this vertex the "centre" of the tree. If instead s is an odd number $2n + 1$ then all paths of length $s = 2n + 1$ have the same middle edge, and this edge is called the "axis" of the tree.

The possession of a uniquely determined edge or vertex is an important property of trees, one not extending to graphs in general. It is of great interest in Reconstruction Theory, a branch of Graph Theory that has developed since the publication of König's book.

Textbooks of organic chemistry often tell of how the molecular structure of some substance has been inferred from the structures of its decomposition products. An archaeologist may hope to sketch the appearance of an ancient building after studying what remains of its wreckage. These are examples of theoretical "reconstruction".

The process of reconstruction is idealized in Graph Theory in the following way. Consider a graph G, with vertices enumerated as $v_1, v_2, \ldots, v_n$. For each suffix j let us form from G a graph G_j by deleting the vertex v_j and all its incident edges.

We now suppose a graph theorist to be given drawings of each of the graphs G_j. He is asked to "reconstruct" the graph G from the n drawings. It is important to realize that in these drawings all the vertices have lost their suffixes. The graph theorist is not told which vertices in one drawing correspond to which in another, and his real task is to establish these correspondences.

We need not doubt that the graph theorist, perhaps using elegant arguments and perhaps reduced to trial and error, can succeed in his task. He constructs a graph H of n vertices, whose n vertex-deletions give the n graphs G_j. He then asks "Is this solution unique? Is it, to within an isomorphism, the same graph G that was used by my challenger? And if it does happen to be unique this one time, then would it necessarily be unique for every possible choice of G?" In this last question he propounds the Reconstruction Problem, still unsolved at the time of writing.

Let us refer to the n graphs G_j as the vertex-deleted subgraphs of G. We say that G is "reconstructible" if it is the only graph that gives rise to this particular family of vertex-deleted subgraphs. The Reconstruction Conjecture, also called Ulam's Conjecture, asserts that every finite graph with three or more vertices is reconstructible. To see why the restriction to 3 or more vertices is necessary consider a graph with just two vertices u and v, and with just n edges, each joining u and v. Here n can be any non-negative integer. For each value of n our graph produces the same pair of vertex-deleted subgraphs, a pair of vertex-graphs. Yet different values of n give different initial graphs.

Expositions of Reconstruction Theory may give long lists of those graph-theoretical properties of G that are "reconstructible", that is deducible from the drawings of the G_j. First the number of vertices of G is reconstructible; it is the number of the drawings, and one more than the number of vertices in each drawing. To get the number of loops of G we count the loops of all the graphs G_j and divide by $n - 1$. To get the number of other edges, the number of "links" of G, we count the links of the graphs G_j and divide by $n - 2$. Now for each G_j we know how many loops of G, and how many links, are missing from it. Hence we know the valency in G of the missing vertex v_j. Hence we can determine the number of vertices of G of any specified valency. In particular if G is regular, that is if all its vertices have the same valency, we can "recognize" that regularity. We can also recognize whether or not G is a tree. The necessary and sufficient condition for arboricity is threefold: No G_j must have a circuit, at least two of the G_j must be connected, and there must be at least one G_j that is not an arc of length $n - 2$.

The list of reconstructible properties goes on, even to the number of subgraphs of G isomorphic with a given graph H of fewer than n vertices. It goes on even to the determination of the characteristic and chromatic polynomials of G. At that stage we are hardly able to go on doubting the Reconstruction Conjecture. Yet however many properties of G we determine we are never able to prove that they suffice to determine the structure of G uniquely. Either the Conjecture is false after all or some essential insight still eludes us.

The Reconstruction Conjecture is known to be true for some kinds of graph. Thus regular graphs are reconstructible as well as recognizable. Once we have recognized G as regular we can rebuild it from G_1. First we restore the missing vertex v_1. Then we join it by links to vertices of G_1 so as to bring the valency of each of these vertices up to its known value in G. Finally, if necessary, we put enough loops on v_1 to give it too the right valency.

Less trivially it can be shown that all trees are reconstructible [21]. The essential trick in proving this is to show that in at least some of the graphs G_j we can recognize the original centre or axis of the tree G. Trivial cases can be disposed of trivially: if G is an arc of length $n - 1$, we shall discover that when we determine how many vertices there are of each valency. In non-trivial cases the paths of greatest length in the collection of graphs G_j are the paths of greatest length in G, and their middle vertices or edges must be identified with the centre or axis of G respectively.

König goes on to define another kind of central vertex or edge in a tree. But for that I have no comment or updating to offer.

Finally he touches upon problems of enumeration. There we come upon another sector of graph theory in which there has been explosive development since König's time.

The problem of finding the number of (non-isomorphic) trees satisfying stated restrictions arose in the nineteenth century, and it was successfully investigated by Cayley. It appears in Organic Chemistry in the following form: how many paraffins are there with a given number of carbon atoms in the molecule? In other words, how many non-isomorphic trees are there in which n vertices are quadrivalent and all the other vertices are monovalent? Cayley gave recursion formulae for calculating the required numbers [9]. He observed that the enumeration problem is simpler for "rooted" trees, that is trees in which one vertex is distinguished as the "root".

The new theory of enumeration began with the classic paper of G. Pólya, published in 1937 [26]. After Pólya's methods had become well-established in Graph Theory it was discovered that essentially the same theory had been published before, by J. H. Redfield in 1927 [28]. There is an excellent account of the Pólya theory, written by N. G. de Bruijn, in the book *Applied Combinatorial Mathematics*, published in 1964 [8].

Pólya's theory depends on some results in Group Theory. It starts with the consideration of a group G of permutations of a finite set S. A given element g of G partitions S into disjoint subsets, and permutes the members of each subset cyclically. Let us suppose g to determine in S b_1 cycles of length $1, b_2$ of length 2, and so on. Then we can associate with g the product

$$x_1^{b_1} x_2^{b_2} \cdots x_r^{b_r} \cdots,$$

where x_1, x_2, x_3 etc. are independent variables. The expression always reduces to a finite product with no suffix greater than the cardinality of S.

The "cycle index" of G is the function

$$(10) \qquad P_G(x_1, x_2, \ldots, x_r, \ldots) = |G|^{-1} \sum_{g \in G} (x_1^{b_1} x_2^{b_2} \cdots x_r^{b_r} \cdots).$$

The results of Pólya's theory are stated in terms of such cycle indices.

Pólya's Theorem applies when labels of two or more kinds are attached to the members of S. Perhaps we can call them "colours" here, saying that some members of S are blue, some red, some green, and so on. More formally we could introduce another finite set R and define a labelling of S as a mapping of S into R. The members of R would be our colours.

We may now ask how many different label patterns on S are possible, given that two patterns are counted as distinct only if no member of G transforms one into the other. According to Pólya's Theorem the number of distinct patterns is

$$P_G(k, k, k, \ldots),$$

where $k = |R|$ is the number of colours available.

The relevance of this theorem to Graph Theory can be seen from the following example. Let us ask how many non-isomorphic loopless graphs there are with just q vertices. To answer this we introduce a set Q of q elements and we let Γ be the group of all permutations on Q. The number required is the number of patterns that can be formed by joining some pairs of members of Q by edges. Two such patterns are counted as distinct if and only if no member of Γ transforms one into the other.

We are not yet ready to apply Pólya's Theorem. For that, instead of the set Q of vertices we need the set S of pairs of vertices. Instead of the permutation group Γ on Q we need the permutation group G induced by Γ on S. We must classify the cycles of G in S, and infer the cycle index of G. Then we can say that our desired patterns are formed on S by using the two colours "joined" and "not-joined". Those patterns are the graphs of n vertices. So the number of such graphs must be

$$P_G(2, 2, 2, \ldots, 2).$$

This brief sketch of Pólya's method for enumerating graphs must suffice here. The theorem can also be used to enumerate some special kinds of graphs, such as trees. But the treatment of trees is less direct. Even de Bruijn, in the work mentioned, thinks it best to leave tree-enumeration to another writer. (F. Harary).

The present writer has had some success in the enumeration of planar maps. However in his work the maps are always "rooted". This means that some edge is marked as the "root" and arbitrary directions along and across it are distinguished. This device destroys all symmetry. Therefore it makes theories of the Pólya type unnecessary. Often it gives very simple enumerating formulae. For example, the number of rooted planar maps of n edges, allowing loops and multiple joins is

$$\frac{2.(2n)!3^n}{n!(n+2)!}.$$

The main results of this theory are given in four "Census" papers of 1962-1963 [41, 42, 43, 44]. Recently progress has also been made in the theory of non-rooted planar maps, notably by E. A. Bender, L. B. Richmond and N. C. Wormald. The results are summarized by Bender in the issue of the American Mathematical Monthly dated January 1987 [3, 29]. With due reference to Steinitz's Theorem, a formula is given, as an asymptotic approximation, for the number of non-isomorphic convex polyhedra of $i + 1$ vertices and $j + 1$ faces. The old goal of T. P. Kirkman [4] of finding the number of non-isomorphic convex polyhedra with m vertices and n faces seems now almost within reach.

7. König's Sixth Chapter.

The sixth chapter is about infinite graphs. It is mainly concerned with the "Unendlichkeitslemma", a device by which many theorems already proved for finite graphs can be extended to infinite but locally finite ones.

König's Unendlichkeitslemma, or "infinity lemma", has wide applications. In one example he uses it to prove the Heine-Borel Theorem of mathematical analysis. In another he proves a result in Number Theory.

He follows these examples with a graph theoretical proof of an equivalence theorem in set theory. The theorem asserts that if two sets A and B are each similar to a subset of the other, then A and B are similar. (Two sets are said to be similar if there exists a one-to-one correspondence between them). König adapts a proof of Julius König, his father, putting it into graph theoretical language. He remarks that the translation does not shorten the proof but does seem to make it clearer. He promises more applications of graph theory to set theory in Chapter XIII.

More than once I have heard an eminent graph theorist express the suspicion that many theorems about topological spaces could be reduced to graph theory if some researcher would take enough trouble. But to my knowledge not much more has been done along these lines than has already been achieved by König.

The Unendlichkeitslemma has been a powerful tool for investigating locally finite graphs. In a sense it has been too powerful, causing some of us to lose interest in that kind of graph. Their theory, we say, is but a succession of exercises on König's Lemma. Some graph theorists feel that for genuinely new theorems on infinite graphs we have to study graphs in which one or more vertices may be of infinite valency.

There is for example an infinite acyclic graph, what most of us would call an "infinite tree". It is a connected acyclic graph in which the valency of each vertex is countably infinite. Call it T. If we delete one vertex and all its incident edges we obtain a graph with a countably infinite set of components, each of which is another T. We get the same result if we start with a graph of two components, each a T. We deduce the intriguing result that the Reconstruction Conjecture is not true for infinite graphs. So here is one result on infinite graphs that does not depend on the infinity lemma.

8. König's Seventh and Eighth Chapters.

We come now to the chapters primarily concerned with directed graphs. Here I see no need to update König's general theory. I should indeed remind the reader of the BEST Theorem, and of the determinantal formula for the number of spanning directed trees converging to a given vertex. But those matters have already been discussed in this Commentary.

There is a book on directed graphs entitled *Structural Models,* by F. Harary, R. Z. Norman and D. Cartwright [19]. One of its authors is a psychologist, and one of its messages is that directed graphs have applications to the social sciences. In theory of course they have applications wherever binary relations occur — for what is a directed graph but a binary relation?

Chapter VII is mainly concerned with an application to logic, with reformulating axiomatic theory in graph theoretical terms. But it is nevertheless the introductory general theory of directed graphs. In Chapter VIII we hear again of this logical application. After that the application to Game Theory is discussed in some detail.

As an example, consider the game known in some parts of the English-speaking world as Noughts and Crosses, and in others as Tic Tac Toe. We can describe the usual version of this game as played on a 3 × 3 square board between two players, Black and White. Black plays first, putting a piece on any one of the nine squares. Then White and Black play alternately, using no square twice. The object of each is to secure a row of three pieces of his own colour, either parallel to an edge of the board or along one of the two diagonals. The first player to achieve such a row is the winner. But if the board is filled without producing any such row, then the game is drawn.

There is a 3-dimensional version of the game, more interesting to adults, played in a 4 × 4 × 4 cube. This time each player tries to get a row of 4 pieces of his own colour, either parallel to an edge of the cube, or parallel to a face-diagonal, or along one of the main diagonals. A. H. Stone has experimented with a 4-dimensional version, played in a 5×5×5×5 hypercube. Stone thought it necessary to forbid Black, the first player, to occupy the central hypercube with his first move. In practice the board consists of five cubes arranged in a row, each representing one slice of the hypercube.

In any of these three versions we can consider that the game is played on a directed graph G. The vertices of G are the possible positions, the possible arrangements of pieces on the board that can occur in the course of a game. In such a position either there are an equal number of black and white pieces and it is Black's move, or Black has one piece more than White and it is White's move. The directed edges are the moves; each is directed from the initial position of its move and to the resultant one. In theory it is possible to classify each vertex as (i) a winning position for the mover, (ii) a losing position for him or (iii) a drawn position. In (i) the mover can always win if he plays properly. In (ii) he must always lose if his opponent plays properly. In (iii) he can get a draw if he plays properly, and cannot get a win if his opponent plays properly.

Other board games can be represented by directed graphs in a somewhat similar way, chess for example. Apart from the obvious extra complication of chess, we have to remember that a player's next move is sometimes restricted, not only by the arrangement of the pieces on the board, but by the history of the game. He may not castle, for example, if his king or rook has previously been moved. If he wants to take a pawn *en passant* he must do it immediately. And if 50 moves have gone by with no pawn moved or piece taken, and a king not left in checkmated position, then the game is drawn. I suppose that in specifying a vertex-represented position for that game one must state not only how the pieces are placed on the board but whose move it is and what the historical restrictions are.

A procedure for classifying positions in a game as winning, losing or drawing has been explained by C. A. B. Smith and others [16, 17, 32]. It works backwards from the possible final positions.

In some puzzles, presentable as simple games, the underlying directed graph is easy enough to draw. A solution of the puzzle or a victorious strategy in the game, if such exists, can then be obtained by analyzing the graph. Alas, for any game of interest, the graph is too complicated to be drawn. There simply is not world enough to display it, nor time to construct it. As in other mathematical contexts, one marvels at the vastness of the numbers and sets that are being discussed, and wonders how they can relate to physical reality. Yet the argument goes on without apparent logical contradictions, and sometimes it throws up a result that we can appreciate and verify.

I would be puzzled to tell you of such a result here. The nearest approach that occurs to me concerns 3-dimensional Tic Tac Toe. It is the theorem that the second player, White, cannot force a win. In the 2-dimensional game, we know of course that neither player can force a win, that the the initial position is a drawn one. This can be established by analysis of the directed graph, simplified by an exploitation of the symmetry of the board. I do not know if the graph theoretical method can be applied plausibly to the 3-dimensional problem. The usual argument goes like this. Suppose there is a winning strategy for White, the second player. Then let Black ignore his own first move, pretend to be the second player, and apply that player's winning strategy. This must give him the victory, since an extra piece on the board can never be a disadvantage for him in this game. If the strategy calls for him to put a piece in Cube X which he has already occupied then he plays in some other unoccupied cube, it does not matter which. If no unoccupied cube is left, then Black has already won the game, has he not?

König ends Chapter VIII with an application to Group Theory, by which a group is represented by its Cayley graph, with edges in many colours. I confess to having always looked askance at these Cayley graphs. For most groups they are too complicated to draw conveniently, and I wonder how they are supposed to improve upon the group's multiplication table. But that is the way with applications of Graph Theory, or of any other theory. Some seem helpful and natural, others forced and artificial. But the distinction between the two is subjective; one man's graph is another man's gripe.

9. König's Ninth Chapter.

In Chapter IX König studies linear forms on the edges of a graph, thus carrying on a tradition already well-established in Combinatorial Topology. I learned the procedure from the textbook *Lehrbuch der Topologie* by Seifert and Threlfall [30], published like König's book in Leipzig, but two years earlier. I read of simplicial and other complexes, of oriented simplexes, and of hierarchies of linear forms called 0-chains, 1-chains, 2-chains and so on. I read also of cycles of various dimensions and of homology groups.

A graph, being but a 1-dimensional complex, has only rudiments of that algebraic structure. Using the topological terminology we can call its vertices "0-cells" and its edges "1-cells". We usually do not bother to orient a vertex. But we orient each edge by marking an arrow on it, thus formally making G into a directed graph, or "digraph". Accordingly, König works with directed graphs in this chapter.

It should be emphasized that, for the purposes of Chapter IX, the directions or orientations assigned to the edges of G are not part of the graph's structure. They are merely marks put upon the graph for purposes of reference, like lines of lattitude and longitude on a map of the Earth. If we were discussing an electrical application we might say that some oriented edge E carried a current of J amps, meaning J amps in the direction of the arrow. If we then reversed the arrow, the current in the edge would thereafter be called $-J$ amps. That kind of notational transformation we are to regard as trivial.

Let us enumerate the vertices of G as $v_1, v_2, \ldots, v_m$, and the edges, duly oriented, as $E_1, E_2, \ldots, E_n$. Having decided upon a domain D from which coefficients may be taken we can then discuss such linear forms as

$$\lambda_1 E_1 + \lambda_2 E_2 + \cdots + \lambda_n E_n$$

and

$$\mu_1 v_1 + \mu_2 v_2 + \cdots + \mu_m v_m.$$

Following Seifert and Threlfall we may call these 1-chains and 0-chains respectively of G over D. We may think of them as mappings of the edge-set or vertex-set of G into D. Two 0-chains or two 1-chains are added or subtracted by adding or subtracting corresponding coefficients, as is usual for linear forms. To multiply a chain by an element λ of D we multiply each coefficient by λ.

König indicates that he will usually take his coefficients from the sets of complex, real or rational numbers. But quite often it suffices for D to be the domain I of the integers. However there are other domains D, usually additive Abelian groups, that are in quite common use. For example, D may be the group of residues modulo a positive integer q.

König is interested in 1-chains. There are some especially interesting ones associated with circuits of G. Let us go around a circuit J, assigning the coefficient $+1$ to each edge we traverse in the direction of its arrow and -1 to each edge we traverse against its arrow. Every edge outside J gets a zero coefficient. Let us denote the resulting 1-chain by $K(J)$. If we had gone around J the other way, we would have constructed $-K(J)$ instead.

König calls $K(J)$ a "cycle-form" and he asks, "What is the condition for a given 1-chain on G to be a linear combination of cycle-forms?" He answers this question in his Theorem 1. I propose to state the answer in terms of a symbol ϵ_{ij}. Here the suffix i ranges in value from 1 to m, and j from 1 to n. We give ϵ_{ij} the value $+1$ if v_i is at the tail of the arrow of E_j, the value -1 if v_1 is at the head of this arrow, and the value 0 when v_i and E_j are not incident. Here König assumes that each edge has two distinct ends. But we can extend the symbolism to the case of a loop E_k by postulating that then $\epsilon_{jk} = 0$ for every vertex v_1.

König's condition can be stated as follows: the 1-chain

$$\sum_j \epsilon_{ij} E_j$$

is a linear combination of cycle-forms if and only if

$$(11) \qquad\qquad \sum \epsilon_{ij}\lambda_j = 0$$

for each vertex v_1. If λ_j is called the "current" in E_j we can state the condition verbally as follows: the total current in the edges directed to a vertex v_1 is equal to the total current in the edges directed from v_1. Loops are to be ignored here. Or we can say that each loop contributes equal amounts to each "total current".

If we wish to follow Seifert and Threlfall our procedure and terminology must be different. First we must define the boundary ∂K of an arbitrary 1-chain

$$\sum_j \lambda_j E_j$$

on G. It is the corresponding 0-chain

$$\sum_i \left(\sum_j \epsilon_{ij}\lambda_j \right) v_i.$$

Then we must define a "cycle" of G as a 1-chain K such that $\partial K = 0$, i.e. such that the coefficients in ∂K are all zero. We observe that these cycles are König's linear combinations of cycle-forms by his Theorem 1, and that his cycle- forms are just special cases of cycles. I shall refer to those cycle-forms as the "primitive cycles" of G.

Cycles in a graph G are often called "flows". We think of some fluid circulating in the graph. The current of fluid in any edge E_j is measured by λ_j and the conservation law of Equation (11) is satisfied at each vertex.

Since König's time there has been much interest in integral flows, those flows in which every coefficient is an integer. It is customary to define a k-flow in G as an integral flow in which no coefficient is zero and each coefficient is less than the positive integer k in absolute value. Such k-flows are of interest in connection with some well-known colouring problems.

If a graph G has an isthmus (or "bridge") it is easy to show that no k-flow of G can exist for any k; the isthmus has a zero coefficient in every cycle. Among graphs with no isthmuses there are some with no 4-flows. It has been conjectured that every graph without an isthmus has a 5-flow, but no proof or counterexample for this conjecture has yet been found. But it has been shown that every such graph has a 6-flow [31].

König's "star-forms" are related to the topological concept of a coboundary. The coboundary of the 0-chain

$$H = \sum \mu_i v_i$$

is, by definition, the 1-chain

$$\delta(H) = \sum_j \left(\sum_j \mu_i \epsilon_{ij} \right) E_j.$$

In particular H may have only one non-zero coefficient, say a 1 for v_k. Then we can call δH a "vertex-coboundary" or "the coboundary δv_k of the vertex v_k". We then have

$$\delta v_k = \sum_j \epsilon_{kj} E_j.$$

This is König's "star-form" at v_k. It involves only the links incident with v_k, each taken with coefficient $+1$ or -1 according to whether it is directed to or from v_k.

König is interested in those 1-chains of G that are linear combinations of star-forms. These are simply the coboundaries of the 0-chains.

König proves some basic theorems about coboundaries. One asserts that a 0-chain has a zero coboundary if and only if in each component of G the coefficients are all equal. From this he deduces that the rank of the additive group of coboundaries of G is $\alpha(G) - p_0(G)$ (in the notation of this commentary). The rank of the group of cycles is $p_1(G)$, as König shows by considering a spanning tree of a connected graph G and the associated fundamental set of circuits.

König goes on to relate cycles and coboundaries. His results would now be expounded in terms of the notion of orthogonality. Two 1-chains $\sum \lambda_j E_j$ and $\sum \mu_j E_j$ are said to be orthogonal if

$$(12) \qquad\qquad \sum \lambda_j \mu_j = 0.$$

It can be shown from the algebraic definitions that a 1-chain of G is a coboundary if and only if it is orthogonal to every cycle. Moreover a 1-chain is a cycle if and only if it is orthogonal to every coboundary.

If D is the set of real numbers, or some subset thereof, it is clear that no non-zero chain can be both a circuit and a coboundary, simply because it cannot be self-orthogonal. König extends this result in his Theorem 15.

The set of cycles and the set of coboundaries of a graph G are additive Abelian groups, closed under multiplication by an element of D. They can be regarded as examples of purely algebraic structures that I call "chain groups".

To construct a chain group we need a finite set

$$S = \{A_1, A_2, \ldots, A_n\}$$

of "cells" A_j, and a coefficient domain D. It is sufficiently general to define D as a commutative ring with a unity element. There is then an additive group Γ of all linear forms, or "chains",

$$\lambda_1 A_1 + \lambda_2 A_2 + \cdots + \lambda_n A_n$$

with coefficients in D. By a "chain group on S over D" I mean any subgroup N of Γ that is closed with respect to multiplication by an element of D. In particular S can be the edge set of a graph G and N can be either the cycle group or the coboundary group of G with respect to D.

A chain group N has a detailed structure describable in terms of its "elementary chains". To explain these we first define the "support" of a chain K of N as the set of all cells with non-zero coefficients in K. An elementary chain is then defined as a non-zero chain whose support is minimal in the sense that it contains the support of no other non-zero chain.

The supports of the elementary chains of the cycle group of a graph G turn out to be the circuits of G, or more precisely their edge sets. The supports of the elementary chains of the coboundary group are the "bonds" of G, defined graph-theoretically as follows: If a connected graph G can be represented as the union of two disjoint connected subgraphs H and K plus a set B of edges each with one end in H and one in K, then B is a "bond" of G and H and K are the "end-graphs" of B in G. A bond in a disconnected graph G is a bond of a component of G. Like a circuit, a bond corresponds to a "primitive" chain in which only coefficients $+1, -1$ and 0 appear. That primitive chain arises as the coboundary of the 0-chain in which the vertices of H have coefficient 1 and all the other vertices of G have coefficient 0.

The generalization from graphs to chain groups can be regarded as the first stage in the exposition of the new mathematical discipline called Matroid Theory. The term "matroid" was introduced by Hassler Whitney in 1935 [51]. Whitney considered the collection F of the edge sets of the circuits of a graph G. He observed that they satisfied two simple rules.

(i) *No member of F contains another.*

(ii) *If X and Y are two distinct members of F, with a common edge A, and if B is an edge of X that does not belong to Y, then there is a third member Z of F with the following properties: Z includes B but not A, and Z is contained in the union of X and Y.*

Now these two rules are obeyed also by the collection of bonds of G. Indeed they are obeyed by the collection of the supports of the elementary chains of any chain group. Since the rules are of such common occurrence it is natural to take them as the axioms of a new "Theory of Matroids". A "matroid" is thus any collection F of non-null subsets of a finite set S which obeys rules (i) and (ii). The members of F are often called the circuits of the matroid. Matroid Theory is both a generalization of Graph Theory and an abstract study of the concept of linear dependence.

Whitney gave several equivalent sets of axioms for Matroid Theory. The one constituted by Rules (i) and (ii) is not the only one, or even the one that is now most commonly used. For further information on Matroid Theory the reader could consult the textbook of D. J. A. Walsh [49].

10. König's Tenth Chapter.

In Chapter X König is dealing with a special case of his theory of linear forms, that in which the coefficient domain is the group I_2 of residues mod 2. This special case does bring in some pleasing simplifications, for example, every element of I_2 is its own negative. Accordingly we need not worry about orientations of edges; orientation is irrelevant.

Let $E(G)$ be the edge set of our graph G. Then a chain K on $E(G)$ is now completely defined when its support is given. In the remainder of this section we can identify any such chain with its support.

To add two subsets (or chains) P and Q of $E(G)$ we simply take those members of $E(G)$ that occur in one or other of P and Q but not in both. We call the resulting set the mod 2 sum $P + Q$ of P and Q. Similarly we can add n subsets $P_1, P_2, \ldots, P_n$ of $E(G)$, not necessarily all distinct. Their mod 2 sum is the set of those edges of G that appear in an odd number of the sets P_i. We note the rule that $P + P = 0$, the zero chain or null subset, for each subset P of $E(G)$.

With these rules we see that any mod 2 sum of cycles of G is a cycle and any mod 2 sum of coboundaries of G is a coboundary. A cycle, by Theorem 1 of Chapter IX, is equivalent to a subgraph of G in which the valency of each vertex is even, that is to an Eulerian subgraph.

In order to illustrate the methods of Chapter X I would like to present one post-König proof. It is a proof of Smith's Theorem [35] on the Hamiltonian circuits of a cubic graph G. We need two new concepts, that of a Tait cycle and that of a Tait colouring of G.

A Tait cycle of G is a subgraph which is a union of disjoint circuits, each of even length, and which includes every vertex of G. A Tait colouring of G is a colouring of the edges of that cubic graph in three colours α, β and γ so that each vertex is incident with one edge of each colour. A permutation of the three colours among themselves is not deemed to give a new Tait colouring.

Given a Tait colouring T of G, we observe that the edges coloured α and β make up a Tait cycle $C_{\alpha\beta}$. Similarly we have Tait cycles $C_{\alpha\gamma}$ and $C_{\beta\gamma}$. Since each edge belongs to just two of these three Tait cycles we can write

(13) $$C_{\alpha\beta} + C_{\beta\gamma} + C_{\alpha\gamma} = 0.$$

Now we ask the following question. Given a Tait cycle C, to how many distinct Tait colourings does it belong? Evidently all such Tait colourings can be derived, modulo a permutation of the three colours, by colouring the edges of each component of C alternately α and β, and then assigning the colour γ to all edges not in C. If C has just k components there are 2^k ways of doing this. Since the interchange of the two colours α and β throughout does not alter a Tait colouring we deduce that the number of distinct Tait colourings to which C belongs is 2^{k-1}.

We note that the Hamiltonian circuits of G are the Tait cycles with $k = 1$. Let us sum Equation (13) over all the Tait colourings of G. A Tait cycle with $k > 1$ appears in an even number 2^{k-1} of these and therefore cancels out under the mod 2 addition. But each Hamiltonian circuit, with $k = 1$, appears in just one Tait colouring and therefore persists in the final sum. We thus obtain the result known as Smith's Theorem, that the Hamiltonian circuits of any cubic graph sum to zero mod 2.

11. König's Eleventh and Twelfth Chapters.

König's expression "factor of degree n" is often abbreviated as "n-factor". An n-factor of a graph G is a spanning subgraph of G in which the valency of each vertex is n.

In his Chapter XI König considers the problem of finding an n-factor in a regular graph. He pays special attention to Eulerian and bipartite graphs. For these he records far-reaching results, resolving regular Eulerian graphs into edge-disjoint 2-factors and regular bipartite graphs into edge-disjoint 1-factors. The case of regular graphs of odd valency that are not bipartite he leaves for partial resolution in Chapter XII.

To some later workers restriction to regular graphs seemed unnatural. For whether a graph G is regular or not it is still meaningful to ask if G has a 1-factor or a 2-factor. König himself recognizes that theorems about factors of regular graphs have to be generalized before they can be proved inductively. In his generalizations, however, he deregularizes the factors of G as well as G itself. See his Theorems 5, 6, and 15 of Chapter XI.

The first major theorem on factorization not treated in these chapters is Hall's Theorem, postponed by König to Chapter XIV. It was published by P. Hall in 1935 as a result about sets and subsets [17]. It gives a necessary and sufficient condition for a bipartite graph G, regular or not, to have a 1-factor.

Let $\{U, V\}$ be a bipartition of G, a partition of the vertex-set of G into complementary subsets U and V such that each edge of G has one end in U and one in V. Let us define a U-constriction of G as a subset S of U such that the number of members of V joined to S is less than the number of members of S. In order that G may have a 1-factor it is clearly necessary that G shall have no U-constriction or V-constriction. Hall's Theorem says that this condition is also sufficient.

Let us denote the number of members of a finite set T by $\alpha(T)$. Then if $\alpha(U)$ and $\alpha(V)$ are unequal there is a U-constriction or a V-constriction, and of course G has no 1-factor. We thus have a non-trivial problem only when $\alpha(U) = \alpha(V)$. But then it can be shown that if G has a V-constriction it has also a U-constriction. So in the case of interest Hall's Theorem says that G has a 1-factor if and only if it has no U-constriction.

In 1947 came a 1-Factor Theorem stating a necessary and sufficient condition for an arbitrary finite graph G to have a 1-factor [36]. Consider any set S of vertices of G. Let G_S be the graph obtained from G by deleting the vertices of S and their incident edges. We classify a component of G_S as even or odd according to whether the number of its vertices is even or odd, and let $h(S)$ denote the number of odd components of G_S. It is not difficult to see that if

$$h(S) > \alpha(S)$$

then G can have no 1-factor. We refer to such a set S as a 1-barrier of G. So if a graph G is to have a 1-factor it is necessary that it shall have no 1-barrier. The 1-Factor Theorem asserts that this condition is also sufficient. Thus a graph G has either a 1-factor or a 1-barrier, but not both.

There has been a further generalization of the theory of graph factors. Let f be a function from the vertex set $V(G)$ of a graph G to the set of non-negative integers. Let us define an f-factor of G as a spanning subgraph F of G such that the valency in F of each vertex v of G is the corresponding number $f(v)$. Necessary and sufficient conditions have been obtained for a graph G to have an f-factor [38, 46]. They take their simplest form when G is bipartite. Oystein Ore gave a factor theory for directed graphs, putting restrictions on the numbers of incoming and outgoing edges at each vertex [25]. The theory of f-factors for bipartite graphs derives from this. For a graph with a bipartition $\{U, V\}$ can conveniently be regarded as directed, each edge going from U to V.

Chapter XII is concerned with Petersen's Theorem, published by J. Petersen in 1891 in the guise of a result on factorizing algebraic forms. As usually stated it asserts that every cubic graph without an isthmus has a 1-factor. A stronger form says that if a cubic graph has no 1-factor then it has three isthmuses not lying on one arc. It was a very difficult theorem to prove in 1891 and even in 1936. But its main difficulties have now been incorporated into, and resolved by, the 1-Factor Theorem. Petersen's Theorem can now be proved as a simple exercise on that result [46].

12. König's Thirteenth Chapter.

I look upon this chapter as a continuation of Chapter VI. The infinity lemma is used to extend some of the theorems on factors to infinite regular graphs of finite valency. But the chapter concludes with a theorem on regular graphs of infinite valency.

For a more recent discussion of factor problems for infinite graphs the reader may consult [23], a paper by C. St. J. A. Nash-Williams entitled "Marriage in denumerable societies".

13. König's Fourteenth Chapter.

This chapter leads off with the standard theory of cut vertices and blocks. If a connected graph G can be represented as the union of two subgraphs H and K, each with at least one edge, having only one vertex W (and no edges) in common, then G is said to be "separable" and W is called a "cut point" or "cut vertex" of G. A "block" of G can be defined as a maximal non-separable connected subgraph of G, maximal in the sense that it is contained in no other non-separable connected subgraph of G.

König shows that the blocks of G are edge-disjoint. He resolves G into a tree-like structure of blocks and cut vertices. Two blocks can intersect only in a cut vertex, and in at most one cut vertex. There is no circular sequence alternately of blocks and of cut vertices.

By one definition a "3-connected" graph is a connected non-separable graph G that cannot be represented as the union of two subgraphs H and K, each with at least two edges, having exactly two vertices (but no edges) in common. There is a procedure for decomposing a connected non-separable graph into 3-connected units joined together by "virtual edges". It is somewhat analogous to that for decomposing a connected graph into blocks and cut vertices, but is more complicated. It was published in 1966 ([45] and [46]).

König next considers a graph G with a bipartition $\{U,V\}$. He asks for the maximum number of edges, no two having a common end. He finds that this is the minimum number of vertices required to separate U and V, that is, such that each path from U to V includes one of them. If $\alpha(U) = \alpha(V)$ this result is equivalent to Hall's Theorem. Accordingly, König goes on to a discussion of that result.

To this commentator it seems a little odd that König should treat Hall's Theorem not in a chapter on factorization but in the chapter on separation theorems since he does recognize it as a factor theorem, and applies it in a new proof that any regular bipartite graph has a 1-factor.

He goes on to discuss some matrices associated with graphs and their associated determinants. There is, for example, the "adjacency matrix" $A(G)$. For this we enumerate the vertices of the graph G as $v_1, v_2, \ldots, v_n$. We then define $A(G)$ as the $n \times n$ square matrix in which the element a_{ij} of the ith row and j^{th} column is the number of edges joining v_i and v_j. In a variant of this definition the edges E_j are associated with independent variables x_j, and a_{ij} is the sum of the variables associated with the joining edges. Then of course each variable associated with a link appears twice, once in a_{ij} and once in a_{ji}. In a further modification every such variable x_k is entered as $+x_k$ in one of these places and as $-x_k$ in the other. In that last case loops are ignored, that is, the diagonal elements are made all zero. We then have a skew-symmetric matrix whose determinant, if the matrix is of even order, is the square of a polynomial called a "Pfaffian". The Kirchhoff matrix described in Section 5 is still another variation on the adjacency matrix.

For a graph with a bipartition $\{U, V\}$ we can define a matrix with rows corresponding to the members of U and columns corresponding to those of V. In the intersection of a row and column we enter the number of edges joining the corresponding vertices, or more generally, the sum of variables corresponding to those edges.

Theorems about graphs imply theorems about the corresponding matrices, and König gives examples. In one example he shows that if the rows and columns of a square matrix have all the same positive sum then one of the product terms in its determinant is non-zero. König notes also that the products with non-zero coefficients in the Pfaffian determine the 1-factors of the graph. This remark is of special interest because the first proof of the 1-Factor Theorem [36] was based on the properties of Pfaffians. Moreover in the case of planar graphs P. Kasteleyn has been able to circumvent the intrinsic ambiguities of sign and to express the number of 1-factors directly as a Pfaffian [20].

Finally König discusses Menger's Theorem. Later work on this result has largely taken the form of a search for new proofs, the shorter and simpler the better. Some of these proofs are given in [24].

There is an important form of Menger's Theorem concerned with directed graphs. It asks for the number of internally disjoint directed paths from a vertex u to another vertex v. Two "internally disjoint" paths have no edge or vertex in common other than their two ends. This is found to be the least number of directed edges going from a vertex set U containing u but not v to its complementary set V (containing v but not u). Since König's time this theorem has been expanded into a mathematical discipline known as Transportation Theory, of great theoretical and practical importance. See the textbook *Flows in Networks* by Ford and Fulkerson [13].

REFERENCES

[1] T. van Aardenne-Ehrenfest and N. G. de Bruijn, *Circuits and trees in oriented linear graphs*, Simon Stevin 28 (1951), 203–217.

[2] W. W. Rouse Ball and H. S. M. Coxeter, *Mathematical Recreations and Essays*, 12th Edition, University of Toronto Press, 1974.

[3] E. A. Bender, *The number of Three-dimensional Convex Polyhedra*, Amer. Math. Monthly, **94** (1987), 7–21.

[4] N. L. Biggs, *T. P. Kirkman, Mathematician*, Bull. London Math. Soc., **13** (1981), 97–120.

[5] N. L. Biggs, E. K. Lloyd and R. J. Wilson, *Graph Theory, 1736-1936*, Oxford 1976.

[6] R. L. Brooks, C. A. B. Smith, A. H. Stone and W. T. Tutte, *The Dissection of Rectangles into Squares*, Duke Math. J., **7**, (1940), 312–340.

[7] W. G. Brown, ed. *Reviews in Graph Theory*, Amer. Math. Soc., 1980.

[8] N. G. de Bruijn, *Pólya's Theory of Counting*, Chapter 5 of *Applied Combinatorial Mathematics*, E. F. Beckenbach ed., New York 1964.

[9] A. Cayley, *On the theory of the analytical forms called trees*, Mathematical Papers, Vol. 3, 242–246 and Vol. 4, 112–115.

[10] B. Descartes, *The expanding Unicurse*, on page 25 of *Proof Techniques in Graph Theory*, F. Harary, ed., New York 1969.

[11] G. A. Dirac, *Some Theorems on Abstract Graphs*, Proc. London Math. Soc., **(3), 2** (1952), 69–81.

[12] P. Erdös, T. Grünwald and E. Vázsonyi, *Über Euler-Linten un-*

endlicher Graphen, J. Math. and Phys., **17** (1938), 59–75.

[13] L. R. Ford and D. R. Fulkerson, *Flows in Networks,* Princeton 1962.

[14] E. Ya. Grinberg, *Plane Homogeneous Graphs of Degree Three without Hamiltonian Circuits,* (Russian), Latvian Math. Yearbook, **4** (1968), 51–58.

[15] B. Grünbaum, *Convex Polytopes,* New York 1967.

[16] P. M. Grundy and C. A. B. Smith, *Disjunctive Games with the Last Player Losing,* Proc. Cambridge Phil. Soc., **52** (1956), 527–533.

[17] R. K. Guy and C. A. B. Smith, *The G-Values of Various Games,* Proc. Cambridge Phil. Soc., **52** (1956), 514–526.

[18] P. Hall, *On Representations of Subsets,* J. London Math. Soc., **10** (1935), 26–30.

[19] F. Harary, R. Z. Norman and D. Cartwright, *Structural Models,* New York 1965.

[20] P. Kasteleyn, *The Statistics of Dimers on a Lattice,* Physica **27** (1961), 1209–1225.

[21] P. J. Kelly, *A Congruence Theorem for Trees,* Proc. Cambridge Phil. Soc., **65** (1969), 387–397.

[22] J. W. Moon, *Topics on Tournaments,* New York 1968.

[23] C. St. J. A. Nash-Williams, *Marriage in Denumerable Societies,* J. Comb. Theory A, **19** (1975), 335–366.

[24] C. St. J. A. Nash-Williams and W. T. Tutte, *More Proofs of Menger's Theorem,* J. Graph Theory, **1** (1977), 13–17.

[25] O. Ore, *Studies on Directed Graphs, I,* Ann. of Math. **2**, 63 (1956), 383–406.

[26] G. Pólya, *Kombinatorische Anzahlbestimmungen fur Gruppen, Graphen und chemische Verbindungen,* Acta Math. **68** (1937), 145–254.

[27] L. Pósa, *On the Circuits of Finite Graphs,* Publ. Math. Inst. Hung. Acad. Sci **8** (1963) A3, 355–361.

[28] J. H. Redfield, *The Theory of Group-reduced Distributions,* Amer. J. Math., **49** (1927), 433–455.

[29] L. B. Richmond and N. C. Wormald, *The Asymptotic Number of Convex Polyhedra,* Trans. Amer. Math. Soc., **273** (1982), 721–735.

[30] a. H. Seifert and W. Threlfall, *Lehrbuch der Topologie*, Leipzig 1934.

b. H. Seifert and W. Threlfall, English translation by Michael Goldman, edited by Joan Birman and Julian Eisner, *A Textbook of Topology*, (Pure and Applied Mathematics), 1980.

[31] P. Seymour, *Nowhere-zero 6-flows*, J. Comb. Theory B, **30** (1981), 130–135.

[32] C. A. B. Smith, *Graphs and Composite Games*, J. Comb. Theory, **1** (1966), 51–58.

[33] C. A. B. Smith and W. T. Tutte, *On Unicursal Paths in a Network of Degree 4*, Amer. Math. Monthly, **48** (1941), 233–237.

[34] E. Steinitz, *Polyeder und Raumentetlungen*, Encyclopädie der Mathematischen Wissenschaften, III AB 12 (1922), 1–139.

[35] W. T. Tutte, *On Hamiltonian Circuits*, J. London Math. Soc., **21** (1946), 98–101.

[36] W. T. Tutte, *The Factorization of Linear Graphs*, J. London Math. Soc., **22** (1947) 107–111.

[37] W. T. Tutte, *The Dissection of Equilateral Triangles into Equilateral Triangles*, Proc. Phil. Soc., **44** (1948), 463–482.

[38] W. T. Tutte, *The Factors of Graphs*, Can. J. Math., **4** (1952), 314–328.

[39] W. T. Tutte, *A Theorem on Planar Graphs*, Trans. Amer. Math. Soc., **82** (1956), 99–116.

[40] W. T. Tutte, *A Non-Hamiltonian Planar Graph*, Acta. Math. Acad. Sci. Hung., **11** (1960), 371–375.

[41] W. T. Tutte, *A Census of Planar Triangulations*, Can. J. Math., **14** (1962), 21–38.

[42] W. T. Tutte, *A Census of Hamiltonian Polygons*, Can. J. Math., **14** (1962), 402–417.

[43] W. T. Tutte, *A Census of Slicings*, Can. J. Math., **14** (1962), 708–722.

[44] W. T. Tutte, *A Census of Planar Maps*, Can. J. Math., **15** (1963), 249–271.

[45] W. T. Tutte, *Connectivity in Graphs*, Toronto 1966.

[46] W. T. Tutte, *Graph Theory*, New York 1984.

[47] H. Walther, *Ein kubischer planarer zyklisch fünffach zusammenhängender Graph, der keinen hamiltonkreis besitzt*, Wiss.

Z. Tech. Hochsch. Ilmenau, **11** (1965), 163–166.

[48] H. Walther and H.-J. Voss, *Uber Kreise in Graphen*, Berlin 1974.

[49] D. J. A. Welsh, *Matroid Theory*, London 1976.

[50] H. Whitney, *The Coloring of Graphs*, Ann. of Math., **33** (1932), 688–718.

[51] H. Whitney, *On the Abstract Properties of Linear Dependence*, Amer. J. Math., **57** (1935), 509–533.

[52] R. J. Wilson, *An Eulerian Trail through Königsberg*, J. Graph. Theory, **10** (1986), 265–275.

By Dénes König

Theory of Finite
and Infinite Graphs

Foreword

Graph theory can be conceived from two different standpoints. First, like the theory of one dimensional complexes, it forms the first part of general topology. Secondly, it can be conceived, if one abstracts from its continuous - geometrical content, as a branch of combinatorics and abstract set theory. This book takes the second standpoint mainly because we do not ascribe to the elements of graphs, points and edges, any geometrical content at all: the points (vertices) are arbitrary distinguishable elements, and an edge is nothing other than a collection of its two endpoints. This abstract conception, which Sylvester emphasized (1873), will be strictly adhered to in our presentation, with the exception of several examples and applications. The geometrical notation, which we nevertheless use, gives a very convenient terminology without assuming any geometrical view or geometrical axioms. This graph theoretical terminology has a great heuristic value: it furnishes "natural" problems and connects quite abstract things with clear ideas, whereby new connections among concepts and problems seemingly distant from one another often come to light. Also this abstract - combinatorial conception is of advantage for geometrical topology, as is shown by the great development that multidimensional combinatorial (Poincaré - Veblen) topology has experienced in recent times. Certainly for complexes and manifolds of more than one dimension the avoidance of the "continuous" is connected with greater difficulties. For this reason we restrict ourselves, in order to preserve the quite elementary character of this book, to absolute graph theory, which considers graphs in themselves, while

"graphs in surfaces and spaces" (relative graph theory) stay out of
consideration. (§5 of Chapter XII forms an exception.) Furthermore,
the material is restricted by the fact that with few exceptions problems
concerning number determinations are not included in the book.

Just like most newer branches of mathematics graph theory has not
been created as an end in itself but in connection with older parts of
mathematics and the natural sciences. To name here the most important
works concerning graphs: the work of Kirchhoff (1847) owes its origin
to the theory of electricity, that of Petersen (1901) to the theory of
invariants, and that of Menger (1927) to the new theory of curves.
Through a work of P. Hertz (1922) formal logic was also at the source of
graph theoretical research. For Cayley and Sylvester chemistry was for
the most part the starting point of their research in graphs. Some
authors studied graphs for their own sake; along with G. Brunel and A.
Sainte - Laguë we mention especially H. Whitney, whose systematic
studies have taken place in recent years and are still not finished.

Perhaps graph theory owes more to the contact of mankind with
himself than to the contact of mankind with nature. The practical -
social problems of combinatorics, including intellectual games, are
chiefly treated in the literature of recreational mathematics, which
through this is being turned into a rich storehouse for graph theory.
Thus Euler (1736) was led to the problem of the Königsberg bridges,
which historically was the first treatment of graph theory to be
published. Edouard Lucas was the first, both mathematically and
historically, to bring recreational mathematics together in a truly
scientific work, his Récréations Mathématiques (1882-1894). So it is
only natural that he is one of those who recognized the importance of
the concept of graph (réseau) and sought to study and apply this
concept. The material collected and processed by Lucas was complemented

-- especially in an historical - bibliographical respect -- in an

exemplary work by Ahrens. The extraordinarily abundant and reliable

biographical information in Ahrens' Mathematischen Unterhaltungen was of

great use to me when I was collecting the very scattered material about

graphs. In this respect I should also mention the article by Dehn and

Heegaard concerning analysis situs in the Encyklopädie der

mathematischen Wissenschaften. There the history of "line systems" up

to the beginning of this century is assembled on seven pages in a

clearly arranged way. This presentation was able to serve me as an

outline. Also I was able to use some biographical information from the

memorial article of Sainte-Laguë concerning graphs.

I have put emphasis in my book on references to the literature and

have striven for completeness. Numbers appearing in brackets in the

text after authors' names refer to the bibliography at the end of the

book. For every place in the literature in which I have found a

concept, a theorem, or a proof that was of use to me in my presentation,

I have given exact references. There is an exception only in the first

chapter; here it seemed impossible to determine the first appearance of

the basic concepts and facts, since in the graphical interpretation of

graph theory, at least in what concerns finite graphs, this chapter

contains for the most part only the obvious. But as for the other

parts, I believe I am permitted to assert that my book contains in it

the history of graph theory. It is not difficult to determine which

results of the book the author considers as his own results. In what,

in particular, concerns infinite graphs I could cite only my own works.

The book has grown out of lectures which I gave often at the

Technische Hochschule in Budapest. For an understanding of its main

contents a certain background in mathematical thinking is necessary, but

not a knowledge of mathematical theorems and concepts, at least as far as the treatment of finite graphs is concerned. With some of the material concerning infinite graphs the elements of abstract set theory are assumed. In some places what is used of elementary number theory, linear algebra, theory of determinants, projective geometry, group theory, etc. will scarcely exceed the amount which can be expected of a would-be mathematician at the beginning of his university studies.

I must remember here with gratitude the late founder of this series, Prof. E. Hilb, who welcomed my book into the series when a short plan of the book was first ready. During the many years in which I worked on this book, I could always enjoy the valuable interest of my highly respected teacher, Prof. J. Kürschák, who is likewise meanwhile deceased. He read the greatest part of my manuscript and accompanied this with remarks, which were of the greatest value for the book. Also I was able to discuss some of my material with some other Hungarian colleagues of mine in Budapest and in Szeged. They are Messrs. E. Egerváry, P. Erdös, L. Fejér, T. Grünwald, G. Hajós, A. Haar (deceased), L. Kalmár, J. von Neumann, T. Schönberger, St. Valko, and P. Veress. I am indebted to them for their many suggestions and contributions, which are mentioned at the appropriate places of the book. Messrs. Hajós, Kalmár, and Veress have also read nearly all the proof-sheets, and with critical care. Mr. H. Nehrkorn (Hamburg) has very kindly helped me by reading the galley proofs. Mr. T. Grünwald has rendered me great service by producing the figures. To all those named I owe hearty thanks for the many suggestions for improvement and also to the Akademische Verlagsgesellschaft (Academic Publishing Society) for its coöperation in the printing of this book.

Budapest (Technische Hochschule) September, 1935 Dénes König.

Translator's note: The translator wishes to thank the Word Processing

Center of Loyola College in Maryland for their painstaking efforts in processing this manuscript.

June, 1989

Contents

Chapter I
Foundations

Chapter II
Euler trails and Hamiltonian cycles

Chapter III
The Labyrinth Problem

Chapter IV
Acyclic Graphs

Chapter V
Centers of Trees

Chapter VI
Infinite Graphs

Chapter VII
Basis Problems for Directed Graphs

Chapter VIII
Various Applications of Directed Graphs
(Logic - Theory of Games - Group Theory)

Chapter IX
(Directed) Cycles and Stars and the
Corresponding Linear Forms

Chapter X
Composition of Cycles and Stars

Chapter XI
Factorization of regular finite graphs

Chapter XII
Factorization of regular finite graphs of degree 3

Chapter XIII
Factorization of regular infinite graphs

Chapter XIV
Separating Vertices and Sets of Vertices

CHAPTER I

Foundations

§1. The Basic Concepts

Let {A, B, C ...} be a set of "points." If certain pairs of these
points are connected by one or more "lines", the resulting configuration
is called a graph. Those points of {A, B, C ...} which are connected
with at least one point are called vertices of the graph. (Vertices
which could be called "isolated" are therefore excluded.) The lines
involved are called edges of the graph[1]. An edge which connects A and
B, i.e. whose endpoints are A and B, and which goes to A (and B), we
shall designate by AB. It is possible that several edges are designated
as AB. If A is an endpoint of edge k, we shall also say that A and k
are incident to each other. If the set of vertices and the set of edges
of a graph are both finite, the graph is called finite, otherwise
infinite. An infinite graph has infinitely many edges but possibly only
finitely many vertices (e.g., two vertices can be connected by
infinitely many edges.)

For an intuitive interpretation of graph theory (especially in the
case of finite graphs) the vertices and edges can be considered to be
embedded in a metric (or possibly only topological) space. For example,
we can choose two - or three - dimensional Euclidean space, and the
edges can be interpreted as Jordan arcs. But this is by no means
necessary. We assume that the vertices are arbitrary elements and that
the edge AB has the single defining property that it determines its

1

endpoints, A and B.

The single "axiom" which we shall require is, in its most general form, as follows:

If M is an arbitrary finite or infinite set and if to every (unordered) pair (A, B), which can be formed from the elements of M, a finite or infinite cardinal number $m_{AB} = m_{BA}$, which can also be zero, is assigned such that for each A at least one m_{AB} is not zero, then there is a graph, which has as vertices the elements of M and in which every two vertices A and B are connected by m_{AB} edges.

For the theory of finite graphs this axiom is needed only in the case where M and the m_{AB}'s are finite. In the general theory of graphs neither the m_{AB}'s nor the cardinality of M is subjected to any restriction, and M can be, for instance, of cardinality higher than that of the continuum. In this case, of course, the graph cannot be considered to be embedded in a Euclidean space. Geometrical nomenclature, however, can be introduced purely formally even in this case without using a geometrical approach -- as is always done in this book. The great heuristic value of graph theory lies in the fact that its abstract concepts can be pictured by concrete spatial relations. But from a logical point of view graph theory has nothing to do with a spatial conception.

Outside of certain examples and applications no geometrical significance should be ascribed to the concepts of vertex, edge, and graph; in particular, edges are not to be thought of [2] as (infinite) point sets. Graph theory, as presented in this book as "absolute" graph theory, is pure combinatorics and pure (abstract) set theory. "Relative" graph theory, which considers graphs as subsets (of surfaces, spaces, etc.), and in particular, the treatment of "graphs on surfaces" lie outside the realm of this book (see, however, Chapter XII, §5).

In order to avoid exceptional cases we shall introduce in addition to proper graphs, the null (empty) graph, which has no vertices and no edges at all, just as the number zero is introduced in number theory and the null (empty) set in set theory.

We shall be a little imprecise sometimes in omitting the word "proper", but naturally only when it is clear from the context whether the null graph is excluded or not.

Two graphs, G and G', are considered to be identical in case their sets of vertices are identical, as well as their sets of edges. They are called disjoint if they have no vertex in common and no common edge. If the vertices of the graph G' are at the same time vertices of G and the edges of G' are also edges of G then G' is called a subgraph of G; and G' is called a proper or improper subgraph of G, depending on whether G' is different from G or not, respectively. We shall often designate a graph by giving its edges. It is always to be understood that we consider the endpoints of these edges, and only these endpoints, as the vertices of the graph. If, for example, we say that G' is obtained from G by removing the edge k from G, then the vertices of G' are the endpoints of the remaining edges of G.

The number (cardinality) of edges, which go to a vertex P of the graph G, is called the degree of P in G. A vertex of degree 1 is called an endpoint of the graph. An edge, which ends in an endpoint of the graph, is called an end edge of the graph. If a finite number of edges go to every vertex of a graph, the graph is said to be of finite degree. Infinite graphs of finite degree play an important role in that, as we shall see, some properties of finite graphs, which lose their validity (or even their meaning) for arbitrary infinite graphs, can be extended to graphs of finite degree. An infinite graph of finite degree, of course, always has infinitely many vertices.

3

By the definition of a graph we have assumed that every edge connects two distinct vertices with each other. Sometimes it is advisable to consider graphs in the broader sense, i.e. to allow also edges which connect a vertex P with itself. Such an edge is called a loop[3] and can be designated by PP. We remove the loop PP and introduce in place of it a new vertex Q and two new edges PQ (the loop is divided by Q into two edges, see Fig. 1). If this is done for each loop, then a graph G' without loops is obtained from the original graph G. Most properties of graphs, which will concern us, hold for both G and G'. For this reason, unless the opposite is explicitly stated, we will restrict ourselves to graphs in the narrower sense (i.e., without loops). Where it appears to be of interest, it will be mentioned

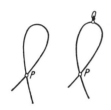

Fig. 1.

whether or not the theorem in question holds for graphs in the broader sense.[4]

An edge PQ can be assigned a direction in two different ways: P can be taken as the initial point and Q the terminal point, or vice versa. So the directed edge $\overrightarrow{PQ}$ or $\overrightarrow{QP}$ can be obtained from PQ, and we say the direction is towards Q or towards P, respectively. If a definite direction is assigned to every edge of a graph, which, if α denotes the number (cardinality) of edges, can happen in 2^{α} ways, a directed graph results. In the illustrative representation of a directed graph the direction of an edge can be indicated by an arrow.

For a given graph G let π denote the set of vertices of G, and let K denote the set of edges. Let π' and K' have corresponding meanings for a graph G'. If there is a one-to-one correspondence between the sets π and π' , on the one hand, and between K and K', on the other hand, in such a way that incident elements correspond to incident

4

elements, then graphs G and G' (regarded as undirected graphs) are said
to be isomorphic. We shall say that by stating the two one-to-one
correspondences with the property mentioned a (mutual) mapping of the
graphs on each other is defined. If, in particular, G is identical to
G', we speak of a mapping of G into itself. As long as disjoint graphs
are considered, only the structure of the graph is of interest, and
isomorphic graphs can be considered as identical: they are different
only in the designation of their vertices and edges. As for directed
graphs we shall agree that two directed graphs can be called isomorphic
if to the initial point of each edge the initial point of the
corresponding edge corresponds. We consider the set $M = \{\psi_\alpha\}$ of
mappings of a graph G into itself. A mapping ψ_α is determined by two
permutations (of the vertices, edges, respectively) in such a way that
the product $\psi_\alpha \psi_\beta$ of two elements ψ_α, ψ_β of M has a definite meaning and
likewise represents a mapping of G into itself. The inverse ψ^{-1} is also
a mapping of G into itself. Consequently M is a group, which as an
abstract graph can be called the group of the graph G. In this
connection the following problem can be studied (which we shall not do
in this book). When can a given abstract group be interpreted as the
group of a graph and if this is the case, how can the corresponding
graph be constructed? This same question can be asked for directed
graphs.

Next to isomorphism there is a second relationship between graphs
which plays an important role (which, in a corresponding generalization,
forms a basis for all topology). In order to explain it, a new
definition is introduced. If one of the edges, PQ, of the graph G is
removed and a new vertex R is introduced, as well as two new edges PR
and RQ (as already considered above for loops) we shall say that the new
graph G' obtained arises as a result of a <u>subdivision</u> of the edge PQ of

5

G. Now we wish to designate two graphs G and G' as homeomorphic to each other provided that they are either isomorphic to each other or can be transformed by arbitrarily many subdivisions of edges into isomorphic graphs. This concept of homeomorphism will not concern us very much; it is, however, important in the continuous topological conception of graphs, where the edges are interpreted as continua -- as topological images of straight lines.

From what has been said above, we wish to emphasize particularly the following:

Theorem 1 A set of elements Π and a set of elements K can be the set of vertices and set of edges, respectively, of a (unique) graph if and only if the endpoints of each edge of K belong to Π and every element of Π is an endpoint of an edge of K.

§2. Walks

If all the edges of a (finite) graph can be listed in the form

$$AB, \; BC, \; CD, \; \ldots, \; KL, \; LM \qquad\qquad (1)$$

where each vertex and each edge can occur arbitrarily (finitely) often, then the graph is characterized as a walk. The walk is called open or closed depending on whether $A \neq M$ or $A = M$, respectively. In the former case we say that the vertices A and M are joined with each other, or also that A and M are the endpoints of an open walk. The remaining vertices, as well as every vertex of a closed walk, are called interior vertices. If an edge occurs in (1) n times, we call n the multiplicity of this edge with respect to the walk. The following theorem is obvious.

Theorem 2 The sum of the multiplicities of the edges going to

6

the same vertex P of a walk is even or odd, depending on
whether P is an interior vertex or an endpoint of the walk,
respectively.

The theorem also holds, of course, for graphs in the broad sense,
provided that loops are counted twice.

If no edge occurs twice in (1), i.e. if every edge is of
multiplicity 1,[5] the walk is called a (closed or open) _trail_. If all
vertices A, B, ..., L, M are distinct from one another the walk is
called a _path_; if A = M, but A, B, ..., L are distinct from one another
the closed walk is called a _cycle_.

If the number of edges of a path is even, the path has a middle
vertex, and if the number is odd, it has a middle edge. The simplest
path has one edge alone. The simplest cycle is formed (if loops are
excluded) by two edges, which join the same two vertices with each
other. Such a cycle is called a _2-cycle_. A cycle with n edges is
called an _n-cycle_. If one edge AB is removed from a closed walk (cycle)
the remaining edges form an open walk (path), which joins A with B. If
two paths join the same two vertices and if they have no other common
vertices and no common edges, then the edges of these two paths form a
cycle. Also, vice versa, if P and Q are two arbitrary vertices of a
cycle, then the cycle consists of the edges of two paths, which join P
and Q and which have no other common vertices and no common edges. If n
edges, which pairwise have no common vertex, are removed from a cycle,
then the remaining edges form n paths, which are pairwise disjoint from
one other. If P is an endpoint (PQ an end edge) of a graph G, this
holds also with respect to every subgraph G' of G in which P (PQ) is
contained. A closed trail has neither endpoints nor end edges. So
neither an endpoint nor an end edge can be contained in a closed trail
of a graph.

We prove now

> Theorem 3 If certain edges of a graph form an open walk which
> joins the vertices A_1 and A_n with each other, then there are
> edges of this walk which form a path, which also joins A_1 with
> A_n.

Proof Let A_1A_2, A_2A_3, ..., $A_{n-1}A_n$ or more briefly written

$$A_1A_2 \cdots A_n \qquad\qquad (A)$$

be formed from certain edges of the walk in such a way that the
resulting walk joining A_1 with A_n has a minimum number of edges. Now
if, for $i < k$, it could be that $A_i = A_k$ then there would result the walk

$$A_1A_2 \cdots A_iA_{k+1} \cdots A_n$$

joining A_1 with A_n which would have fewer (by $k-i$) edges. So if
$A_i = A_k$ (i, $k = 1, 2, \ldots, n$) then it follows that $i = k$. But then
$A_iA_{i+1} \neq A_kA_{k+1}$ for $i \neq k$. Otherwise $A_i = A_k$ (and $A_{i+1} = A_{k+1}$) which
contradicts the assumption that $i \neq k$; or $A_i = A_{k+1}$ and $A_{i+1} = A_k$, from
which it follows that $i = k + 1$ and $k = i + 1$, which is impossible. In
(A) neither a vertex nor an edge can occur twice, and hence (A) is a
path.///

In a similar way the following can be proved.

> Theorem 4 If certain edges of a graph form a closed trail,
> which contains the vertex A, then there is a cycle which can be
> formed from certain edges of this trail, which also contains A.

In this theorem the word "trail" cannot be replaced by
"walk".[6]

> Theorem 5 If A, B, and C are three different vertices of a
> graph and there is a (open) walk F_1, which joins A with B, and
> a (open) walk F_2, which joins B with C, then there are certain
> edges, which are in F_1 or F_2, which form a path, which joins A
> with C.

<u>Proof</u> Let the edges of F_1 be followed by the edges of F_2, so that there results a walk F, which joins A with C. Now W does not need to be a trail, much less a path, since the same edge can occur more than once. But by Theorem 3 certain edges of F form a path joining A with C.///

<u>Theorem 6</u> If there is a path W = $P_1 P_2 P_3 \ldots$ which joins P_1 with an endpoint of the edge k then there is a path

$P_1 Q_2 \ldots Q_{n-1} Q_n$ where $Q_{n-1} Q_n = k$.

<u>Proof</u> If in the path W = $P_1 P_2 P_3 \ldots P_r$ is the first vertex which is an endpoint of k then we get a path of the desired kind by adjoining to the subpath $P_1 P_2 \ldots P_r$ the edge k at the end.///

<u>Theorem 7</u> If W_1 and W_2 are two distinct paths, which join the vertices P and Q with each other, then there are certain edges, which belong either to W_1 or to W_2, which form a cycle.

<u>Proof</u> Let $W_1 = A_1 A_2 \ldots A_n$, $W_2 = B_1 B_2 \ldots B_m$, $A_1 = B_1 = P$, $A_n = B_m = Q$, $n \le m$. (See Fig. 2). There is a first edge $A_\alpha A_{\alpha+1}$ in W_1, which is distinct from $B_\alpha B_{\alpha+1}$, since otherwise W_1 and W_2 would be the same path, if n=m, or, if n<m then we would have that $A_n = B_n = B_m = Q$, which would contradict the definition of path (for W_2). Let A_β be the first among the vertices $A_{\alpha+1}$, $A_{\alpha+2} \ldots$, A_n, which is the same as one B_γ of the vertices $B_{\alpha+1}$, $B_{\alpha+2}$, $\ldots$, B_m. (In the extreme case where β=n we have that $A_n = B_m$). Now the edges

$A_\alpha A_{\alpha+1}$, $A_{\alpha+1} A_{\alpha+2}$, $\ldots$, $A_{\beta-1} A_\beta$,
$B_\gamma B_{\gamma-1}$, $B_{\gamma-1} B_{\gamma-2}$, $\ldots$, $B_{\alpha+1} B_\alpha$

form a closed walk. But this is also a cycle, since the A's are distinct from one another and the B's are distinct from one another and also

$$A_i \ne B_k \text{ for}$$
$$i = \alpha + 1, \alpha + 2, \ldots, \beta - 1;$$
$$k = \alpha + 1, \alpha + 2, \ldots, \gamma$$

9

since otherwise such an A_i would have been chosen instead of A_β . (The simplest case is $\beta = \alpha + 1 = \gamma$, because then the two distinct edges, which join $A_\alpha = B_\alpha$ with $A_{\alpha+1} = B_{\alpha+1}$, form the desired cycle.)///

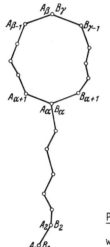

Fig. 2.

Similar reasoning leads us to

Theorem 8 If certain edges of a graph G are distinguished from the remaining edges in such a way that every cycle of G contains an even number of distinguished edges, then the number of distinguished edges in each of two paths of G, which join the same two vertices, always has the same parity.

Proof If the theorem were not true, then there would be a path W_1 with a minimum number of edges with the property that there could be found a path W_2 joining the same two vertices, P and Q, for which the parity of the number of distinguished edges is different from that of W_1. In order to show that this is impossible we use the notation of the preceding proof and distinguish two cases.

Case 1. $\alpha > 1$.

So $A_1 A_2 = B_1 B_2$, $A_2 = B_2$. The paths $A_2 A_3 \ldots A_n$ and $B_2 B_3 \ldots B_m$ join the same vertices $A_2 = B_2$ and $A_n = B_m = Q$. The first of these paths has one less edge than W_1. The parity of the number of distinguished edges is therefore by induction assumption the same for these two paths: but then this must hold also for W_1 and W_2 as well, whether $A_1 A_2 = B_1 B_2$ is a distinguished edge or not.

10

Case 2. $\alpha = 1$.

Then the parities for the paths

$A_\beta A_{\beta+1} \cdots A_n$ and $B_\gamma B_{\gamma+1} \cdots B_m$ are the same by induction

assumption. On the other hand this also holds for the paths

$A_1 A_2 \cdots A_\beta$ and $B_1 B_2 \cdots B_\gamma$ (since the edges of these paths form

a cycle). But then this must also be the case for W_1 and W_2.///

Theorem 9 If in a graph two distinct cycles K_1 and K_2 both

contain the same edge PQ, then a new cycle can be formed from

certain edges of these cycles, which does not contain the edge

PQ.

Proof If the edge PQ is removed from both K_1 and K_2, two different

paths are obtained which join P with Q. According to Theorem 7 there is

then a cycle of the desired type.///

Theorem 10 If Z_1 and Z_2 are two closed trails, which do not

contain a common edge but do contain at least one vertex P in

common, then all the edges, which belong to Z_1 or Z_2, form a

closed trail.

Proof It is clear how both trails need to be "cut up" and "put

together" again, in order to obtain a closed trail.///

Theorem 11 If in a graph there are a cycle K_1, which contains

the edges k_1 and k_2, and a cycle K_2, which contains the edges

k_2 and k_3, then there is also a cycle, which contains k_1 and

k_3.

Proof Let $k_3 = RS$; then the cycle K_2 is of the form

$R_\mu \cdots R_2 R_1 R S S_1 S_2 \cdots S_\nu R_\mu$, where $k_2 = S_\nu R_\mu$. Let R_α be the first

vertex in the walk $R R_1 R_2 \cdots$ which belongs to K_1. Let S_β be the first

vertex in the walk $S S_1 S_2 \cdots$ which belongs to K_1. The cycle K_1 is

divided by R_α and S_β into two paths, one of which contains k_1.

This forms with the path $R_\alpha R_{\alpha-1} \cdots R_1 R S S_1 S_2 \cdots S_\beta$ a cycle of the

11

desired type.///

Two more concepts are needed for the study of infinite graphs, which we shall introduce now. An infinite set of edges $P_i P_{i+1}$ (i = 0, 1, 2, ..., ad inf.) and the graph formed by them is called a singly infinite path provided $P_i \neq P_j$ for $i \neq j$. Under the same condition the edges $P_i P_{i+1}$ (i = 0, $\pm$ 1, $\pm$ 2, . . . , ad inf.) form a doubly infinite path. (We emphasize that the designation "path" will always be used in its meaning already defined and therefore will always mean a finite graph. Also a walk, unless otherwise indicated, will always be finite.)

If K is the set of edges of a graph G, we shall be interested later in subsets K' of K, which have the property that every vertex of G has one and only one edge of K' going to it. In this case we shall say that the subgraph G' of G consisting of the edges of K' is a 1 - factor of G (or factor of degree 1 of G).[7] The following theorem concerns the existence of a 1 - factor (factor of degree 1) for the simplest graphs.

Theorem 12   If G is a graph which is a cycle with an even number of edges or a singly infinite path or a doubly infinite path, then G has a 1 - factor.

Proof Let K be the set of edges of G. A subset K' of K with the desired property can be obtained in the following way. First of all, if G is a cycle $P_1 P_2$ . . . $P_{2n} P_1$ then K' can be taken to be $\{P_1 P_2, P_3 P_4, \ldots, P_{2n-1} P_{2n}\}$. Secondly, if G is a singly infinite path $P_1 P_2 P_3$ . . . then K' can (and must) be taken to be $\{P_1 P_2, P_3 P_4, \ldots\}$. Finally if G is a doubly infinite path $\ldots P_{-2} P_{-1} P_0 P_1 P_2$ . . .   K' can be taken to be $\{\ldots P_{-3} P_{-2}, P_{-1} P_0, P_1 P_2, P_3 P_4, \ldots\}$. (In the first and third cases above the complementary set of E to the one given can also be chosen. No other choice, however, is possible.)///

12

§ 3. Connected Graphs

If for any vertices, P and Q, in a graph G there is a path which
joins P and Q, then G is said to be connected. In order to avoid having
to assume that P ⊧ Q, we shall agree that in a graph every vertex can be
joined with itself by a path; this convention allows us to avoid
exceptional cases. We shall call a graph disconnected if it is not
connected. As examples of connected graphs there are, according to
Theorem 3, open and closed walks and also infinite paths. The following
four theorems are simple consequences of the definition of "connected"
and are often used.

> **Theorem 13** If in a connected graph G there is a path W which
> joins the endpoints A and B of an edge AB and W does not
> contain this edge AB then the graph G', obtained by removing
> the edge AB from G, is also connected.

Proof We must show that any two vertices, P and Q, of G' are joined by
a path in G'. Let V be a path in G which joins P with Q. If the edge
AB is not contained in V then V is the desired path of G'. But if V
contains AB, we can replace AB by the edges of the path W; we thus
obtain an open walk of G', which joins P with Q. According to Theorem 3
there is then a path of G', which joins P with Q.///

> **Theorem 14** If an end edge or an edge of one of its cycles is
> removed from a connected graph G, and the resulting graph is
> called G', then G' is also connected.

Proof Let A be an endpoint and let AB be an end edge of G. So every
path of G which contains AB has the form AB P_1 P_2 . . . P_n and joins A
with a vertex of G. Since A does not belong to G', any path of G which
joins two vertices of G' can not contain AB. Such a path is therefore
contained in G'. If AB belongs to a cycle of G, the edges of this cycle

other than AB form a path of G, which joins A with B. Then according to
Theorem 13, G' is connected.///

> Theorem 15 Let G' be a disconnected subgraph of the connected
> graph G. Then there is in G an edge not belonging to G' with
> the property that every cycle of G which contains this edge
> also contains at least one other edge of G not contained in G'.

Proof Let A_1 and P be two vertices of G' which cannot be joined by a
path in G'. There must be a path in G from A_1 to P. Let $A_1 A_2 \ldots A_n$
(where A_n = P) be such a path, and let it be chosen in such a way that
the number ν of its edges not belonging to G' is minimal. In this path
there is certainly an edge $A_i A_{i+1}$ not belonging to G'. This is the
edge we are seeking. Otherwise there would be in G a cycle

$$A_i A_{i+1} B_1 B_2 \ldots B_m A_i$$

in which, except for $A_i A_{i+1}$, there are edges only of G'. But then the
walk $A_1 A_2 \ldots A_{i-1} A_i B_m B_{m-1} \ldots B_1 A_{i+1} A_{i+2} \ldots A_n$
and hence the path (Theorem 3) from A_1 to A_n=P contained in this walk
contains fewer than ν edges which do not belong to G'. And this
contradicts the minimal property of ν .///

> Theorem 16 If G' is an arbitrary subgraph of the connected
> graph G and P is any vertex of G not belonging to G' then there
> is a path W = P $P_1 P_2 \ldots P_{n-1} P_n$ of G with the property that
> P_n belongs to G' but neither the other vertices of W nor its
> edges belong to G'.

Proof Let Q_ν be an arbitrary vertex of G' and let P $Q_1 Q_2 \ldots Q_\nu$ be a
path in G which joins P with Q_ν. If Q_i ($i \le \nu$) is the first vertex
belonging to G' in the sequence $Q_1, Q_2, \ldots, Q_\nu$, the subpath
P $Q_1 Q_2 \ldots Q_i$ of this path fulfills the conditions of the theorem.
This is clear for the vertices; but also an edge of P $Q_1 Q_2 \ldots Q_i$
cannot belong to G', since then both of its endpoints would also belong

14

to G', whereas Q_i is the only vertex of P Q_1 . . . Q_i belonging to
G'.///

A metric element is introduced into the theory of connected graphs
by defining a "distance" (as a natural number) between any two vertices
of a connected graph. (The graph may be finite or infinite.)

We define the length of a path as the number of edges it
contains. We designate the length of the shortest path joining two
vertices as the distance between these vertices. The use of the word
"distance" is justified by the the fact that the triangle inequality is
true, which we can formulate as follows:

> Theorem 17   If the distance between vertices a and b, b and c,
> and c and a are d_1, d_2, and d_3, respectively, then
> $d_1 + d_2 \geq d_3$.

Proof If the path W_1 of length d_1 joins the vertices a and b with each
other and the path W_2 of length d_2 joins the vertices b and c with each
other, then by Theorem 5 there is a path joining vertices a and c, which
contains only edges from W_1 and W_2. So the length of this path is, of
course, $\leq d_1 + d_2$. So the length of the shortest path joining a and c
is then $\leq d_1 + d_2$.///

The greatest distance between vertices of a graph, if it exists, we
call the diameter of the graph. This is a "maximum minimorum" and is
for finite graphs a well defined natural number, since finite graphs
have only finitely many paths.

§4. Components

If a graph is not connected, it can be "divided into well-defined
parts." The sense and correctness of this assertion, at least as far as
finite graphs are concerned, is intuitively clear immediately. This

15

assertion should now be explained and proved without any geometric intuition, in order that both finite and infinite graphs can be dealt with.

If there is a path in the graph G which joins the vertices A and B with each other we shall for the time being express this by

A R B.

By doing this a binary relation R for the set π of vertices of G is completely defined. It has the following three properties:

α) A R A for all A in π. (Reflexivity)

β) A R B implies B R A. (Symmetry)

γ) A R B and B R C imply A R C. (Transitivity)

The last property is a consequence of Theorem 5. It is well known that from these three properties (which characterize the relation R as a so-called equivalence relation) it follows that π can be uniquely partitioned into pairwise disjoint classes in such a way that two elements, A and B, of π belong to the same class if and only if A R B.[8] Let $\pi = \Sigma \, \pi_\alpha$ be this partition, where the set of summands as well as the summands themselves can be infinite (of arbitrary cardinality).

If π_α is an arbitrary summand, we shall let K_α denote the set of those edges of G which join two vertices of π_α. If an endpoint of an edge k of G belongs to π_α, then the other endpoint of k also belongs to π_α (k is itself a path); every edge of G thus belongs to one and only one of the sets K_α. According to Theorem 1 the vertices of π_α and the edges of K_α form a subgraph G_α of G. We show that each G_α is connected. If A and B are two arbitrary vertices of G_α (and therefore of π_α), there is by definition of π_α a path $W = A \, P_1 \, P_2 \, \ldots \, P_n \, B$ in G which joins A with B. Then $A \, P_1 \, P_2 \, \ldots \, P_i$ (i = 1, 2, . . ., n) are also paths in G, so that the vertices of W and thus also the edges of W belong to G_α. Hence G_α is connected.

The connected proper subgraphs G_α of G introduced in this way we shall designate as <u>components</u> of G. The following theorems are obvious:

Theorem 18 Two different components of a graph are mutually disjoint.

Theorem 19 Every vertex and every edge belong to one and only one component. Two vertices (two edges) belong to the same component if and only if there is a path to which they belong.

Theorem 20 A graph is connected if and only if it has only one component; this is then the graph itself.

Here we shall introduce the concept of <u>addition of graphs</u>. Let $\{G_\alpha\}$ be an arbitrary set of graphs. According to Theorem 1, those vertices and edges which are contained in at least one of the graphs G_α form a graph G. If the graphs G_α are pairwise disjoint, we designate G as the <u>sum of graphs</u> G_α and write $G = \Sigma\, G_\alpha$. The assumption that the G_α's are disjoint from one another is essential here; only in this case can we say that a graph splits into subgraphs G_α and only in this case can we write equations of the form $G = \Sigma\, G_\alpha$. It should be noted that not every subgraph G' of G gives rise to an equation $G = G' + G''$. By the way that we have introduced the concept of a graph, the endpoints of the edges must be vertices of the graph; thus not every subgraph G' of G can have a "complementary" subgraph G" with the property that $G = G' + G''$: those vertices and edges of G, which do not belong to G', do not generally form a graph.

Let $G = \Sigma\, G_\alpha$ and let $W = P_1\, P_2\, \ldots\, P_n$ be a path of G. We assume that P_1 belongs to the summand G_1; then the edge $P_1\, P_2$ must likewise belong to G_1, since otherwise (Theorem 1) its endpoint P_1 would have to belong to another summand. But then the other endpoint P_2 must also belong to G_1. Continuing in like manner we get step by step that $P_2\, P_3$, P_3, $P_3\, P_4$, $\ldots$, $P_{n-1}\, P_n$, P_n also belong to G_1. From this we have

17

Theorem 21   If $G = \Sigma\, G_\alpha$ and two vertices belong to different

summands, then there is no path in G which joins these two

vertices with each other.

Theorem 18 and the first part of Theorem 19 can now be expressed as

follows:

Theorem 22   If $\{G_\alpha\}$ is the set of components of the graph G

then $G = \Sigma\, G_\alpha$.

So every graph splits into connected subgraphs. We show now that

this is always possible in only one way. Let $\{H_\alpha\}$ be a set of pairwise

disjoint connected graphs and let $G = \Sigma\, H_\alpha$. On the other hand

let $\{G_\alpha\}$ be the set of components of G. Our assertion will be proved if

we show that every H_α is a G_α and vice versa.

Let H_α be an arbitrary element of $\{H_\alpha\}$ and let P be a vertex

of H_α. If there is a path of G going from P to Q, then by Theorem 21

the vertex Q also belongs to H_α. If there is no such path, Q belongs to

another summand since H_α is connected. Therefore H_α contains exactly

the same vertices as that component of G --- call it G_α --- which

contains P. But the edges of H_α also are the same as the edges of G_α.

Since, according to Theorem 1, every edge of G belongs to that

summand H_α to which its endpoints belong, H_α must --- exactly as G_α does

by definition --- contain exactly the same edges which join two vertices

of H_α (i.e. of G_α). Therefore $H_\alpha = G_\alpha$.

On the other hand an arbitrary component G_α is identical to one of

the summands H_α, since otherwise the vertices of G_α would be contained

in none of the H_α. By this reasoning we have obtained the following

result:

Theorem 23   Every graph splits in one and only one way into

connected subgraphs, namely, into its components.

18

§5. Further Conclusions

In order to characterize connected graphs in another way we prove

Theorem 24 A graph is connected if and only if it is not the

sum of two graphs.

Proof If $G = G_1 + G_2$, a vertex of G_1 and a vertex of G_2 cannot be

joined by a path in G by Theorem 21. If, on the other hand, G is not

connected, it has by Theorem 20 at least two components; if one of these

is G_1 while the sum of the remaining ones is G_2, then $G = G_1 + G_2$.///

The property expressed in the theorem just proved is also suitable

as a definition of a connected (indecomposable) graph. The same holds

for the next two theorems, which also express a characteristic property

of a connected graph.

Theorem 25 If all the vertices of a connected graph are

partitioned in an arbitrary way into two or more nonempty

pairwise disjoint classes, then there is an edge whose

endpoints belong to different classes. If, however, the graph

is disconnected it is possible to partition its vertices into

at least two nonempty classes such that every edge joins

vertices of the same class.

Proof In order to prove the first part of this theorem, let

$P_1 P_2 \ldots P_n$ be a path which joins the vertices P_1 and P_n belonging to

two different classes. If in the sequence P_2, P_3, . . ., P_n the vertex

P_r is the first one which belongs to the same class as P_n then the edge

$P_{r-1} P_r$ has the desired property. As for the second part of the

theorem, let the graph be disconnected, so that by Theorem 20 it has at

least two components. If the vertices are distributed to different

classes, according to whether they belong to different components, we

have a partition of the desired kind.///

Theorem 26   A graph is connected, if and only if it can be a
subgraph of a sum of graphs only when it is a subgraph of one
of the summands.

Proof   If, first of all, the graph G is disconnected then it is by
Theorem 24 the sum of two proper subgraphs: $G = G_1 + G_2$, and G is a
subgraph of itself without being a subgraph of G_1 or G_2.

If, secondly, G is a subgraph of $H = \Sigma H_\alpha$, without being a
subgraph of one of the summands, then there are two different summands,
which each contain a vertex of G. Then by Theorem 21 these two vertices
cannot be joined by any path in H, much less in G. So G is
disconnected.///

It is easy to recognize in Theorems 23,24 and 26 an analogy with
the first theorems of elementary number theory; connected graphs
correspond to prime numbers.

With regard to components of the simplest graphs we now prove

Theorem 27   If at most two edges go to any vertex of a graph G,
then every component of G is either

	1.	a path
or	2.	a cycle
or	3.	a singly infinite path
or	4.	a doubly infinite path.

Proof   Let P be an arbitrary vertex of G. There is in G an edge $P\ P_1$,
possibly another uniquely determined edge $P_1\ P_2$, then possibly a
uniquely determined edge $P_2\ P_3$ distinct from $P_1\ P_2$, etc. In the
sequence P_1, P_2, P_3, . . . none of these vertices can appear a second
time, since otherwise there would be three edges going to the same
vertex. Now there are two cases possible. First, one of these vertices
can be identical to P, and then these edges form a cycle. In the second
case, which we would now like to consider, they form a finite or singly

infinite path according to whether the sequence P_1, P_2, . . . stops or

not. Now there is possibly an edge PQ_1 distinct from P P_1, possibly an

edge Q_1 Q_2, distinct from P Q_1 etc. (these edges, in case they exist,

are also uniquely determined). Also in the sequence Q_1, Q_2, Q_3, . . .

no vertex can appear a second time. For the same reason none of the

vertices Q_1, Q_2, . . . can be identical to any of the vertices

P, P_1, P_2, . . . and also not identical to the possibly last vertex P_i

of this sequence, since otherwise this sequence could not stop with

P_i. Also the edges P Q_1, Q_1 Q_2, Q_2 Q_3, . . . form in this second case

either a finite or a singly infinite path, which has no vertex, except

for P, in common with the finite or infinite path P P_1 P_2 . . . and

therefore also no common edge. The combined set of edges

$$. . ., Q_3\ Q_2,\ Q_2\ Q_1,\ Q_1\ P,\ P\ P_1,\ P_1\ P_2,\ P_2\ P_3, . . .$$

(which is either finite or of either ordinal number ω or $^{*}\omega + \omega$) forms

either a finite or singly infinite or doubly infinite path. So we have

the following result:

For every vertex P of G a subgraph G' of G containing the vertex P

can be uniquely [9] determined, which belongs to one of the four types

mentioned in the theorem. In cases 1 and 3 no second edge of G runs to

the endpoints (or endpoint, respectively) of G'.

There remains only to show that in all four cases G' is the

component G_p of G containing the vertex P. By Theorem 26 we have that

G' is a subgraph of G_p. We assume that there is an edge or a vertex ---

and so also an edge RS --- of G_p, which does not belong to G'. We may

assume that one endpoint R of this edge belongs to G'. (If the latter

were not the case, then the edge $P_n\ P_{n-1}$ of Theorem 16, where we replace

G now by G_p and P by R, would be such an edge.) And this is impossible,

since if R is not an endpoint of G' then RS would be a third edge going

to R. But if R is an endpoint of G' (which is possible only in Case 1

21

or Case 3) we have seen that no second edge of G can go to R.

By the above we have completely proved Theorem 27, which appears to be almost intuitively obvious.///

If exactly two edges run to each vertex then neither a path nor a singly infinite path can occur among the components of the graph. So we have

> Theorem 28 If exactly two edges go to every vertex of a graph, then the graph is the sum of cycles and doubly infinite

> paths. For finite graphs there can, of course, be only cycles.

Like Theorem 2, Theorems 27 and 28 retain their validity also for graphs in the broad sense, provided that loops are counted twice. The rest of the theorems of this chapter are easily seen to be valid for graphs in the broad sense.

Notes on Chapter I

[1] The word <u>graph</u> was probably first used in this sense in 1878 by Sylvester [3, p. 149] As a short, precise designation workable in all languages it appears to us to be the most suitable. In the same or similar sense the following expressions have been used: <u>Linearkomplexion</u>, <u>Liniensystem</u>, <u>Streckenkomplex</u> (line complex), <u>Kantenkomplex</u> (edge complex), <u>Netz</u> (net). The French use mostly: <u>réseau</u>.

For <u>vertex</u> there is in English: <u>node</u>, <u>knot</u>, <u>vertex</u>; in French: <u>sommet</u>, <u>noeud</u>, <u>carrefour</u> have been introduced and for <u>edge</u> in English: <u>branch</u>, <u>edge</u>; in French: <u>chemin</u>, <u>arête</u>, <u>lien</u>.

[2] For this concept the following words of Sylvester [1, p. 23] are characteristic: "The theory of ramification is one of pure colligation, for it takes no account of magnitude or position; geometrical lines are used, but have no more real bearing on the matter than those employed in genealogical tables have in explaining the laws of procreation."

[3] In French the designations <u>boucle</u>, <u>impasse</u>, <u>autoliaison</u> have been introduced for this.

[4] For some reduction processes (for example, for the one which will be treated in Chapter XII, §§2 and 3) it would be advisable to introduce graphs in a still broader sense, namely, also to admit edges without endpoints, which would appear in the graphic representation as isolated circles. In this book we do without them.

[5] It is also possible in this case that an edge symbol PQ appears more than once in (1), but then PQ must denote distinct edges, which join P with Q.

[6] But the theorem also holds for a walk, if it has at least one edge going to A of odd multiplicity. We omit here the proof of this assertion, which I owe to a kind communication from Mr. P. Veress.

23

[7] Factors of higher degree will be introduced later (Chapter XI).

[8] In most mathematical disciplines which are concerned with infinite sets, this theorem is important and, in fact, indispensable. Most of the time it is used implicitly. A clear presentation is found in Hasse, Higher Algebra, I, 1926, pp. 17-18. Like Zermelo's Axiom of Choice this fruitful theorem can be attacked and doubted (even with the addition of uniqueness) from the standpoint of those who accept an existence proof only if it is constructive; cf. Lusin, Sur les ensembles analytiques, §65 (l'axiome du partage), Fundamenta Mathematicae, 10, 1927, p. 83. In this book it is used, just like the Axiom of Choice, unhesitatingly.

[9] Certainly the edge PP_1 was not uniquely determined, if a second edge going to P, PQ_1, exists. But it can be immediately seen that the same subgraph G' is reached, if PQ_1 is chosen instead of PP_1.

CHAPTER II

Euler Trails and Hamiltonian Cycles

§1. The Euler Theorem and Related Theorems

Some questions, which relate to certain mathematical games and to which graph theory and especially topology owe their first interest, lead to the question of when a graph can be represented as a closed trail. If a closed trail contains all the edges of the graph, then it can be called an Euler trail of the graph. In the above mentioned question the so-called Euler graphs play an important role. A graph is called an Euler graph if it is finite and all its vertices are of even degree.[1]

For such graphs we have the following:

Theorem 1 Every vertex Q_1 of an Euler graph is contained in a cycle in the graph.

Proof If there is an edge $Q_1 Q_2$, there is also an edge $Q_2 Q_3$, possibly a third edge $Q_3 Q_4$, etc. The trail $Z = Q_1 Q_2 \ldots Q_n$, in case $Q_n \neq Q_1$, can be continued with an edge $Q_n Q_{n+1}$ since in Z, according to Theorem I.2,[2] an odd number of edges go to Q_n, while in an Euler graph there is an even number of such edges. Since the number of edges is finite, the sequence $Q_1 Q_2$, $Q_2 Q_3$, $\ldots$ must end, and so finally we have $Q_m = Q_1$; hence $Q_1 Q_2 \ldots Q_{m-1} Q_1$ is a closed trail. Then according to Theorem I4 there is a cycle which contains Q_1.///

As an answer to the question posed above we now prove a theorem[3] which originated with Euler [1]:

25

Theorem 2   All the edges of a graph can be in a closed trail if and only if the graph is a connected Euler graph.

Proof   This condition is necessary since a closed trail is finite and connected and, by Theorem I.2, the degree of every vertex is even.

In order to show that the condition is also sufficient, let G be an arbitrary connected Euler graph and let Z be a closed trail in G with a maximum number of edges. We must show that Z contains all the edges of G.

We assume that this is not the case. If the edges of Z are removed from G then the remaining edges form a proper subgraph G' of G. G' is an Euler graph, because by Theorem I.2 there is an even number of edges going to each vertex. Z and G' contain a common vertex Q, for otherwise we would have that G = Z + G', and then by Theorem I.24 G could not be connected. Theorem 1 asserts that there exists a cycle K of G', which contains the vertex Q. According to Theorem I.10 there is a closed trail of G which contains not only the edges of Z but also those of K. This contradicts the assumption that Z contains the maximum possible number of edges.///

With the help of Theorem 1 it can be seen that an Euler graph is a cycle if and only if it contains no proper Euler subgraph. The concept of cycle (circuit) is defined by Veblen and Alexander [1, §7] by this property.

In order to give an example for the Euler Theorem 2, we consider a finite number ν of straight lines in the projective plane. If we consider the points of intersection as vertices and the (finite or infinite) line segments, into which these lines are divided by the points of intersection, as edges, then this system of lines can be represented as a graph (possibly in the broad sense). It is obviously always a connected Euler graph and therefore is a closed trail.

26

If $\nu = 4$ and no three lines have a common point then we get Fig. 3.   If the infinite line segments are replaced by "finite" edges, Fig. 4 is obtained, where only the numbered points are considered vertices.[4]

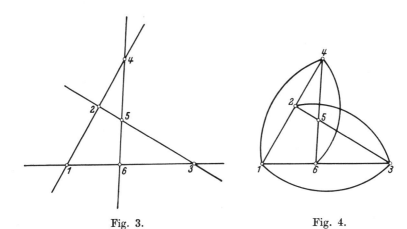

Fig. 3.   Fig. 4.

The graph is a closed trail as can be seen by the sequence 1 2 4 5 6 4 1 3 2 5 3 6 1.

In order to be able to answer similar questions for graphs which also contain vertices of odd degree, we now prove a theorem, which was first proved by Euler [1, §17].

   Theorem 3   In every finite[5] graph the number of vertices of odd degree is even.

Proof Let ρ_n be the number of vertices of degree n.   So, since every edge has two endpoints, double the number of edges satisfies the following equation:

$$2\alpha_1 = \rho_1 + 2\rho_2 + 3\rho_3 + 4\rho_4 + 5\rho_5 + 6\rho_6 + 7\rho_7 + \ \cdot \ \cdot \ \cdot$$

If we subtract from this even number the even number

$$2\rho_2 + 2\rho_3 + 4\rho_4 + 4\rho_5 + 6\rho_6 + 6\rho_7 + \ \cdot \ \cdot \ \cdot$$

the resulting difference

$$\rho_1 + \rho_3 + \rho_5 + \cdots$$

is also an even number. But this number

$$\rho_1 + \rho_3 + \rho_5 + \cdots$$

is the number of vertices of odd degree. This theorem can also be proved by induction on the number of edges, by simply noticing (Lucas [I, Vol. I, p. 239]) that if an edge is removed the number

$$\rho_1 + \rho_3 + \rho_5 + \cdots$$

is changed by -2, 0, or 2, and thus its parity does not change.///

We now prove a theorem which was first stated[6] by Listing [1, p. 60] without proof and proved by Lucas [1, Vol. I., p. 239].

Theorem 4   If a finite connected graph G contains exactly 2p (>0) vertices of odd degree, then there is a system

$$Z = (Z_1, Z_2, \ldots, Z_p),$$

consisting of p open trails, which has the property that every edge of G is contained in one and only one of the trails Z_i.

Any system with this property must contain at least p trails.

Proof   Let the vertices of G of odd degree be arbitrarily paired off:

$$P_1, P_1', P_2, P_2', \ldots, P_p, P'_p.$$

We introduce p new edges $P_1 P_1'$, $P_2 P_2'$, $\ldots$, $P_p P_p'$, not contained in G, and we call the resulting connected graph H.

Now all the vertices of H are of even degree, and so by Theorem 2 H constitutes a closed trail. If the p new edges are again removed, there results from this trail, since the new edges pairwise have no common endpoint, p open trails Z_i. The system $\{Z_1, Z_2, \ldots, Z_p\}$ is easily seen to have the desired property.

Let $\{Z_1, Z_2, \ldots, Z_r\}$ now be an arbitrary system with the desired property. If the vertex P is distinct from the endpoints of the Z_i (i = 1, 2, $\ldots$, r), then by Theorem I.2, the degree of P with respect to each Z_i, in which P is contained, is even. The sum of these degrees, which is the degree of P with respect to G, is therefore also even. Since the Z_i's have, in all, at most 2r endpoints, the number 2p of vertices of odd degree is at most 2r. So $r \geq p$.///

As an example the graph of Fig. 5, since it has eight vertices of odd degree, cannot be traversed in fewer than four trails, if every edge may[7] be traversed only once.

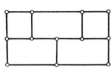

Fig. 5.

It can be seen immediately that Theorem 4 also remains valid for disconnected graphs, if every component of the graph contains vertices of odd degree.

The following interpretation can be given[8] to the problem solved by Theorem 4. Let the network of a trolley car rail system be given. The trolley car company wants to set up its "lines" in such a way that one and only one line travels on any given rail. How many lines, at least, must be set up?

The following theorem was stated by Euler [1, §18] and reduces to Theorem 2 in the following way.

Theorem 5   Every finite connected graph G can be considered a closed walk where every edge is traversed exactly twice.

Proof   Corresponding to every edge k of G we introduce a new edge k' not contained in G which joins the same two vertices with each other that k does. So there results a new (connected) graph H which has twice as many edges as G. H is, of course, an Euler graph. By Theorem 2   H can be represented as a closed trail. If in this trail every new edge e' is replaced by its corresponding edge e, then there results a closed walk with the desired property.///

In this sequence of theorems belongs a theorem due to Veblen [1, §3 and 2, Chap. I, §22], in which the graph does not have to be assumed to be connected.

29

<u>Theorem 6</u> For a graph G there exists a finite system

$S = \{K_1, K_2, \ldots, K_n\}$ of cycles with the property that every

edge of G belongs to one and only one cycle K_i if and only if G

is an Euler graph.

<u>Proof</u> If there is such a system and p is a vertex of G contained

in ν cycles of the system, then the degree of p is 2ν . If, on the

other hand, G is an Euler graph, G has a cycle K_1 by Theorem 1. If the

edges of K_1 are removed from G then the resulting graph is again an

Euler graph, as we have already seen in the proof of Theorem 2. This

new graph likewise contains a cycle, K_2, etc. Since the number of edges

in G is finite the sequence $\{K_1, K_2, \ldots \}$ must end and forms a system

S with the desired property.///

§2. The Bridge Problem and the Domino Problem

The preceding material and some questions connected with it have a

role in recreational mathematics, where diverse applications are to be

found,[9] as well as the origins of some of the concepts.

Euler [1] was led to Theorem 2 by the Königsberg bridge problem,

which was communicated to him as a well known problem. It asks that the

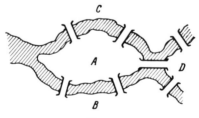

Fig. 6.

seven bridges in Fig. 6 be crossed, each one exactly once. If A, B, C,

and D are the four regions separated by the river and if we let a vertex

correspond to each region and we join any two vertices by as many edges

as there are bridges connecting the corresponding regions, then we get
the graph of Fig. 7. Then our question is simply whether this graph can
be represented as a closed or open trail. Since the graph has four
vertices of odd degree, Theorems 2 and 4 tell us that neither a closed
nor an open trail is possible. So the Königsberg bridge problem is
unsolvable and also unsolvable in the case where one must return to the

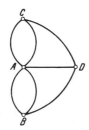

Fig. 7.

starting point.

We consider now the graph G_7, which has $\alpha_0 = 7$ vertices (0, 1, 2,
3, 4, 5, 6) and in which the edges join each pair of vertices exactly
once, so that the graph has $\alpha_1 = 21$ edges (Fig. 8). Each vertex is of
degree 6, so that by Theorem 2 the graph G_7 can be represented as a
closed trail. A cyclic assignment of the 21 edges, which contains every
edge exactly once, is, for example, as follows:

(01) (13) (36) (60) (02) (25) (56) (61) (14) (45) (50)
(03) (34) (46) (62) (23) (35) (51) (12) (24) (40).

If the pair (ik) is interpreted as a domino piece which bears the numbers i and k, then this cyclic assignment solves the following domino problem:

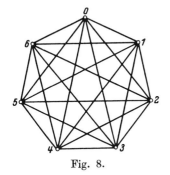

Fig. 8.

All the pieces of a domino game in which the numbers on the pieces go from 0 to 6 and from which all double numbered pieces have been removed (so that 21 pieces remain) are to be placed next to one another in a continuous closed chain (without branches) in such a way that halves of adjacent pieces always show the same number.

That this problem, as we have already seen, can be identified with a problem in graph theory was discovered by Terquem [1, pp. 72-74] in 1849 for the general game of dominoes, in which the numbers on the pieces go from 0 to α_0 - 1. The corresponding graph is then the general finite complete[10] graph G_{α_0}, where a graph is called complete if every two of its vertices are connected by one and only one edge. In G_{α_0} every vertex is of degree α_0 - 1, so that by Theorem 2 our problem always has a solution if α_0 is odd; but for even α_0 (>2) there is no solution.

In the work cited above Terquem gave a procedure by which, if α_0 is odd, G_{α_0} can be described as a closed trail. (The cyclic assignment given above for α_0 = 7 comes from Terquem's method.) In the same work Terquem posed the question, in the special case of α_0 = 7 , of how many ways there are of describing the complete graph as a closed trail; i.e. how many complete chains can be formed from the 21 domino pieces. Although this question was put repeatedly (1852, 1863) in Nouvelles Annales de Mathématiques as a problem a solution did not appear

32

(for α_0 = 7) until 1871 in a work of M. Reiss. In 1886 G. Tarry gave a much simpler solution.[11]

We mention that the case where double-numbered pieces are not removed can also be handled using graphs. A loop in that case must be added to each vertex of G_{α_0} .

For any given connected Euler graph the number of ways can be calculated by a recursion process.[12]

§3. Hamiltonian Cycles

With the Euler Theorem (Theorem 2) we obtained necessary and sufficient conditions for a graph to be interpreted as a closed trail. A similar question is as follows: when is there a cycle in a graph, which contains every vertex of the graph? Such a cycle --- which can, of course, exist only in the case of connected graphs --- is called a Hamiltonian cycle. The similarity between these two problems rests on a purely external analogy, but the second problem is difficult and unsolved. Nothing in general is known[13] concerning questions of existence of a Hamiltonian cycle. Here much deeper properties of graphs must be involved than in the question of an Euler trail. This can be

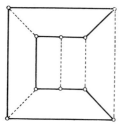
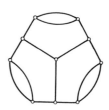

Fig. 9.   Fig. 10.

illustrated by the examples of Figures 9 and 10. Both graphs shown in these figures have 10 vertices and 15 edges and both have only vertices of degree 3. Yet the graph in Fig. 9 has a Hamiltonian cycle (indicated by the lines in heavy type), while the graph in Fig. 10 has no

33

Hamiltonian cycle, as the reader can quickly convince himself.

We mention only the two most studied examples of Hamiltonian cycles.

In 1859 Hamilton, excited by a remark by T. P. Kirkman, published a game which among other things had as object the discovery of a Hamiltonian cycle for the system of edges of a dodecahedron. A solution is given[14] in Fig. 11 by the lines in heavy type.

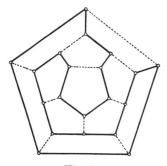

Fig. 11.

The ancient and often treated problem of the knight's move on the chessboard is equivalent[15] to the discovery of a Hamiltonian cycle for a special graph. The problem is as follows. By a continuous sequence of knight's moves one should move from an initial square of the chessboard around to all squares and back again to the initial square[16], hitting each square exactly once. We define a graph G now. To each of the 64 squares of a chessboard we let correspond a vertex and join two vertices by an edge if and only if a knight's move is possible between the corresponding squares. The problem of the knight's move is then equivalent to finding a Hamiltonian cycle for this connected graph (with 64 vertices and 168 edges). There the graph theoretic interpretation does not seem very well suited for knight's moves. We give a solution, originating with Euler, in the usual representation (Fig. 12) and we refrain from picturing the graph and one of its Hamiltonian cycles, since this would not be very helpful[17].

The problem of Hamiltonian cycles has particular interest in the case of graphs with only vertices of degree 3. Tait [2, §16] expressed the conjecture, with reference to Kirkman, that the system of edges of a "true polyhedron" always has[18] a Hamiltonian cycle in case it contains only vertices of degree 3. By the way the concept of "true polyhedron" is defined, this conjecture is still unproved today. The problem is connected with the problem of map colorings (see Chapter XII, §5). Fig. 10 shows that a finite connected graph with only vertices of degree 3 does not necessarily have to have a Hamiltonian

58	43	60	37	52	41	62	35
49	46	57	42	61	36	53	40
44	59	48	51	38	55	34	63
47	50	45	56	33	64	39	54
22	7	32	1	24	13	18	15
31	2	23	6	19	16	27	12
8	21	4	29	10	25	14	17
3	30	9	20	5	28	11	26

Fig. 12.

cycle. Fig. 13 and Fig. 14 give examples[19] of this and, in fact,

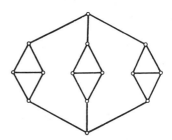

Fig. 13.   Fig. 14.

examples where there is no 2-cycle. These graphs have the property that an edge (Fig. 13)[20] or a pair of edges (Fig. 10 and 14) can be removed in such a way that the graph loses[21] its connectedness. We shall later (Fig. 90, 91, 96 in Chap. XII) see an example of the non-existence of a Hamiltonian cycle in a graph which has none of these properties (not even a 2-cycle).

The problem of a Hamiltonian cycle can be extended and facilitated by seeking a system of cycles $(K_1, K_2, \ldots, K_\nu)$, with ν arbitrary,

35

such that every vertex of the graph is contained in one and only one of the cycles K_i. Similar difficulties appear here. As Figures 10, 13, and 14 show this problem can have a solution (with $\nu = 2$) if no Hamiltonian cycle exists, but it must be that $\nu > 1$. This problem will concern us in Chapters XI and XII for a special class of graphs (which we shall call regular). There we shall give examples of the unsolvability of this problem, (see Fig. 68 and 70) including the case where all vertices are of degree 3.

§4. Transition to Directed Graphs

Theorems 2, 5, and 6 can be carried over to directed graphs. If the edges of an open or a closed trail $P_1 \, P_2 \, . \, . \, . \, P_{n-1} \, P_n$ (where either $P_n = P_1$ or $P_n \neq P_1$) are so directed that $P_i \, P_{i+1}$ is always directed to P_{i+1} (or always to P_i) ($i = 1, 2, . . ., n-1$) the trail is said to be directed. In like manner a singly infinite or doubly infinite path can be directed. This concept can be carried over to walks, where two directions are possible, the same edge having different directions for different "places" in the walk. A (directed) cycle can be defined for a directed graph. It is clear that in the case of a closed trail or doubly infinite path each vertex appears as many times as the initial vertex of a directed edge as it appears as a terminal point of a directed edge.

Corresponding to Theorem 2 we have for directed graphs the following

> Theorem 7 In a directed graph there exists a closed trail going through all the edges in their given directions if and only if the graph is finite and connected and every vertex appears exactly as many times as an initial vertex of a directed edge as it appears as the terminal vertex of a

directed edge.

We have already seen that the condition is necessary. That it is sufficient comes out of the proof we have given for Theorem 2 by a translation into the language of directed graphs (and by application of the corresponding translation of Theorem 1).

In like manner, by translation of the proof given above for Theorem 6, we get the following

Theorem 8   For a finite directed graph G there is a system
$S = \{C_1, C_2, \ldots, C_\nu\}$ of (directed) cycles with the property
that every edge of G is contained in one and only one the
cycles C_i -- and indeed with its original direction -- if and
only if all vertices of G appear exactly as many times as
initial vertices as they do as terminal vertices of edges.

As a complement to these theorems let it be mentioned that every undirected Euler graph can be transformed into a directed graph in such a way that every vertex is an initial vertex of a directed edge exactly as many times as it is the terminal vertex of a directed edge.

This follows immediately from Theorem 2:   each component of the graph is transformed into a (directed) closed trail.

Corresponding to Theorem 5 we have

Theorem 9   Every finite connected graph G can be represented as
a (directed) walk in such a way that every edge is traversed
exactly twice, once in each direction.

Proof   If a direction is assigned arbitrarily to the edges of G and the direction to each new edge k' in the proof of Theorem 5 which is opposite to the corresponding old edge k, then the so constructed directed graph H satisfies the conditions of Theorem 7.   H can therefore be represented as a (directed) trail.   If the edges k and k' are then again identified, there results a representation of G with the desired

37

property.///

It is perhaps best to mention here a theorem of L. Rédei [1]. In order to be able to express it briefly, we introduce another term, which will also prove useful later (Chapter XII).

If every edge of a path $\overrightarrow{P \ldots Q}$ of the directed graph G has the same direction which it has as an edge of G, the path is called a directed path of G; it "goes from P to Q".

We now state a theorem due to L Rédei.

Theorem 10   If in a finite directed graph G every pair of vertices, P and Q, are joined either by the edge $\overrightarrow{PQ}$ or $\overrightarrow{QP}$, then G has a (directed) path which contains[22] all the vertices of G.

Proof   The theorem holds, of course, for the case of two vertices. We can apply induction and assume that the theorem holds for that graph G' obtained from G (with n + 1 vertices) by removing the vertex P and all edges which begin or end at P. G' therefore has a path $\overrightarrow{Q_1 Q_2 \ldots Q_n}$ which contains all the vertices of G'. If PQ_n is directed towards P, then $\overrightarrow{Q_1 Q_2 \ldots Q_n P}$ is a path of G of the desired kind. We may therefore assume that one of the edges PQ_1, PQ_2, ..., PQ_n is directed towards the Q-vertex. If in the sequence PQ_1, PQ_2, ..., PQ_n the first such edge is PQ_i then $\overrightarrow{Q_1 Q_2 \ldots Q_{i-1} P Q_i Q_{i+1} \ldots Q_n}$ is a path of the desired kind. (For i=1 the path is, of course, $\overrightarrow{PQ_1 Q_2 \ldots Q_n}$.)///

The theorem just proved (which can be applied[23] to advantage by introducing a simply ordered finite set) can also be formulated in the following way:

In a chess tournament if every participant has played a game against every other and if none of the games is undecided, then after the tournament all the participants can be arranged in a sequence A_1, A_2, ..., A_n in such a way that A_1 has won the game with A_2, A_2

with $A_3, \ldots, A_{n-1}$ with A_n.

§5. Transition to Infinite Graphs

The problems of Euler trails and Hamiltonian paths can be carried
over to infinite graphs, and we now consider undirected graphs again.

A doubly infinite walk of the type

$\ldots, P_{-2}\, P_{-1}, P_{-1}\, P_0, P_0\, P_1, P_1\, P_2, P_2\, P_3, \ldots$ in which no edge
occurs twice is called a doubly infinite Euler trail of the graph if
every edge of the graph is contained in it. A doubly infinite walk, in
which $P_i \neq P_k$ for $i \neq k$ (which is therefore a doubly infinite path), is
called a doubly infinite Hamiltonian path of the graph, provided that
all the vertices of the graph are contained among the vertices P_i. At
this point the problem of the existence of an Euler trail encounters
difficulties. Just as in the case of a finite Euler trail of a finite
graph it is also clear here that a doubly infinite Euler path can exist
only if all the vertices of finite degree are of even degree. But this
condition is now not sufficient, as the example of the graph, which is
indicated in Fig. 15, shows. This graph consists of vertices P_i and Q_i

(i = 0, ± 1, ± 2, ..., ad inf.)

where $P_0 = Q_0$ but all the other P_i and
Q_i are distinct from one another; the
graph consists also of the edges

$P_i\, P_{i+1}$ and $Q_i\, Q_{i+1}$

(i = 0, ± 1, ± 2, . . . , ad inf.).

Here $P_0 = Q_0$ is of degree 4 and the
rest of the vertices are of degree 2.
So all vertices are of even degree. As

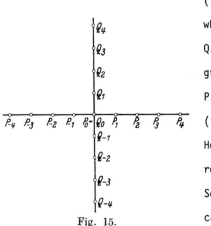

Fig. 15.

can be easily seen, the graph has no
doubly infinite Euler trail. On the other hand the infinite square

lattice has a doubly infinite Euler trail, as Fig. 16 shows. It can
also be shown for the three-dimensional analogue of this formation
(which represents an infinite graph with vertices only of degree 6) that
it can be traversed by a doubly infinite trail.[24]

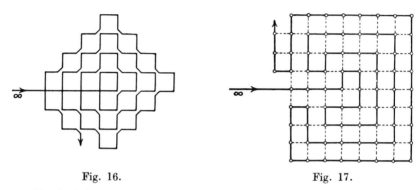

Fig. 16.   Fig. 17.

The infinite square lattice also has a doubly infinite Hamiltonian
path, as Fig. 17 shows. The question can be asked whether this is also
the case for the corresponding formations of higher dimension.[25]

The knight's move problem can also be extended to an infinite (in
each direction) chessboard. It can be shown that the corresponding
infinite graph -- which is still connected -- has a doubly infinite
Hamiltonian path, so that it is possible to traverse the whole
chessboard with the knight in a doubly infinite trail in such a way that
every square is entered[26] exactly once.

An infinite Euler trail or Hamiltonian path can, of course, exist
if and only if the graph is connected and has only countably many edges
and vertices.

As has already been indicated, it is not known in what way
Theorem 2 (and also, of course, Theorems 4 and 5) can be extended to
infinite graphs. On the other hand, Theorem 6 can be generalized in the
following way:

Theorem 11 For a graph G of finite degree there is a system

40

S = {G_α} of cycles and doubly infinite paths of G which has the property that every edge of G is contained in one and only one subgraph G_α, if and only if every vertex of G is of even degree.

Proof It can be seen by the proof of Theorem 6 that the condition is necessary. The following now shows that it is also sufficient. To any vertex P^λ of G there go $2\nu_\lambda$ edges. We split these arbitrarily into ν_λ pairs:

$(a_1{}^\lambda, b_1{}^\lambda)$, $(a_2{}^\lambda, b_2{}^\lambda)$, ..., $(a_{\nu_\lambda}{}^\lambda, b_{\nu_\lambda}{}^\lambda)$.[27] Now we define a new graph H. To each of these pairs $(a_\rho^\lambda, b_\rho^\lambda)$ we let correspond a vertex P_ρ^λ (ρ= 1, 2, ..., ν_λ). Corresponding to each edge $P^\lambda P^\mu$ of G we introduce an edge $P_\rho^\lambda P_\sigma^\mu$, where we choose P_ρ^λ and P_σ^μ from the vertices $P_1{}^\lambda$, ..., $P_{\nu_\lambda}{}^\lambda$ and $P_1{}^\mu$, ..., $P_{\nu_\mu}{}^\mu$, respectively, so that $P^\lambda P^\mu$ is on the one hand one of the edges a_ρ^λ and b_ρ^λ and on the other hand one of the edges a_σ^μ and b_σ^μ. The vertices P_ρ^λ and edges $P_\rho^\lambda P_\sigma^\mu$ thus introduced determine a graph H, in which exactly two edges go to each vertex.[28] According to Theorem I.28, H is therefore the sum of cycles and doubly infinite paths. And again cycles and doubly infinite paths, respectively, in G correspond to them. It is easily seen that their totality has the desired property of S.///

The proof just completed is copied after the one which Veblen gave in the first edition of his book [2, Chap. I, §22] for finite graphs (for Theorem 6). On the other hand the proof given above for Theorem 6 which we borrowed from the second edition of Veblen's book and which involves induction cannot be made use of for infinite graphs. As for finite graphs (see Theorem 8) Theorem 11 can be carried over to infinite directed graphs.

Likewise as we have already shown above for finite graphs, we can get from Theorem 11 for infinite graphs the following

Theorem 12   If every vertex of a graph of finite degree is of even degree, it can be changed into a directed graph in such a way that every vertex appears as an initial point of an edge as many times as it appears as a terminal point.

Proof We only need to make the cycles and doubly infinite paths in the system S of Theorem 11 continuously directed. (Since this continuous directing can always occur in two ways, the Axiom of Choice is needed in the case where S is infinite.)///

All the results of this chapter also remain valid for graphs in the broad sense, as long as, in calculating the degree of a vertex, each loop is counted as two edges. This follows immediately, if every loop is split into two edges. (see Fig. 1).

Notes on Chapter II

[1] In the terminology of Veblen and Alexander [1, §7], who introduced it for the topology of n-dimensional manifolds, Euler graphs are called <u>closed</u>. This designation was also used sometimes for graphs without endpoints.

[2] Theorems will be designated this way from now on; the Roman numerals always stand for the numbers of the chapters of this book. If the Roman numeral is missing, then the theorem is from the current chapter.

[3] The formulation by Euler refers not to vertices and edges but to "islands" and "bridges" (cf. also the Königsberg bridge problem below). Euler proves only that the condition in Theorem 2 is necessary but not that it is also sufficient. This gap was frequently overlooked (for example, by Dehn and Heegaard [1, p. 174]). Hierholzer [1] gave the first complete proof for Theorem 2, without seemingly having known of Euler's work. Our proof agrees essentially with Hierholzer's proof.

[4] In the graphical -- plane -- representation of graphs points of intersection must, of course, often appear which do not correspond to vertices of the graph represented. Because of this, from now on the points in our figures which are supposed to correspond to vertices of the graph represented will be identified by small circles.

[5] On the other hand, for example, the unique endpoint of a singly infinite path is its only vertex of odd degree.

[6] For p = 1 the theorem is found in Euler [1, §19] and was first proved by Hierholzer [1]. The first part of the proof following here comes from Brunel [3, p. 63].

[7] This "theorem" was already expressed by Clausen [1] in 1844. With reference to Clausen it was frequently asserted (for example, Lucas [1, Vol. IV, p. 133 and 2, p. 102], Dehn and Heegaard [1,

p. 174], Sainte-Lague [4, p. 12]) that Clausen had expressed here a
general theorem like Theorem 4. This is not the case; Clausen's
statement refers only to the special Fig. 5. It would probably be
difficult to determine what Clausen understood by an "analytical" proof
of his assertion.

[8] See Rademacher and Toeplitz [1, §2].

[9] See Lucas [1, vol. I, pp. 19-55, 238-240; Vol. II, pp. 37-71,
199-229; Vol. IV, pp. 123-151, 205-223], Rouse Ball [1, Vol. II,
pp. 198-266], Ahrens [2, Vol. II, pp. 170-210, 216-276], to name only
the best known collections.

[10] This designation (complet) stems from Sainte-Lague
[2, p. 4].

[11] More about this with exact historical details is to be found
in Ahrens [2, Vol. II, pp. 264-276]. In this book number questions (how
many ways can something be done) will not be dealt with further.

[12] Lucas [2, pp. 103-109]. The calculation of this number for
many special connected Euler graphs is found in this reference.

[13] Brunel [2] was concerned -- without touching the question of
existence -- with counting all the Hamiltonian cycles of a given graph.
Sainte-Lague [2] studied for the simplest graphs of certain classes of
graphs whether they are "cercle" , i.e. whether they have a Hamiltonian
cycle. For special graphs, which can be drawn in the plane (on the
surface of a sphere) (cf. Chapter XII, §5), this question has recently
been treated by Whitney [1].

[14] For more on this see, for example, Ahrens [2, Vol. II,
pp. 196-210]. A novel solution is given by A. Kowalewski [1], cf. also
G. Kowalewski [2, pp. 86-98].

[15] This was noticed for the first time by Tait [2, §§11 and
16]. In the graph theoretical interpretation Brunel [2, p. 177] was

concerned with this problem.

[16] Sometimes this last requirement is not included ("unclosed"
moves of the knight).

[17] See Figure 38a in G. Kowalewski [2, p. 63].

[18] That this theorem was not proved, Tait explains by saying
that "habitual stargazers are apt to miss the beauties of the more
humble terrestrial objects."

[19] Since the graph of Fig. 14 can be drawn in the plane (on the
surface of a sphere), this graph yields a simple counterexample for an
often recurring false assertion in the literature of the Four Color
Theorem, see for example 1930: Jahresbericht der deutschen
Mathematiker- Vereinigung, 39, p. 51.

[20] The occurrence of such an edge makes in itself the existence
of a Hamiltonian cycle impossible.

[21] If a graph has an edge or a pair of edges with this property
(and so the graph in the terminology of Sainte-Laguë [4, p. 6] has the
puissance 1 or 2, respectively), then it can be easily shown that it, as
Tait noted [2, §16], cannot form the system of edges of a polyhedron
with planar surfaces. And, of course, it also cannot if it contains a
two-cycle.

[22] Rédei proves loc. cit. not only the existence of such a
directed path but also shows that the number of such directed paths is
always odd. This theorem plays a role in his investigations into the
class number of quadratic number fields; see his Hungarian treatment in
Matematikai és Természettudományi Értesíto , 48, 1932, pp. 648-682.

[23] Cf. Chapter VIII, §2.

[24] Mr. Tibor Grünwald communicated a proof for this to me.

[25] This question was answered affirmatively for three
dimensional space by Mr. Tibor Grünwald .

[26] Kürschák [1]. - In another paper Kürschák [2] showed that this is also possible in a trail infinite in only one direction.

[27] Since this is supposed to happen for all vertices of G at the same time, the possibility of this division is assured only by the Axiom of Choice.

[28] Expressed in a graphical way, we get H from G by replacing every vertex P^λ of G by ν_λ vertices P_1^{λ}, P_2^{λ}, ..., $P_{\nu_\lambda}^{\lambda}$ and letting the $2\nu_\lambda$ edges, which go to P^λ , instead of going to P^λ , go to the vertices P_ρ^λ in such a way that two edges go to each of the vertices P_ρ^λ . -- In generalized form this idea was used by König [5, p. 465].

CHAPTER III

The Labyrinth Problem

§1. Formulation of the Problem

Wiener's Solution

Theorem II.5 is closely connected with the Labyrinth problem[1],
which will be treated in this chapter. Since the width of the passages
does not matter, a labyrinth can be identified with a graph and, in
fact, with a finite, connected graph; in the labyrinth the branching
places and the endpoints of the dead ends correspond to the vertices of
this graph. So, for example, the graph of Fig. 19 (or the graph of Fig.
20) corresponds to the labyrinth of Fig. 18. It is now required to give
a method by which a definite spot in the labyrinth, usually

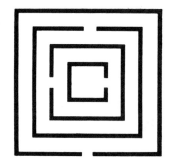
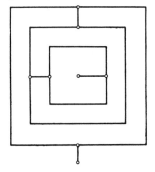

Fig. 18.   Fig. 19.

designated as the center of the labyrinth, can be reached. Since it is
quite indefinite which spot is considered the center, the problem to be
solved is to give a method by which every spot of the labyrinth is
reached. If the map of the entire labyrinth is known, then there is, of
course, no difficulty, since there is only a finite number of
possibilities. But we do not wish to assume that we have a map of the

47

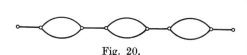

Fig. 20.

labyrinth. Instead, it will be assumed only that if in walking in the labyrinth a certain branching place X (or endpoint X of a dead end) is reached, the labyrinth (the graph) is known in the "neighborhood of X", that is, the passages (edges) going to X are known as well as certain properties of the path already traversed, insofar as they relate to the neighborhood of X.

We wish first to assume that if in walking in the labyrinth a vertex P is reached, then for every edge going to P it is known whether it was traversed earlier or not; moreover, we assume that it is possible at any time to trace back the entire path already traversed in the opposite direction, with the same repetitions, (just like Theseus with the help of Ariadne's thread). The problem with these assumptions was solved by Wiener [1][2], who was the first to treat the ancient labyrinth problem as a mathematical problem. Wiener's method (in a somewhat specialized form) goes as follows:

Beginning at an arbitrary vertex A as starting point, the edges AB, BC, CD, . . . are chosen quite arbitrarily until an endpoint Q of the graph is reached or such a vertex Q, which has already been passed. Having reached Q, we turn around and retrace in the opposite direction the path already traversed, until a vertex R is reached which has an edge going to it which has not yet been traversed, Having reached R, we turn off and follow any edge RS not yet traversed, and then we continue walking over as yet untraversed edges in an arbitrary manner until we reach an endpoint or such a vertex which was already traversed; here we turn back again, etc.

We now prove that the sequence of edges ABC . . . Q . . . RS . . .

1. leads back to A and

2. then contains every edge of the graph.

If there is no turning off, then we, of course, come back to A. But if there is a turning off, then, since with every turning off a new edge is added, there must be a last turning off because of the finite number of edges. Since the vertices are also finite in number, after this turning off we must reach an endpoint or a vertex Z already traversed. By the directions we must, since no more turnings off are possible, traverse this whole sequence of edges F = AB . . . Z in the opposite direction, by means of which we reach A.

We see that every edge of the original graph G was traversed in the following way. We suppose that this is not the case; then the edges not traversed form a proper subgraph G" of G, while the edges traversed form the (connected) subgraph G' of G. G' and G" have no common edge; but they have at least one common vertex P, since otherwise G = G' + G" (Theorem I.24) would not be connected. Since P belongs to G', P is a vertex of the sequence of edges F = AB . . . . Z. Since P also belongs to G", there is an untraversed edge ending in P. In the above mentioned traversing of G in the direction of A we must therefore turn off at P, and this contradicts the fact that the last turning off already took place.

§2. Tremaux' Solution

The process outlined above of Wiener's cannot be called economical, since the edges must be, in general, traversed quite often. This is to be noticed even more when -- by Theorem II.5 -- a walk covering all edges of the graph exists, in which no edge is crossed more than twice. We shall show that such a solution can also be found if the whole graph does not lie before us as known, but only certain local properties of the walk begun are known at any given moment.

49

We do not, however, wish to assume now that we are always able to retrace the path traversed in the opposite direction, but instead of Wiener's assumption we make the assumption (partly stronger and partly weaker) that if we reach in our walk a certain vertex P, then for every edge going to P it is known whether it was already traversed and how often (the directions in which the edges were traversed are not assumed as known). Under this assumption the labyrinth problem was solved[3] in an elegant manner by Trémaux. His solution is as follows:

If, beginning at an arbitrary starting vertex A_1, we reach by an edge PQ a vertex Q which is not an endpoint of the graph and which is reached for the first time, then we continue the walk with an arbitrary other edge QR. But if Q is either a vertex traversed earlier or is an endpoint, and PQ was traversed for the first time, then we turn back, that is, we continue the walk with the same edge PQ (in the opposite direction QP). If both Q and PQ were traversed earlier, then we continue on some other edge QR not yet traversed or, if there is no such edge, we continue on an edge QR, which was traversed only once.

If these rules are followed, then each edge is traversed at most twice; because of the finiteness of the number of edges the walk must come to an end, that is, we must finally reach a vertex K, such that all edges going to K have already been traversed twice. Now K must be A_1, since we otherwise would have a contradiction with Theorem I.2 (for the endpoint K of an <u>open</u> sequence of edges). Now we wish to prove Trémaux' Theorem, which states that at the stopping of the process at an[4] arrival at A_1 every edge was traversed twice and, in fact, once in each direction. We would like to describe our somewhat complicated proof quite thoroughly, since no complete proof of Trémaux' Theorem appears in the literature.[5]

Let

$$A_1 \; A_2, \; A_2 \; A_3, \; \ldots, \; A_{\rho-1} \; A_\rho, \; A_\rho \; A_{\rho+1}, \; \ldots, \; A_n \; A_1 \qquad (1)$$

be the edges, in order, which are traversed in a proper (that is,

satisfying Trémaux' rules) walk in the finite, connected graph G. First

of all, we prove that there must be an edge in (1) such that,

immediately after we have traversed it, this same edge is traversed

again (in the opposite direction). Let A_μ be the first element in the

sequence A_1, A_2, A_3,..., which has already come up, say as A_ν ($\nu < \mu$) .

Now two cases are possible:

    1) The edge $A_{\mu-1} \; A_\mu$ has not come up earlier; then we must turn

    back here, since A_μ has already come up earlier as A_ν, that is,

    $A_{\mu-1} \; A_\mu$ has the desired property.

    2) $A_{\mu-1} \; A_\mu$ has already come up, that is, there is a $\rho < \mu$ such

    that $A_{\mu-1} \; A_\mu = A_{\rho-1} \; A_\rho$; but $A_{\mu-1}$ has not come up earlier and

    therefore cannot be $= A_{\rho-1}$ since $\rho-1 < \mu-1$, and consequently

    $A_{\mu-1} = A_\rho$; but, since $A_{\mu-1} \neq A_\rho$ for $\rho < \mu-1$, this can happen only

    in the case $\rho = \mu-1$. If this value of ρ is substituted we get

    $A_{\mu-1} \; A_\mu = A_{\mu-2} \; A_{\mu-1}$, and consequently $A_{\mu-2} \; A_{\mu-1}$ now has the desired

    property. (It can easily be shown that case 2 happens if and only

    if $A_{\mu-1}$ is an endpoint of G.)

So let $A_\rho \; A_{\rho+1}$ be an edge which is identical to the edge

immediately preceding it in (1); and we have that $A_{\rho+1} = A_{\rho-1}$. If this

edge $A_\rho \; A_{\rho+1}$ ($=A_{\rho-1} \; A_\rho$) is removed from G, the remaining edges form a

subgraph G' of G. We prove that the graph G' is also connected.

If A_ρ is an endpoint of G, then this follows from Theorem I.14.

If A_ρ is not an endpoint, then turning back at A_ρ can be proper only

if A_ρ was already passed, and then a subsequence of the set (1) of edges

gives an open sequence of edges joining $A_{\rho-1}$ and A_ρ and not containing

the edge $A_{\rho-1} \; A_\rho$. By Theorem I.3 there is also a path in G not

containing the edge $A_{\rho-1} \; A_\rho$ which joins these two points. But then

Theorem I.13 says that G' is actually connected.

If $A_{\rho-1} A_\rho$ and $A_\rho A_{\rho+1}$ in (1) are deleted, then, since $A_{\rho-1} = A_{\rho+1}$, we get a sequence of edges:

$$A_1 A_2, A_2 A_3, \ldots, A_{\rho-2} A_{\rho-1}, A_{\rho+1} A_{\rho+2}, \ldots, A_\mu A_1. \qquad (2)$$

This represents a walk in G', since $A_{\rho-1} A_\rho$ (as with every other edge) can occur at most twice in (1); so $A_{\rho-1} A_\rho$ is not contained in (2). We prove now that (2) gives a proper walk in G', that is, if A_σ has been reached by any edge $A_{\sigma-1} A_\sigma$ (where σ is distinct from ρ and from ρ+1), the choice of the edge following immediately after $A_{\sigma-1} A_\sigma$ in (2) corresponds[6] to Trémaux' rules.

This is clear if A_σ is distinct from both A_ρ and $A_{\rho+1}$ $(=A_{\rho-1})$ and also if σ < ρ-1, since in these cases Trémaux' rules allow the same possibilities for G' as for G. Since σ = ρ and σ = ρ+1 are excluded, only the following three cases need to be taken care of:

1) σ = ρ-1; 2) σ > ρ+1 and $A_\sigma = A_\rho$; 3) σ > ρ+1 and $A_\sigma = A_{\rho-1}$ $(= A_{\rho+1})$.

Case 1) We must show that in crossing over from $A_{\sigma-1} A_\sigma = A_{\rho-2} A_{\rho-1}$ to $A_{\rho+1} A_{\rho+2}$ the rules are not violated. First let (Fig. 21) $A_{\rho-2} A_{\rho-1} = A_{\rho+2} A_{\rho+1}$ (and so $A_{\rho-2} = A_{\rho+2}$). Then crossing over to the identical edge (a turning back) is certainly proper in G', if $A_{\rho-1}$ is either an endpoint of G' or was already traversed. One of the two is surely the case, since otherwise there would have to be a third edge, $A_{\rho+1}$ K, still not traversed, going to $A_{\rho+1}$, and then we would have had, in the walk in G to $A_\rho A_{\rho+1}$, to choose not the edge $A_{\rho+1} A_{\rho+2}$, which was already traversed, but this edge $A_{\rho+1}$ K.

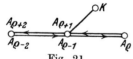
Fig. 21

Second, let $A_{\rho+1} A_{\rho+2} \neq A_{\rho-2} A_{\rho-1}$ (Fig. 22). After traversing $A_{\rho-2} A_{\rho-1}$ in G' we do not have to turn back, since $A_{\rho-1}$ is not an endpoint of G' and since otherwise we would have had to turn back in the walk in G, and so $A_{\rho-2}\,A_{\rho-1}$ would not have been followed by $A_{\rho-1}\,A_\rho$. In crossing from $A_{\rho-2}\,A_{\rho-1}$ to $A_{\rho+2}\,A_{\rho+1}$ in G' we would violate the rules only if $A_{\rho+1}\,A_{\rho+2}$ had already been traversed and a third edge $A_{\rho+1}$ K were still not traversed. But then in the walk in G not $A_{\rho+1}\,A_{\rho+2}$ but $A_{\rho+1}$ K would have had to follow $A_\rho\,A_{\rho+1}$.

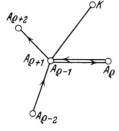

Fig. 22

Case 2) We prove that we are obeying the rules if we let the edge $A_\sigma\,A_{\sigma+1} = A_\rho\,A_{\sigma+1}$ follow the edge $A_{\sigma-1}\,A_\sigma = A_{\sigma-1}\,A_\rho$ in the walk in G', assuming that $\sigma > \rho + 1$ and that therefore the double traversing of $A_{\rho-1}\,A_\rho$ has already happened. Since A_ρ is not an endpoint of G (there are at least two edges going to A_ρ : $A_{\rho-1}\,A_\rho$ and $A_{\sigma-1}\,A_\rho$), A_ρ must already have been passed once, since otherwise this double passing could not be proper. So if, first, $A_{\sigma-1}\,A_\rho = A_{\sigma+1}\,A_\rho$ and consequently $A_{\sigma-1} = A_{\sigma+1}$, then this turning back is also proper in the walk in G'. But secondly if $A_{\sigma-1}\,A_\rho$ and $A_\rho\,A_{\sigma+1}$ are distinct (Fig. 23), then $A_{\sigma-1}\,A_\rho$ must already have been traversed twice (since otherwise we would have had to turn back); so we can have violated the rules only if there were a fourth edge A_ρ K, still not traversed, and $A_\rho\,A_{\sigma+1}$ had only been traversed once; but then we would have violated the rules by the same crossing over already in the walk.

Case 3) We show that the crossing from $A_{\sigma-1}\,A_{\rho-1}$ to $A_{\rho-1}\,A_{\sigma+1}$ is proper in the walk in G', in case this was the case for G and $\sigma > \rho + 1$. First of all, let these two edges be identical; so $A_{\sigma+1} = A_{\sigma-1}$. This turning back is proper, since $A_{\rho-1}$ was passed earlier in the walk in G'. Secondly,

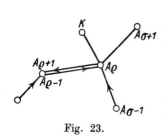

Fig. 23.

53

if $A_{\sigma-1} \, A_{\rho-1} \neq A_{\rho-1} \, A_{\sigma+1}$ (Fig. 24) then $A_{\sigma-1} \, A_{\rho-1}$

must already have been traversed, and we can draw a similar conclusion to that of the second case.

So it is proved that (2) represents a proper walk in G'.

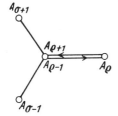

Fig. 24.

Now we can prove Trémaux' Theorem by induction on the number of edges, since the theorem is valid for the graph with a single edge, as can be seen immediately. If in the walk in G given by (1), which satisfies Trémaux' rules, not every edge of G is traversed once in each direction, then, of course, in the walk (2) in G' not every edge of G' would be traversed, once in each direction. But since we have proved that G' is connected and that walk (2) in G' satisfies Trémaux' rules, this is impossible; so we have assumed the validity of the theorem for the connected, finite graph G', which has one fewer edges than G.

So Trémaux' Theorem is proved.

§3. Tarry's Solution

The labyrinth problem can be solved in a somewhat easier way, if, instead of making the assumptions which form the basis of Trémaux' solution, we make the following assumptions. For every arrival at a vertex P it ought to be known for every edge PK going to P, whether or not the edge was already traversed in the direction towards K, and it ought always to be possible to determine that edge, on which P was reached for the first time. ("Entrance edge" of P).

The problem was solved by Tarry [1][7] for this case. Tarry's solution runs as follows. When we come on an edge PQ to a vertex Q, we should continue the walk with any edge QR which was still not traversed

in the direction towards R; but we should choose the entrance edge of Q
for QR only if all other edges QK going to Q were already traversed in
the direction towards K.

By this rule it is assured that every edge is traversed in each
direction at most once, and so is traversed in all at most twice.
Likewise, as above for the Trémaux walks, it turns out here that the
walk must finally stop and that this can happen only if the initial
vertex A_1 of the walk has been reached. We now wish to prove Tarry's
Theorem, which says that at the end of the walk every edge was traversed
exactly twice, once in each direction.[8] Let A_1, A_2, A_3, ... be the
vertices, in order, which are met on the walk. It is clear that all
edges A_1K which go to the initial vertex A_1, were already traversed in
the direction towards K, since otherwise the walk could not stop at
A_1. Since A_1 was reached as often as it was left, every edge A_1K must
also have been traversed in the direction towards A_1. We show also that
the edges which end in the remaining passed vertices A_n were traversed
in both directions. If this were not the case, then there would be a
first vertex A_n in the sequence A_1, A_2, A_3, ..., such that an
edge $A_n K$ was not traversed twice. Since also A_n was left as often as
reached, there must be an edge A_nK_1, which was not traversed in the
direction towards K_1 and also an edge A_nK_2, which was not traversed in
the direction towards A_n. But now surely $\alpha < n$ for the entrance edge
$A_\alpha A_n$ of A_n, and so $A_\alpha A_n$ (as all edges going to A_α) was also traversed
in the direction towards A_α. But this, by Tarry's rules, is proper only
if all edges going to A_n were traversed in the direction towards K. And
this contradicts the fact that A_nK_1 was not traversed in the direction
towards K_1.

We now need only to show that every vertex of the graph is one of
the vertices A_i, and so is passed on the walk. We assume that a vertex

55

B is not passed. We join B with A_1 by a path

A_1 B_1 B_2... B_ν (B_ν = B). Let B_α be the first vertex in the

sequence B_1, B_2, ..., which is not passed on the walk. Then $B_{\alpha-1}$ is

passed (for α = 1 we set $B_{\alpha-1}$ = A_1), so that, as we saw, $B_{\alpha-1} B_\alpha$ was

traversed (in fact, twice). Thus B_α was also passed, in contradiction

to our assumption.

By this Tarry's theorem is proved.

§4. Connection between Trémaux' and Tarry's Solutions

An interesting connection exists between Trémaux' and Tarry's

walks, which is expressed by the following theorem:

Any Trémaux walk is at the same time a Tarry walk.[9]

In other words, we wish to prove the following:

If in a Trémaux walk a certain vertex P is arrived at, which is

distinct from the initial vertex of the walk, then the walk is continued

on the entrance edge of P if and only if all other edges PK, which end

in P, were already traversed in the direction towards K.

This is clear if P is reached for the first time and P is an

endpoint of the graph; so we can assume that this is not the case. We

wish to designate by ρ_ν the number of those edges going to P, which,

after P was passed for the nth time (n=1, 2, 3,...), were traversed

exactly once. We have ρ_1 = 2. If P is arrived at for the nth time

(n>1) on an edge which was traversed for the first time, then it is

necessary to turn back, and consequently ρ_n = ρ_{n-1}. But if this edge

was also traversed earlier, then it now ceases to be an edge traversed

exactly once. So we have ρ_n= ρ_{n-1} - 2 or ρ_n = ρ_{n-1}, according to

whether the edge on which we continue the walk -- after P was reached

for the nth time -- was already traversed once or not. So ρ_1 = 2

implies ρ_2 = 0 or 2, and ρ_3 = 0 or 2, etc.: every ρ_n is either 0 or 2 (if $\rho_{\kappa} = 0$, then we have, of course, $\rho_{\kappa+1} = \rho_{\kappa+2} = \ldots = 0$).

We now assume that after the nth arrival at P on the edge Q_nP $(n > 1)$ the entrance edge Q_1P of P was chosen for the continuation of the walk ($Q_nP \neq Q_1P$ since otherwise, because $n > 1$, Q_1P would have been traversed three times). After the n-1 st passing of P, Q_1P and Q_nP were both already traversed. For Q_1P this is clear, since we have that $n > 1$. For Q_nP it follows from this that we otherwise would have had to turn back on Q_nP, instead of choosing PQ_1. On the other hand these two edges were traversed only once before the nth arrival at P, since otherwise the traversing of these edges at the nth passing of P would not follow the Trémaux rules. So, since ρ_{n-1} cannot be greater than 2, we have $\rho_{n-1} = 2$. Before the nth arrival at P, of all edges KP only Q_1P and Q_nP were traversed exactly once, while the remaining edges going to P were traversed twice, since, if one were not traversed, the choice of the edge PQ_1 already traversed once after the nth arrival at P would not be according to the rules. According to Trémaux' theorem, however, no edge can be traversed twice in the same direction; so, with the exception of PQ_1, all edges PK were traversed in the direction towards K. At the nth arrival at P, the edge PQ_1 is the only edge PK which was still not traversed in the direction towards K.

Here we add a remark concerning Tarry's solution. If Tarry's rules were modified so that, instead of requiring that no edge be traversed twice in the same direction, it is required only that no edge be traversed three times, then the labyrinth problem with this new rule would not be solved. As the walk in Fig. 25 shows, it is possible that edges remain untraversed.

So if, as was proved above, Trémaux' rules come from Tarry's through specialization, a broader result is obtained using Trémaux' solution to the labyrinth problem than using Tarry's when the requirement of opposite directions is, so to speak, automatically fulfilled in Trémaux' walks and the direction in which an edge was traversed does not have to be assumed as known at the return to an endpoint of the edge.

§5. Reverse Walks

Finally we would like to make two observations about the reversing of a walk. I owe both to J. Kürschák's kind disclosure.

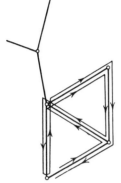

Fig. 25.

If
$$A_1 A_2 \ldots A_{\mu-1} A_\mu A_{\mu+1} \ldots A_1 \tag{I}$$
is a Tarry walk, then the reverse walk
$$A_1 \ldots A_{\mu+1} A_\mu A_{\mu-1} \ldots A_2 A_1 \tag{II}$$
is also one.

It is clear that the (ended) Tarry walks are characterized by the following three properties:

1. No edge is traversed twice in the same direction.

2. The entrance edge of any vertex P is identical to the exit edge of P, on which P is last left (otherwise this entrance edge would be traversed only once in contradiction to Tarry's theorem).

3. All edges are traversed in both directions. (this is Tarry's theorem.)

Now obviously (II) has any of these three properties, if they are already properties of (I).

A corresponding statement also holds for the Trémaux walks. We now prove the theorem:

If (I) is a Trémaux walk, then so is the reverse walk (II).

We need to show that in the walk (II) the crossing from $A_{\mu+1} A_{\mu}$ to $A_{\mu} A_{\mu-1}$ follows the Trémaux rules. If A_{μ} is passed for the first or last time at this crossing, then this crossing follows the Trémaux rules if it follows the Tarry rules (even if A_{μ} is an endpoint of the graph). But this is certainly the case, since (I) -- as a Trémaux walk -- and therefore also (II) -- as the reverse of a Tarry walk -- are Tarry walks. So we still need only to consider the case where, in crossing from $A_{\mu+1} A_{\mu}$ to $A_{\mu} A_{\mu-1}$ the vertex A_{μ} comes up neither for the first nor the last time. Here two cases are to be distinguished. First of all, let $A_{\mu+1} A_{\mu} = A_{\mu} A_{\mu-1}$. Then, since every edge is traversed only twice, this edge was not traversed earlier. So turning back on this edge follows the Trémaux rules. Secondly, let $A_{\mu+1} A_{\mu} \neq A_{\mu} A_{\mu-1}$. In the corresponding crossing from $A_{\mu-1} A_{\mu}$ to $A_{\mu} A_{\mu+1}$ in (I) A_{μ} is then reached neither for the first nor the last time. So in this crossing 1) $A_{\mu-1} A_{\mu}$ is already traversed for the second time (since otherwise we would have had to turn back at A_{μ}) and 2) $A_{\mu} A_{\mu+1}$ is traversed only for this first time. If $A_{\mu} A_{\mu+1}$ had been traversed for the second time then, in passing A_{μ}, the number ρ of edges going to A_{μ} and traversed exactly once would be decreased, and so -- as we saw above -- would become equal to 0, and this is impossible, since -- as a consequence of the fact that the Trémaux walk is at the same time a Tarry walk -- the entrance edge of A_{μ} remains only once traversed up to the last passing of A_{μ}. But now 1) and 2) imply that $A_{\mu+1} A_{\mu}$ is traversed for the second time and $A_{\mu} A_{\mu-1}$ for the first in the crossing under consideration in (II). And this follows the Trémaux rules.

Notes on Chapter III

[1] Almost all works on recreational mathematics concern themselves with labyrinths. Outside of the already mentioned places in the books of Lucas, Rouse Ball, and Ahrens we mention also: H. E. Dudeney, Amusements in Mathematics, London 1917, where many labyrinths (on pp. 127-137) are to be found.

[2] Here we are dealing with the second part of this note. The first part (where Wiener tacitly restricts himself to a plane labyrinth) is actually without mathematical content, since the problem solved there, to give a method of finding one's way back to the starting point, has a trivial solution: one remains stationary.

[3] See Lucas [1, Vol. I, pp. 47-51].

[4] The walk does not necessarily have to stop, if we get back to A_1 for the first time.

[5] Trémaux , as it seems, did not publish his proof. Lucas' proof [1, Vol. I, pp. 47-51] is -- as he says -- a slight modification of Trémaux' proof. This proof of Lucas' is not complete, because, first of all, the case where the edges, denoted in the proof by ZA and ZY, are identical (a case, which can actually happen) is left out of consideration and, secondly, an essential assertion at the end of the proof is taken care of by the short remark "on demontrera de même ". In Ahrens [2, Vol. II, pp. 192-194] both of these critical points are dealt with in more detail, but his proof also contains an essential gap, to which we shall return shortly (see the following footnote). -- Let it be noted here that neither Lucas nor Ahrens mentions that the edges are traversed twice in opposite directions.

[6] In the proof of Ahrens' mentioned above of Trémaux' theorem this theorem (on pp. 193 and 194) is used without proof and without

60

being formulated.

[7] With reference to this work Rouse Ball [1, p. 207] gives a still simpler "solution," but it is not correct. Simple examples show that it is possible in Rouse Ball's process to get stuck before all the edges have been traversed. In the same place Rouse Ball restricts the treatment of the question expressly to two dimensional labyrinths. This is quite superfluous. Everything remains true also, for example, for catacombs, where the corridors can also run under and over one another in an arbitrary way.

[8] The first part of the proof following here stems from Tarry.

[9] See the "remarque" of Tarry [1, p. 190].

CHAPTER IV

Acyclic Graphs

§1. Trees

In a way the simplest graphs are those that contain no cycle and
therefore (see Theorem I.4) no closed trail. A graph without a cycle is
called acyclic. If an acyclic graph is finite and connected then it is
called a tree.[1] (see Figure 26.)

We shall now take a closer look at trees. In order to avoid
unnecessary repetitions we shall consider acyclic graphs, even if they
are infinite or disconnected.

The concept of tree was introduced using another definition (see,
for example, de Polignac [1]); in order to show the equivalence of the
two definitions, we prove

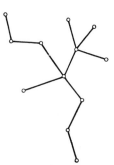

Fig. 26.

Theorem 1 A finite graph is a tree if and only if
for any two of its vertices there exists exactly
one path which joins them.

Proof If between any two vertices such a path exists
then the graph is clearly connected. It can contain
no cycle $A_1 A_2 \ldots A_n A_1$ $(n > 1)$, since otherwise the
two different paths $A_1 A_2 \ldots A_i$ and
$A_i A_{i+1} \ldots A_n A_1$ $(1 < i \leq n)$ would join the vertices
A_1 and A_i with each other.

Conversely, if a finite graph is a tree then it is connected and so
any two vertices are joined by a path. Two different paths, however,

62

can never join the same two vertices since then by Theorem I.7 there
would be a cycle.///

(The above proof holds only for graphs in the narrow sense. As
Fig. 27 shows, a graph with loops can also fulfill the condition of the
theorem, and such a graph is, of course, not a tree.)

The same chain of reasoning leads also to the
following general result:

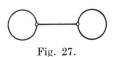

Fig. 27.

Theorem 2   A graph is connected and acyclic if
and only if for any two vertices there exists
exactly one path which joins them.

The following theorem states an essential property of acyclic
graphs:

Theorem 3   If G_1 and G_2 are two connected subgraphs of the
acyclic graph G, then their intersection (i.e., the edges which
are contained in G_1 and G_2 at the same time) is a connected
graph G'.

Proof   If P and Q are arbitrary vertices of G'(and therefore are
contained in both G_1 and G_2), then there is a path W_1 of G_1 and a path
W_2 of G_2, both of which join P with Q. By Theorem 2 $W_1 = W_2$, so that
this path is in G'. Therefore G' is connected.///

Is the converse of this theorem true? In other words, if the graph
G contains a cycle $A_1 A_2 \ldots A_n A_1$, are there always two connected
subgraphs G_1 and G_2 of G with the property that the common edges of G_1
and G_2 do not form a connected graph? This is always the case if $n > 3$,
since then the edges $A_1 A_2$ and $A_3 A_4$, which are common edges of the
paths $A_1 A_2 A_3 A_4$ and $A_3 A_4 \ldots A_n A_1 A_2$ do not form a connected
graph. The case, however, where G is itself a 2-cycle (a cycle with 2
edges) or a triangle (a cycle with 3 edges) is an exception[2] to this
converse. Now we prove

63

__Theorem 4__   Every tree has at least two endpoints.

__Proof__   If $A_0 A_1$ is an arbitrary edge then either A_1 is an endpoint or there is another edge $A_1 A_2$, where $A_2 \neq A_0$, since otherwise these two edges would form a cycle. Then in turn either A_2 is an endpoint or there is an edge $A_2 A_3$, where $A_3 \neq A_0$ and $A_3 \neq A_1$. Continuing this process we keep obtaining new vertices, but since their number is finite the process must end, at which point we obtain an endpoint A_n. As far as the second endpoint is concerned, either A_0 is such a one or there is an edge $A_0 A_{-1}$ which is distinct from $A_0 A_1$; if A_{-1} is not an endpoint then there is another edge $A_{-1} A_{-2}$, etc. Since there is no cycle, each of the vertices A_n, A_{n-1}, ... , A_1, A_0, A_{-1}, A_{-2}, ... is distinct from all the preceding vertices. Since the number of vertices is finite, a second endpoint must be reached.///

The following two theorems hold not only for trees but also for infinite acyclic graphs; they are therefore stated and proved in this more general way.

__Theorem 5__   If $P_0 P_1 P_2 .. P_\nu$ and $Q_0 Q_1 Q_2 ... Q_\mu$ are two paths of an acyclic graph, which have the common endpoint $P_0 = Q_0$, where $P_1 \neq Q_1$, then all the edges of these paths form a path $P_\nu ... P_1 P_0 Q_1 ... Q_\mu$ of the graph.

__Proof__   According to the assumption two P-vertices are distinct from each other, as are two Q-vertices. With the exception of the case $i = j = 0$, $P_i \neq Q_j$, since otherwise certain edges of the paths $P_0 P_1 P_2 ... P_i$ and $Q_0 Q_1 Q_2 ... Q_j$, which are different because $P_1 \neq Q_1$, would form a cycle of the graph by Theorem I.7. Since the vertices P_ν, ..., P_1, P_0, Q_1, Q_2, ..., Q_ν are pairwise distinct, the walk $P_\nu ... P_1 P_0 Q_1 Q_2 ... Q_\mu$ is a path.///

This theorem will be used especially in the case where $\mu = 1$.

__Theorem 6__   If P is a vertex and AB an edge of a connected acyclic graph G and if $W_1 = P\ Q_1\ Q_2\ \ldots\ Q_m\ A$ and

$W_2 = PR_1\ R_2\ \ldots\ R_n\ B$ are the unique (see Theorem 1) paths which

join P with A and B, respectively, then AB is either the last

edge Q_m A of W_1 or the last edge R_nB of W_2 (and therefore

Q_m = B, R_n = A, respectively).

<u>Proof</u> If Q_m A $\neq$ BA then $Q_m \neq$ B, since two edges never join the same two

vertices in an acyclic graph. So by Theorem 5 (μ = 1) $PQ_1\ Q_2\ \ldots\ Q_m$ AB

is a path. By Theorem 1 this must be identical to W_2; and so

AB = R_n B.///

    The following also holds, of course.

        <u>Theorem 7</u> Every finite acyclic graph is the sum of trees.

    If a vertex P is an endpoint of all the edges of a tree it is

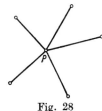

called a <u>star</u>[3] (see Fig. 28) and P is called its

center. The other endpoints of the edges are

naturally pairwise distinct; so all edges of a star

are end edges. Conversely,

<u>Theorem 8</u> If all the edges of a tree are end

Fig. 28   edges, then the tree is a star.

<u>Proof</u> If the tree has only one edge, then the theorem is true. (One

edge by itself is a star; in this one case the star has two centers.)

We therefore assume that the tree has two different edges, AB and CD;

they are, according to assumption, end edges: let A and C be endpoints

of the tree; so A $\neq$ C. Now we join A and C by a path. Since A and C

are endpoints, this path must be of the form AB ... DC, i.e. the first

edge of this path is AB, the last DC. This path cannot contain a third

edge, since only the end edges of a path can be end edges of a graph, so

that this path consists only of the two edges AB and DC. Therefore B=D,

and this vertex B=D is not an endpoint of the tree. We have thus shown

that if AB is any edge of our tree, one endpoint of the edge (A) is an endpoint of the tree, and the other one (B) is not. Furthermore, the latter vertex (B) is the endpoint of any second edge (CD) of the tree. So the tree has been shown to be a star.///

Now we prove a theorem first stated by Listing.

Theorem 9 For every tree the number α_0 of vertices is 1 more than the number α_1 of edges.

Proof If the tree has no other vertices than the two endpoints (see Theorem 4), it has a single edge and therefore $\alpha_0 = 2$ and $\alpha_1 = 1$ and the theorem is proved. By using induction we can assume that the theorem is true if $\alpha_0 = n$ and then prove the theorem for a graph G which has $\alpha_0 = n+1$ (>2) vertices (and α_1 edges). Let A be an endpoint and AB an end edge of G. If this end edge is removed (along with the endpoint A), what remains is again a tree (since by Theorem I.14 the connectedness of the graph is not lost); this new tree has $\alpha_0 - 1 = n$ vertices and $\alpha_1 - 1$ edges. According to our assumption, then,

$$(\alpha_0 - 1) - (\alpha_1 - 1) = \alpha_0 - \alpha_1 = 1.///$$

From the theorem just proved comes an immediate generalization: if a finite graph is the sum of ν trees, then $\alpha_0 = \alpha_1 + \nu$. This result can be extended to infinite graphs. We prove

Theorem 10 If α_0, α_1, and ν are the numbers (or cardinalities) of the vertices, edges, and components, respectively, of an arbitrary acyclic graph G, then $\alpha_0 = \alpha_1 + \nu$.

Proof It suffices to prove the theorem for connected graphs ($\nu = 1$). If subsequently the equations $\alpha_0^{(\gamma)} = \alpha_1^{(\gamma)} + 1$ relative to the individual components $G^{(\gamma)}$ are added, then the desired equation $\alpha_0 = \alpha_1 + \nu$ is obtained for the original graph $G = \sum G^{(\gamma)}$. So let G be taken as connected and let P be a vertex of G. Let us call the set of remaining vertices π_p. We shall designate the set of edges of G by K. The equation $\alpha_0 = \alpha_1 + 1$ will be proved if we show that π_p and K are equivalent. Let AB be any element of K; depending on whether (see Theorem 6) the last edge $Q_m A$ of the path P Q_1 Q_2 ... $Q_m A$ or the last edge ($R_n B$) of the path P R_1 R_2 ... R_n B is AB, we wish to assign the vertex A or B to the edge AB. If one of the vertices A and B, say A, is P, then the other vertex B is assigned to the edge AB, so that an element of π_p is assigned to every edge. In order to prove that this correspondence is one-to-one, we need only to show that every element M of π_p is assigned to one and only one edge. If P Q_1 Q_2 ... Q_m M is the path which joins P with M, then M was assigned to the edge Q_m M and only to this edge, since otherwise a second path would join P with M. The theorem is thus proved.///

If the graph has infinitely many components, the Axiom of Choice is used to choose the vertices P.

The one-to-one correspondence defined here for the sets π_p and K has the property that to every edge there corresponds one of its endpoints. We are therefore led to the following result:

> Theorem 11 If P is an arbitrary vertex of an acyclic connected graph G, then there is a one-to-one correspondence between the set of all vertices of G distinct from P and the set of all edges of G, which assigns to each edge one of its endpoints.

We shall see later (Theorem VI.4) in the case where G is infinite but of finite degree that we can be freed from the exceptional vertex P.

§2. Trees as Subgraphs-- Connectivity.

We wish now to prove a theorem which has as its origin the basic work of Kirchhoff [1] (1847), who is also responsible for the points following the theorem.[4]

> Theorem 12 Every connected finite graph G contains a subgraph which is a tree and which has the same vertices as G.

Proof   If G is a tree, then it is itself a subgraph with the desired

property.   If this is not the case then it contains a cycle. We remove

an edge of this cycle, and the graph (Theorem I.14) remains connected.

If a tree results (and from the following it turns out that this is the

case if and only if $\alpha_0 = \alpha_1$) then we are done; if not, we again remove

an edge of a cycle, etc.   Because of the finiteness of the number of

edges this process must come to an end, i.e.: we must reach a graph G'

which contains no cycle.

   Since we have been removing only edges of a cycle, the

connectedness of the graph cannot be lost, and for any vertex P of G not

all edges going to P can be removed. So G' is a subgraph with the

desired property.///

   In the next paragraph we shall extend this theorem to infinite

graphs.

   If a finite connected graph with α_0 vertices is given, edges can be

removed in different ways so that the resulting graph is a tree G' with

the same vertices. But the number μ of edges to be removed is always

the same. The tree G' contains α_0 vertices and $\alpha_1 - \mu$ edges. By theorem

9 $\alpha_1 - \mu = \alpha_0 - 1$, and therefore

$$\mu = \alpha_1 - \alpha_0 + 1.$$

So we have the following

> Theorem 13   If G is a finite graph with α_0 vertices
>
> and α_1 edges, and a tree with α_0 vertices results by removing
>
> certain edges from G, then the number of removed edges is
>
> always

$$\mu = \alpha_1 - \alpha_0 + 1.$$

   Actually this was proved only for connected graphs. This

condition, however, can be omitted, since a graph must be connected if

it contains a connected subgraph with the same vertices.

68

The number

$$\mu = \alpha_1 - \alpha_0 + 1$$

is called the connectivity of the connected finite graph.[5] If the

finite graph G is not connected[6] and it has components G_1, G_2, ..., G_ν,

which have connectivities μ_1, μ_2, ..., μ_ν, respectively, then the

connectivity of G is defined to be

$$\mu = \mu_1 + \mu_2 + \ldots + \mu_\nu = \sum_{i=1}^{\nu} (\alpha_1^{(i)} - \alpha_0^{(i)} + 1) =$$

$$\sum_{i=1}^{\nu} \alpha_1^{(i)} - \sum_{i=1}^{\nu} \alpha_0^{(i)} + \nu = \alpha_1 - \alpha_0 + \nu$$

(here $\alpha_0^{(i)}$, $\alpha_1^{(i)}$ are the number of vertices and edges, respectively, of

G_i for i = 1, 2, ..., ν.)

Now we have

Theorem 14   The connectivity μ of any finite graph is ≥ 0. We

have $\mu = 0$ if and only if the graph is the sum of trees and

therefore contains no cycle.

Proof   The first part of the theorem follows directly from Theorems 13

and 12 for connected graphs. We then also have, for disconnected

graphs, $\mu = \sum \mu_i \geq 0$. Theorems 9 and 13 tell us that if G is connected

then $\mu = 0$ if and only if G is a tree. If G is the sum of trees we

must likewise have $\mu = \sum \mu_i = 0$. If, on the other hand, $\mu = \sum \mu_i = 0$,

then every $\mu_i = 0$, since $\mu_i \geq 0$, and therefore G is the sum of trees.

For connected graphs the first part of the theorem proved above tells us

the following:

Theorem 15   A finite connected graph with α_0 vertices contains

at least $\alpha_0 - 1$ edges.

An important property of connectivity is stated in the following

theorem.

Theorem 16   If in a finite graph G an edge is removed from a

cycle[7], then the connectivity of the resulting graph is 1 less
than that of G.

Proof Let G_1, G_2, ... be the components of G and let the removed edge
PQ belong to G_1. Let us designate by G_1' the graph resulting from G_1 by
the removal of PQ. By Theorem I.14, G_1' is also connected. Since PQ
belongs to a cycle of G, P and Q belong to G_1'. So G_1' has the same
number of vertices as G_1, but one less edge. Hence, the connectivity of
G_1' is 1 less than that of G_1. Since the removal of PQ does not change
G_2, G_3, ..., the theorem is proved.///

It should be emphasized that it is essential that the removed edge

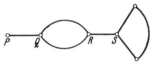

belongs to a cycle. If, for example, PQ or
RS is removed from the graph of Figure 29,
the connectivity remains 2.

Fig. 29.

We now prove four theorems which are
in a way converses of Theorem 13.

Theorem 17 If G is a connected finite graph with connectivity
μ and if a connected subgraph G' is obtained from G by removal
of μ edges, and if the vertices of G' are the same as the
vertices of G, then G' is a tree.

Proof Let the symbols α_0, α_1, and μ refer to the original graph and
α_0', α_1', and μ' to the subgraph. We have $\alpha_0' = \alpha_0$ and $\alpha_1' = \alpha_1 - \mu$ and
therefore, since $\mu = \alpha_1 - \alpha_0 + 1$, we have $\alpha_1' = \alpha_0 - 1$. If these values
of α_0' and α_1' are substituted in, then

$$\mu' = \alpha_1' - \alpha_0' + 1 = (\alpha_0 - 1) - \alpha_0 + 1 = 0,$$

and by Theorem 14 the subgraph, since it was assumed connected, is
consequently a tree.///

The following theorem was stated by Kirchhoff [1, §5] and was
proved by him using a physical assumption (see Chap. IX, §7).

Theorem 18 In order to obtain an acyclic graph from a finite graph with connectivity μ, at least μ edges must be removed.

Proof First assume that the graph G is connected. We remove μ' edges from G and assume that the resulting graph G' contains no cycles and therefore (Theorem 7) is the sum of trees:

$$G' = B_1 + B_2 + \ldots + B_k.$$

If $\alpha_0^{(i)}$ and $\alpha_1^{(i)}$ are the number of vertices and edges, respectively, of B_i (i = 1, 2, ..., k) then the connectivity of G is:

$$\mu = \alpha_1 - \alpha_0 + 1 = (\mu' + \sum_{i=1}^{k} \alpha_1^{(i)}) - (\rho + \sum_{i=1}^{k} \alpha_0^{(i)}) + 1$$

where ρ denotes the number of vertices of G which are not vertices of G'. On the other hand it follows from $\alpha_1^{(i)} - \alpha_0^{(i)} = -1$ (Theorem 9), if one sums over i = 1, 2, ..., k, that

$$\sum_{i=1}^{k} \alpha_1^{(i)} - \sum_{i=1}^{k} \alpha_0^{(i)} = -k;$$

if this is substituted in the expression obtained above for μ , we get

$$\mu = \mu' - \rho - k + 1.$$

But since $\rho \geq 0$ and $k \geq 1$ we have $\mu \leq \mu'$.

If G is not connected, then $G = \Sigma\, G_i$ and μ_i is the connectivity of the component G_i of G; so at least μ_i edges must be removed from G_i in order that no cycle remains. Therefore at least $\Sigma\, \mu_i = \mu$ edges must be removed.///

> Theorem 19 In order to obtain from a finite graph G with connectivity μ a connected subgraph G' with the same vertices, at most μ edges may be removed.

Proof The graph is, of course, connected if it has a connected subgraph with the same vertices. If μ' edges are removed and the resulting connected graph G' has connectivity μ'' (≥ 0), then μ'' more edges (by

Theorems 12 and 13) can be removed, so that the resulting graph is a tree with the same vertices. But then by Theorem 13 $\mu = \mu' + \mu''$ and so $\mu' \leq \mu$.///

By considering the just proved property of μ, we see how connectivity can be defined as a certain maximum number. Jordan [1, P.185] and Skolem [1, P. 302] defined it in this way. In a similar way connectivity can also be defined as a certain minimum number, corresponding to the property stated in Theorem 18; this is Kirchhoff's definition [1, P. 23].

> Theorem 20   If the resulting graph G', obtained by
> removing μ edges of the finite graph G with connectivity μ, is
> acyclic, then G' contains all the vertices of G.

Proof   If the vertex P of G were not contained in G', then if any edge PQ of G ending in P was added to the edges of G', a graph G" which also contains no cycles would result, since the end edge PQ of G" cannot belong to any cycle. But now G" is obtained from G by removal of $\mu - 1$ edges, while according to Theorem 18 at least μ edges must be removed in order to eliminate all cycles.///

§3.   Frames and Fundamental Systems.

A subgraph G' of an arbitrary (finite or infinite) graph G is called a frame[8] of G, if it has the following two properties:

>   1.   G' contains no cycle;
>   2.   if any edge of G not contained in G' is added to G', then
>   the resulting graph contains a cycle.[9]

For an acyclic graph the graph itself is the only frame of the graph.

For example, for the graph pictured in Fig. 30 the solidly drawn edges form a frame. For the plane quadratic lattice, infinite in all directions, all the vertical edges and the edges of a single horizontal line form a frame (see Fig. 31).

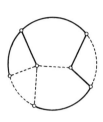

Fig. 30.   Fig. 31.

<u>Theorem 21</u> Every frame of a graph G contains the same vertices

as G.

The proof of this theorem is contained in the first sentence of the

proof which we gave for Theorem 20.///

<u>Theorem 22</u> Every frame G' of a connected graph G is itself

connected.

<u>Proof</u> Supposing the theorem is not true, there would be by Theorem I.15

an edge of G not contained in G', which would not form a cycle together

with the edges of G', and this contradicts the second defining property

of a frame. The frame of a finite connected graph is therefore a

tree.///

<u>Theorem 23</u> If a subgraph G' of G has the three properties:

1. G' is connected;
2. G' has the same vertices as G;
3. G' contains no cycle;

then G' is a frame of G.

<u>Proof</u> It is sufficient to show that any edge PQ of G not contained in

G' forms, together with certain edges of G', a cycle of G. Since P and

Q are contained in G' and G' is connected, there is a path in G', which

joins P with Q. This path, together with PQ, generates a cycle of G.///

The following theorem comes directly from the definition of a

73

frame:

> Theorem 24   If $G = \Sigma\, G_\gamma$, and S_γ is a frame of G_γ, then $\Sigma\, S_\gamma$ is a frame of G. And conversely every frame of G comes about by determining a frame S_γ for each component G_γ and summing them.

Now we prove

> Theorem 25   Every frame of a finite graph G with connectivity μ can be obtained by removing μ edges from G.

Proof   Let α_0 and α_1 be the number of vertices and edges, respectively, of G. If G is connected, then by Theorem 22 the frame S is a tree which contains (Theorem 21) α_0 vertices and therefore (Theorem 9) $\alpha_0 - 1$ edges. Therefore in S there are $\alpha_1 - (\alpha_0 - 1) = \mu$ edges of G missing. If, on the other hand, G is disconnected and $G_1, G_2, \ldots, G_\nu$ are its components: $G = \sum_{\gamma=1} G_\gamma$, then (Theorem 24) $S = \sum_{\gamma=1}^{\nu} S_\gamma$, where S_γ is a frame of G_γ, is therefore a tree. If the number of vertices of S_γ is denoted by $\alpha_0{}^{(\gamma)}$ and the number of edges by $\alpha_1{}^{(\gamma)}$, then $\alpha_1{}^{(\gamma)} = \alpha_0{}^{(\gamma)} - 1$. Summing, we get that $\sum_{\gamma=1}^{\nu} \alpha_1{}^{(\gamma)} = \sum_{\gamma=1}^{\nu} (\alpha_0{}^{(\gamma)} - 1) = \alpha_0 - \nu$ for the number of edges of S. Therefore

$\alpha_1 - (\alpha_0 - \nu)$ edges of G are missing from S. And this is again the connectivity of G.///

If a frame of G is obtained by removing μ edges, then the theorem, just proved above, states that if G is finite, then the connectivity of G must be μ. The question of whether the equation

$$\mu = \alpha_1 - \alpha_0 + \nu$$

also holds for infinite graphs with ν components has no meaning, since in general subtraction is not (uniquely) defined for infinite cardinal numbers. But we can ask if the equation

$$\alpha_0 + \mu = \alpha_1 + \nu$$

74

always holds. We wish to show that this is actually the case, and that therefore the following theorem holds:

> Theorem 26 Let α_o, α_1, and ν be, respectively, the number (or cardinal number) of vertices, edges, and components of an arbitrary graph G, and let S be a frame of G. If μ is the cardinal number of the set of those edges of G which are not contained in S^{10}, then
>
> $$\alpha_o + \mu = \alpha_1 + \nu.$$

Proof Let λ be the cardinal number of the edges of S; so we have that $\lambda + \mu = \alpha_1$ and therefore $\lambda + \nu + \mu = \alpha_1 + \nu$. According to Theorem 22 it follows from Theorem 24 that S also has ν components and from Theorem 21 that S also has α_o vertices. But then Theorem 10 tells us that $\lambda + \nu = \alpha_o$. If this is substituted into our equation, we get

$$\alpha_o + \mu = \alpha_1 + \nu. \ ///$$

As for the existence of a frame, Theorems 23 and 12 yield the result that a finite connected graph always has a frame. But by Theorem 24 this also holds for disconnected finite graphs. For the entire theory of infinite graphs it is significant that the existence of a frame can also be proved for arbitrary infinite graphs. The first proof we would like to give for this assertion is based on Zermelo's Well Ordering Principle. Let the totality of edges of G be ordered as a well ordered sequence:

$$k_1, k_2, \ldots, k_\omega, k_{\omega+1}, \ldots, k_\alpha, \ldots \tag{1}$$

Those of the edges k_α which taken together with certain edges k_β of smaller index form a cycle form a subsequence

$$k_{i_1}, k_{i_2}, \ldots, k_{i_\omega}, k_{i_{\omega+1}}, \ldots, k_{i_\alpha}, \ldots \tag{2}$$

The edges of G not contained in (2) form the complementary subsequence

$$k_1, k_{j_2}, \ldots, k_{j_\omega}, k_{j_{\omega+1}}, \ldots, k_{j_\alpha}, \ldots \tag{3}$$

75

We prove that the subgraph G' of G formed by the edges of (3) is a frame of G.

The first defining property of a frame is fulfilled: G' contains no cycle, since the edge of greatest index of any cycle of G belongs to (2) and not to (3) (among the finitely many edges of a cycle there is one of greatest index.)

It is therefore necessary only to show that any edge from (2), along with certain edges of (3), produces a cycle. If this were not the case, then there would be in (2) an edge k_β of smallest index with the property that every cycle of G which contains k_β contains at least one more edge from (2). We consider the cycle K which -- according to the definition of the sequence (2) -- contains outside of k_β only edges of index smaller than β. If k_γ is an edge of K, distinct from k_β, which belongs to (2), then $\gamma < \beta$, and so, by the definition of k_β and the sequence (2), the edge k_γ is contained in a cycle K_γ which contains outside of k_γ only edges from (3). The edges of K_γ distinct from k_γ form a path W_γ, which joins the endpoints of k_γ with each other and does not contain k_γ, and therefore contains only edges from (3). If the edge k_γ in K is replaced by this path, and if the corresponding replacement is carried out for all edges belonging to (2), with the exception of k_β, while the edges of K belonging to (3) are left alone, there results a walk consisting of edges belonging to (3) which joins the endpoints A and B of k_β. But then by Theorem I.3 certain edges of (3) form a path, which joins A and B. If k_β = AB is added to this path, a cycle results which contains the edge k_β as the only edge from (2). And this contradicts the definition of k_β[11].

So in general we have

Theorem 27. Every graph has a frame.

We wish to give a second proof[12] for this theorem which uses only the Axiom of Choice (and not the Well Ordering Principle). By Theorem 24 we may assume that G is connected, so that for any two vertices there exists a definite distance between the points. Let 0 be an arbitrary vertex of G, which we fix once and for all. An arbitrary vertex A then has a distance $d(A)$ from 0 (where we naturally set $d(0) = 0$.) If there are several paths from A to 0 of length $d(A)$, one of them should be chosen (by the Axiom of Choice). The edge of this path leading to A should be designated as the edge assigned to vertex A. Thus to every vertex A distinct from 0 is assigned an edge leading to A. These assigned edges form the subgraph S of G; we prove that S is a frame of G.

If the edge AB is assigned to the vertex A, then

$$d(B) = d(A) - 1. \tag{4}$$

So from every vertex A ($\neq 0$) of G there runs an edge of S to a vertex B which has a distance from 0 that is 1 less; likewise there runs from B an edge of S to a vertex C, the distance of which from 0 is again 1 less, etc. In this way after $d(A)$ steps the vertex 0 is reached, so that every vertex of G is joined to 0 by a path of S. So by Theorem I.5, we have that S is connected.

As we have just seen, 0 is a vertex of S; for the rest of the vertices of G this is clear; and so S has the same vertices as G.

Finally we must still show that S contains no cycle. We suppose that $P_1 P_2 \ldots P_n P_1$ is a cycle of S and let P_i be the vertex of this cycle which has maximum distance from 0. (Naturally $P_i \neq 0$.) Then by (4) the edge $P_i P_{i+1}$ cannot be the edge assigned to the vertex P_{i+1}; it is therefore assigned to the vertex P_i. The same holds for the edge

77

P_i P_{i-1} (in the case i = 1, n is to be written for i-1). But this is impossible, since to each vertex there was assigned only one edge.

We have now demonstrated the three properties in Theorem 23 for S; this theorem states then that S is a frame of G.///

Now we prove

> Theorem 28   If S is a frame of G and k is an edge of G which
> does not belong to S, then there is one and only one cycle K_k
> of G which contains the edge k, the other edges of which belong
> to S.

The existence of such a cycle follows from the definition of a frame. If there were two such cycles, K_k and K_k', then by Theorem I.9 a cycle could be formed from the edges of K_k and K_k' which does not contain k, and therefore would be a cycle of the frame.///

We shall say that all the cycles K_k assigned in this way to the edges k of G which do not belong to S form a fundamental system[13] F_s of G belonging to the frame S. By Theorem 27 every graph has a fundamental system. For an acyclic graph the empty set must naturally be considered as the only fundamental system.

From this definition there follows immediately:

> Theorem 29   Every cycle of a fundamental system contains an
> edge which belongs to no other cycle of this same fundamental
> system.

The fundamental system F_s of G was defined as a set of cycles which is equivalent to the set of those edges of G which do not belong to S. If G is finite, we get the following from Theorem 25:

> Theorem 30   Every fundamental system of a finite graph with
> connectivity μ consists of μ cycles.

Some results of this chapter (as well as some subsequent results of

Chapter IX) form the foundation for certain important studies which connect general combinatorial topology (the theory of one dimensional and multidimensional complexes) with group theory. The fact that these studies were not included in this book can be excused by the fact that they involve more group theory than graph theory. We shall here only refer to Reidemeister's [1] treatment of the subject.

Notes on Chapter IV

[1] The development of the theory of trees is due partly to the
fact that trees play a role in some investigations in physics (see
Chapter IX, §7) and chemistry (see Chapter V, §5). As the first ones
Kirchhoff [1, §1] and von Staudt [1, pp. 20-21] happened upon graphs in
the same year, 1847. With Kirchhoff they appeared, when he determined
certain wires (edges) in linear galvanic current branchings, "after
whose removal no closed figure remains"; von Staudt arrived at the
concept of tree with the elegant and exact proof, which he gave for
Euler's polyhedron theorem. Independent of both Kirchhoff and von
Staudt the long series of works of Cayley begins in 1857 [among others
1, 2, 3, 4, 7, 9] concerning the "analytical forms called trees," to
which -- in mutual cooperation with Sylvester -- he was led mostly by
the problem of organic chemistry which seeks to determine the form and
number of certain chemical bonds. Certainly the first work of Cayley's
[1] on trees arose in another way; he wished, in connection with studies
of Sylvester's concerning the transposition of variables in differential
calculus, to form graphically a sequence of differentiation processes
$Aa_x + Ba_y + \ldots$ (where A, B depend on the variables x, y, ...). --
Likewise independent of their predecessors and of each other Listing in
1862 [2, §27] and Jordan in 1869 [1] arrived at the concept of a tree
and indeed out of purely mathematical speculations ("without having any
suspicion of its bearing on modern chemical doctrine" -- writes
Sylvester [1, p. 24] concerning Jordan). We treat the studies of
Jordan, Cayley, and Sylvester in the next chapter. The designation tree
stems from Cayley. Sylvester [4] wished to introduce the designation
ramification instead of this and save the word tree only for such trees
in which a vertex was distinguished as root (for more about this see

80

chapter V, §5). In French, outside of <u>arbre</u> the designations

<u>ramification</u> and -- in Jordan [1] -- <u>assemblage</u> (<u>de</u> <u>lignes</u>)

<u>à</u> <u>continuité</u> <u>simple</u> are also used.

    Since 1921 the concept of tree has also played an important role in

the new theory of curves, see Menger [4, Chapter X] and the literature

(mostly Polish and American) cited there. In this theory the

designations <u>Baumkurve,</u> <u>acyclic</u> (<u>continuous</u> <u>curve</u>), and <u>dendrite</u> are

used.

    [2] The exceptional cases would disappear if we would regard the

edges as continua and indeed as arcs of curves and the graph as a curve

(of the new -- for example, Menger's -- theory of curves). In this

sense Theorem 3, together with its converse, was stated and proved by

Menger [2, p. 573].

    [3] This designation (étoile) stems from Sainte-Lague [4, p. 5].

    [4] Outside of the works still to be named cf. also Reidemeister

[1, Chapter IV, §§4-5] and Steinitz [1, §25].

    [5] The number $\alpha_1 - \alpha_0 + 1$ appears first in von Staudt [1 Pp. 20-

21] and in Kirchhoff [1, p. 23,31]. Listing [2, §21] designated it as

<u>cyklomatische</u> <u>Ordnungszahl</u> (Rang). In Jordan [1] the number 1 greater

is designated as <u>degré</u> <u>de</u> <u>continuité</u> .

    [6] For disconnected graphs, connectivity (cyclomatic number) was

introduced by Veblen [2, pp. 8-9].

    [7] This operation is designated by Listing [2, §9] as <u>Dialyse.</u>

Also only the name stems from Listing; the concept already occurred in

von Staudt and Kirchhoff. "Dialyse" plays here a role similar to the

cutting up of a surface along a nondisintegrating cross section in the

Riemannian topology of surfaces; the elementary surface corresponds here

to the trees.

[8] Translator's note: "Frame" is a translation of the German word "Gerüst," suggested by W.T. Tutte. Professor Tutte pointed out to the translator that the translation of "Gerüst" as "spanning forest" or "spanning tree" is wrong because a "tree" in König is always a finite graph whereas a "Gerüst" is not necessarily so, and much of the chapter is concerned with infinite "Gerüste". "Forest", he pointed out, is Whitney's term and "spanning forest" is, he thought, post-König and therefore undesirable in the translation. He also clarified that components of a forest are trees and thus finite, whereas components of a "Gerüst" can be infinite.

[9] The concept of frame can be characterized most clearly through the concept of basis. Since the concept of basis plays an important role elsewhere in graph theory, we would like to give its definition here. Let M be an arbitrary set; let one or more elements be assigned to certain subsets of M; then a subset B of M is called a basis for M with respect to this correspondence, if 1. each element of M not belonging to B is assigned to a subset of B (first basis property) and 2. no element of B is assigned to a subset of B which does not contain this element (second basis property). The definition given here of a frame is then equivalent to the following: Let G be an arbitrary graph; let K be the set of its edges; an element k of K is assigned to a subset K' of K if k forms with certain edges of K' the edges of a cycle. Now certain edges form a frame of G if they form a basis for the set K of edges with respect to this correspondence.

-- The edges, which are removed from G in order to obtain a frame of G, determine a "group" of edges in the terminology of Ahrens [1, p. 315].

[10] Later (see Chapter IX, end of §3) we shall see that even for infinite graphs this (finite or infinite) μ has the same value for every

frame and so depends only on G.

[11] The set theoretical basic idea of this proof was first used by G. Hamel: Eine Basis aller Zahlen ..., Mathematische Annalen, 60, 1905, pp. 459-462.

[12] I owe this fine proof to the kind communication of Mr. G. Hajós.

[13] This name was introduced by Ahrens [1, §4] for connected finite graphs. The concept stems from Kirchhoff [1].

CHAPTER V

Centers of Trees

§1. Center and Axis of linear dimension.

The concepts of <u>length</u> of a path, <u>distance</u> between two vertices,
and <u>diameter</u> of a graph, which we have introduced in a general way for
arbitrary (finite) connected graphs, play a special role in the theory
of trees. This is based on the fact that in the case of a tree -- in
which two vertices can always be joined by only one path -- the distance
between two vertices, A and B, does not need to be thought of as a
minimum but is simply the length of the path which joins[1] A and B.

Let $W = P_1 P_2 \ldots P_n$ be a path of a tree and let one of the
endpoints of this path, say P_n, not be an endpoint of the tree. Then
there is, besides $P_{n-1} P_n$, a second edge $P_n P_{n+1}$ ending in P_n.
Then $P_1 P_2 \ldots P_{n-1} P_n P_{n+1}$ is (see Theorem IV.5 for $\mu = 1$) also a
path of the tree; its length is 1 greater than that of W. So we have

> **Theorem 1** If P is an arbitrary vertex of the tree B, then the
> vertex of B which is at the greatest distance from P is an
> endpoint of B.
>
> **Theorem 2** A longest path of a tree joins two endpoints of the
> tree.

If all the end edges of a tree B are deleted, the edges remaining
(if there are any) also form a tree B' by Theorems IV.7 and I.14.
Likewise by deletion of the end edges of B' there results a tree B",
etc. Because the number of edges is finite we must (see Theorem IV.4)

end up after a certain r^{th} step with the null graph. This number r we

shall designate as the <u>radius</u> of the tree B. After the next to last

step we have a tree $B^{(r-1)}$, the edges of which are all end edges;

therefore, by Theorem IV.8, $B^{(r-1)}$ is a star. Now there are two cases

to be distinguished, according to whether the number of edges of $B^{(r-1)}$

is 1) equal to 1, or 2) greater than 1. In the first case the unique

edge of $B^{(r-1)}$ is called the <u>axis</u> of the tree B, and the endpoints of

this edge are called <u>central points</u> (bicenters) of B. (The central

points together are called the <u>center</u>.) In the second case the common

endpoint of all edges of $B^{(r-1)}$ is called the center of B. So the tree

of Fig. 32, for example, has a center consisting of C and the tree in

Fig. 33 has a center consisting of C_1 and C_2 and has an axis C_1C_2.

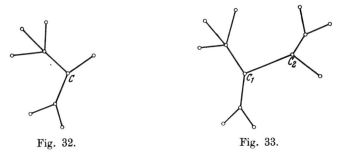

Fig. 32.   Fig. 33.

These concepts were introduced[2] by Jordan (1, pp. 188-189) and, in fact,

by the construction conducted here. In contrast to other centers of

trees (see further §4 below) these centers were designated by Dehn and

Heegard [1, p. 176] also as centers of linear dimension.

The designations B, B', B", . . . $B^{(r-1)}$ will be retained in our

further work. From the given definitions we immediately have

Theorem 3 The trees B, B', B", . . . , $B^{(r-1)}$ all have the

same center and the same axis (the latter if there are two

central points).

Now we prove

85

Theorem 4 If $P_0 P_1 P_2 \ldots P_{d-1} P_d$ is a longest path of the tree B, then $P_1 P_2 \ldots P_{d-1}$ is a longest path of B'.

Proof First it is clear that the path $P_1 P_2 \ldots P_{d-1}$ belongs completely to B'; the vertices $P_1, P_2, \ldots, P_{d-1}$ as inner vertices of a path of B cannot be endpoints of B and therefore the edges $P_1 P_2, \ldots, P_{d-2} P_{d-1}$ cannot be end edges of B. Now let $Q_0 Q_1 Q_2 \ldots Q_k$ be a longest path of B'. Then k is the diameter of B'. According to Theorem 2 the vertices, Q_0 and Q_k, are endpoints of B' but not of B, since otherwise $Q_0 Q_1$ and $Q_{k-1} Q_k$ as end edges of B could not belong to B'. Therefore there is in B an edge RQ_0 distinct from $Q_0 Q_1$ and an edge $Q_k S$ distinct from $Q_{k-1} Q_k$, neither of which belongs to B' (i.e. they are end edges of B). Thus we have obtained (see Theorem IV.5) a path $R Q_0 Q_1 \ldots Q_k S$ of B, which has length k + 2. Since we have designated the diameter of B by d, we have that $k + 2 \le d$, $k \le d - 2$. But now the longest path of B' has length at least the length of the path $P_1 P_2 \ldots P_{d-1}$ and therefore $k \ge d - 2$. Thus k = d - 2 and so $P_1 P_2 \ldots P_{d-1}$ is actually a longest path of B'.///

So we also have the following result:

Theorem 5 The diameter of the tree B is 2 greater than the diameter of B'.

If the diameters of B, B', B'', ..., $B^{(r-1)}$ are designated by d, d', d'', ..., $d^{(r-1)}$, respectively, then d = d' + 2, d' = d'' + 2, ..., $d^{(r-2)} = d^{(r-1)} + 2$ and so $d = d^{(r-1)} + 2r - 2$.

Now it is clear that the diameter $d^{(r-1)}$ of the star $B^{(r-1)}$ is either 1 or 2 depending on whether it has one edge or more than one edge, and so depending on whether the center of B has two vertices or one. If the center of B has two vertices, d = 2r - 1, and if the center of B has one vertex, d = 2r. So we have

86

Theorem 6   If the radius of a tree is r and its diameter is d, then d = 2r if the tree has one central point and d = 2r - 1 if the tree has two central points (and so has an axis).

We have also obtained the following result:

Theorem 7   A tree has one central point if its diameter is an even number. A tree has two central points (and an axis) if its diameter is odd.

Now we wish to prove one more property of the center, which Cayley [3, p. 429] used as a definition of center.[3]

Theorem 8   Let W be a longest path of the tree B. If the number of edges of W is even, the center of B consists of the middle vertex of W; if the number is odd, then the middle edge of W is the axis of B.[4]

Proof   The theorem is clear in the case that the radius r of B is equal to 1 (in which case we have a star). So by using induction we can consider the theorem true for radius r - 1 and assume that r > 1. If both end edges of W are removed, the path W' remaining is a longest path of B' (Theorem 4). Since -- according to the definition of radius -- B' has radius r - 1, the middle vertex (middle edge) of W is by induction assumption the central point (axis) of W'. Now the middle vertices (middle edges) for W and for W' are, of course, identical; and -- according to Theorem 3 -- B and B' have the same center. The theorem is therefore proved.///

§2. Characteristic properties of the center

The following two theorems express characteristic properties of the center.

Theorem 9   If r is the radius and C is a central point of the tree B, then r is the length of the longest path of B which goes to C.

Proof First, let C be the only central point of B. We know (Theorem 8)

that C is the middle vertex of any longest path

$P_r\ P_{r-1}\ \cdot\ \cdot\ \cdot\ P_1\ C\ Q_1\ Q_2\ \cdot\ \cdot\ \cdot\ Q_r$ of B and consequently the path

$P_r\ P_{r-1}\ \cdot\ \cdot\ \cdot\ P_1$ C ending in C has length r. Now let $C\ R_1\ R_2\ \cdot\ \cdot\ \cdot\ R_k$

be any path of B which ends in C. If $C\ R_1 \neq C\ P_1$ (i.e. $R_1 \neq P_1$) then by

Theorem IV.5 $P_r\ P_{r-1}\ \cdot\ \cdot\ \cdot\ P_1\ C\ R_1\ R_2\ \cdot\ \cdot\ \cdot\ R_k$ is a path of length

r + k; but if $C\ R_1 = C\ P_1$, then $C\ R_1 \neq C\ Q_1$ and so the

path $Q_r\ Q_{r-1}\ \cdot\ \cdot\ \cdot\ Q_1\ C\ R_1\ R_2\ \cdot\ \cdot\ \cdot\ R_k$ has length r + k. But now the

diameter of B is d = 2r, and so $r + k \leq 2r$, and from this it follows

that $r \geq k$.

Secondly, let C and C' be the two central points of B. Then by

Theorem 8 every longest path of B has the form

$P_{r-1}\ P_{r-2}\ \cdot\ \cdot\ \cdot\ P_1\ C\ C'\ Q_1\ Q_2\ \cdot\ \cdot\ \cdot\ Q_{r-1}$. Its length

(r - 1) + 1 + (r - 1) = 2r - 1 is the diameter d of B. And

$C\ C'\ Q_1\ Q_2\ \cdot\ \cdot\ \cdot\ Q_{r-1}$ is a path of B ending in C which has length r. We

show again that the length k of any path $C\ R_1\ R_2\ \cdot\ \cdot\ \cdot\ R_k$ of B ending in

C is at most equal to r. If $R_1 \neq C'$, then by Theorem IV.5

$R_k\ \cdot\ \cdot\ \cdot\ R_1\ C\ C'\ Q_1\ Q_2\ \cdot\ \cdot\ \cdot\ Q_{r-1}$ is a path of length k + r. But if

$R_1 = C'$ (and therefore $R_1 \neq P_1$) then $R_k\ \cdot\ \cdot\ \cdot\ R_1\ C\ P_1\ \cdot\ \cdot\ \cdot\ P_{r-1}$ is a

path of length k + r - 1. But now the diameter of B is d = 2r - 1; we

must have in the first case $k + r \leq 2r - 1$ and therefore $k \leq r - 1$; in

the second case $k + r - 1 \leq 2r - 1$, and therefore $k \leq r$. Thus in both

cases $k \leq r$.[5]///

The theorem just proved can be turned around as follows:

> Theorem 10 If A is any vertex of the tree B which is not a
>
> central point, then there is a path of B ending in A whose
>
> length is greater than the radius r of B.

Proof In the case that there is only one central point C let

P_r . . . P_1 C Q_1 . . . Q_r be a longest path of B and let

A R_1 R_2 . . . R_{k-1} C $(k \geq 1)^6$ be the path which joins A with C. If

Theorem IV.5 is again applied, then in the case $R_{k-1} \neq Q_1$, the walk

A R_1 . . . R_{k-1} C Q_1 . . . Q_r is a path, which has length k + r; but if

R_{k-1} = Q_1 and therefore $R_{k-1} \neq P_1$, then A R_1 . . . R_{k-1} C P_1 . . . P_r

is a path of length k + r. In both cases we have found a path ending in

A, whose length, k + r, is greater than r.

In the case that there are two central points C and C', we can

reason in a similar way, and Theorem IV.5 is again applied. Let

P_{r-1} P_{r-2} . . . P_1 C C' Q_1 . . . Q_{r-1} again be a longest path of B. If

A is joined with C by an edge AC, then, since A $\neq$ C',

A C C' Q_1 . . . Q_{r-1} is a path of length r + 1 > r. We may therefore

assume that no edge AC exists and that therefore the length k of the

path A R_1 R_2 . . . R_{k-1} C, which joins A with C, is greater than 1. If

$R_{k-1} \neq$ C', then A R_1 . . . R_{k-1} C C' Q_1 . . . Q_{r-1} is a path of length

k + r > r (in fact, > r + 1). If, on the other hand, R_{k-1} = C' and

therefore $R_{k-1} \neq P_1$, then the path A R_1 . . . R_{k-1} C P_1 . . . P_{r-1}

has length k + r - 1 > r.///

Theorems 9 and 10 can be proved by induction (on the radius). It

is left to the reader to carry out these proofs as exercises.

Theorems 9 and 10 give the solution to the following problem. We

wish, for a tree, to determine those vertices for which the maximal

distance of vertices is at a minimum, and we wish to determine this

minimum.[7] By applying Theorem 6 the diameter can be introduced instead

of the radius, and then our result can be stated as follows:

> Theorem 11 In every tree the maximal distance of a vertex from
>
> the remaining vertices assumes its minimal value only for a
>
> central point. If d is the diameter of the tree, then this
>
> minimum is $\frac{d}{2}$ in the case of a single central point and

$\frac{d+1}{2}$ in the case of two central points.[8]

The problem solved here can be formulated not only for trees but for arbitrary finite connected graphs. But in this general case the problem has little interest, since it can have arbitrarily many solutions. In fact, every vertex can be a solution if, for example, the graph is a cycle.

Now we prove the following theorem:

> Theorem 12 If the tree B has a single central point C, then C
> is the only vertex in B with the following property. There are
> two paths leading to C, which have two distinct edges going to
> C and whose lengths are equal to each other and greater than or
> equal to the length of any third path which goes to C.

Proof That C has this property is clear: a longest path contains, by Theorem 8, C as the middle vertex, and since it has length d=2r by Theorem 6, it has the form $P_r P_{r-1} \ldots P_1 C Q_1 Q_2 \ldots Q_r$; the two paths going to C, $C P_1 \ldots P_r$ and $C Q_1 Q_2 \ldots Q_r$ (where $P_1 \neq Q_1$) both have length r, and by Theorem 9 a longer path cannot go to C. If now $A \neq C$ and $A P_1 P_2 \ldots P_k$ and $A Q_1 Q_2 \ldots Q_k$ were two paths of common maximal length k and with different edges going to A, then by Theorem 10 the inequality k > r would have to hold. But then by Theorem IV.5 $P_k P_{k-1} \ldots P_1 A Q_1 \ldots Q_k$ would be a path; it would have length 2 k > 2r = d, which contradicts the definition of diameter (see Theorem 6).///

Correspondingly there holds for trees with an axis the following

> Theorem 13 If the tree B has an axis s, then s is the only
> edge of B whose endpoints, C and C', have the following
> property. There are two paths of equal length, which go to C
> and C', respectively, which do not contain the edge s and whose
> lengths are greater than or equal to the length of any third

path, which goes to C or C' and does not contain s.

Proof   That s = C C' has this property can be seen by the following

argument. By Theorem 8 a longest path

P_{r-1} P_{r-2} . . . P_1 C C' Q_1 Q_2 . . . Q_{r-1} determines the two paths

C P_1 P_2 . . . P_{r-1} and C' Q_1 Q_2 . . . Q_{r-1} which do not contain C C' and

which have the common length r - 1. If there were a path

C R_1 R_2 . . . R_k (or C' R_1 R_2 . . . R_k) which also did not contain

C C' (and therefore $R_1 \neq$ C' or $R_1 \neq$ C, respectively) and whose length k

satisfied the inequality k > r - 1, then C' C R_1 R_2 . . . R_k or

C C' R_1 R_2 . . . R_k, respectively, would be a path, whose length k + 1

would satisfy the inequality k + 1 > r, which contradicts Theorem 9.

As for the second part of the proof, let k = M N be an edge of the

tree distinct from the axis. We can assume, for instance, that M is

distinct from the central points. Now we assume that there are two

paths M P_1 P_2 . . . P_k and N Q_1 Q_2 . . . Q_k, which do not contain M N

and which have the common maximal length. By Theorem 10 there is a path

M R_1 R_2 . . . R_{r+1} of length r + 1. If $R_1 \neq$ N here, then this path does

not contain the edge M N; but if R_1 = N, then we have the path

N R_2 R_3 . . . R_{r+1} of length r. In any case there is always a path

going to either M or N and not containing M N, the length of which path

is greater than r - 1. Therefore we have k > r - 1 for the maximal

length k. But then the length of the path

P_k P_{k-1} . . . P_1 M N Q_1 Q_2 . . . Q_k is 2k + 1 > 2r - 1 = d, which

contradicts the definition of diameter.///

The concept of center can also be introduced by Theorems 12 and

13. Cayley chose this way [4, p. 599, cf. also 3, p. 429] in connection

with the solution which he gave for a problem posed by Sylvester [2].

[These investigations have an external analogy with certain

investigations of metric geometry, which developed later quite independently from graph theory. In these investigations points of a bounded closed point set are considered instead of vertices of a graph. The point set, for instance, is in the plane, and the word "distance" is understood in its usual metrical sense. The maximal distance of any two points of the set is called the diameter. As in the case of a tree with one central point, there is a uniquely determined point, whose maximal distance from all points of the set has a minimum value. This is the center of the unique "smallest covering circle" of the set. Its center corresponds to the (unique) central point of a tree, and its radius corresponds to the radius of the tree. (Certainly with point sets this radius r is not determined by the diameter d, but there holds in general only the inequality[9] $r \leq \dfrac{d}{3}$.) Further analogies exist, if special point sets are considered, e.g. those point sets which are bounded by a Jordan curve in the plane.]

§3. Branches of Trees.

Height of Vertices

In order to be able to introduce a second kind of central point of trees, a few simple remarks are necessary regarding "branches" of trees.

Let PQ be an arbitrary edge of the tree B. All the edges of the tree which are contained in any path of the form PQX...Z form a connected graph by Theorem I.5, and therefore a tree. This is designated[10] as "the branch (P, PQ) of B coming from P" (it is different from the branch (Q, QP)).

Let PQ_1 PQ_2, ... , PQ_n be the edges of B going to P. It follows directly from Theorem I.7 that two branches (P, PQ_i) and (P, PQ_j) $(i \neq j)$ have no common edge and, except for P, no common vertex. On the other hand every vertex and every edge of B is contained in one of the edges (P, PQ_i) $(i = 1, 2, \ldots, n)$, since B is connected and every path

which joins P with any vertex of B begins with one of the edges PQ_1,

$PQ_2, \ldots, PQ_n$.

It follows immediately from the definition of branch that we have

Theorem 14   If PQ and PR are two different edges of a tree,

then the edge (P, PR) is a proper subgraph of the branch

(Q, QP).

Now we prove

Theorem 15   If $A_1 A_2 \ldots A_n$ is any path of a tree then every

edge of the tree belongs to one of the branches $(A_1, A_1 A_2)$ and

$(A_n, A_n A_{n-1})$.

Proof If an edge k does not belong to $(A_1, A_1 A_2)$, then it belongs, as

we have already seen, to a branch $(A_1, A_1 A_2')$ where $A_2' \neq A_2$.

Therefore there is a path $A_1 A_2' A_3' \ldots A_{m-1}' A_m'$, where $A_{m-1}' A_m'$ is

the given edge k. Since $A_2' \neq A_2$, this path, together with the path

$A_1 A_2 \ldots A_n$, yields a path $A_n A_{n-1} \ldots A_1 A_2' \ldots A_{m-1}' A_m'$. Therefore

$k = A_{m-1}' A_m'$ is an edge of the branch $(A_n, A_n A_{n-1})$.///

For n = 2 the theorem just proved states that if PQ is an arbitrary

edge of a tree, every edge of the tree belongs to one of the edges

(P, PQ) and (Q, QP). This theorem can be supplemented by the following:

Theorem 16   Every edge PQ of a tree is the only common edge of

the branches (P, PQ) and (Q, QP).

Proof If there were a second such edge k, then there would be an

endpoint A of this edge distinct from P and from Q. Since k belongs to

(P, PQ), a path which does not contain PQ leads from Q to A, and since k

also is contained in (Q, QP), a path which also does not contain PQ

leads from P to A. By Theorem I.5 certain edges of these two paths

yield a path from P to Q not containing PQ. This path would produce

with PQ a cycle of the tree.///

If PQ_1, PQ_2, ... , PQ_n are the edges of the tree B going to P, we
shall denote the number of edges of the branch (P, PQ_i) by h_i
(i = 1, 2, ... , n). The largest of the numbers h_1, h_2, ... , h_n is
called the <u>height</u> of the vertex P in the tree B. Concerning heights
there holds the following.

Theorem 17   If a tree contains α_1 edges then the minimal height

q of its vertices is $\leq \dfrac{\alpha_1 + 1}{2}$.   11

<u>Proof</u> Let P be the vertex (or one of the vertices) of minimal height q
and let PP_1, PP_2, ... be the edges going to P. We designate the number
of edges of the branches (P, PP_1), (P, PP_2), ... by p_1, p_2, ...
Without loss of generality let p_1 be the largest of these numbers. So
p_1 = q. Let the edges going to P_1 be P_1P, P_1Q_2, P_1Q_3, ... (Fig. 34),

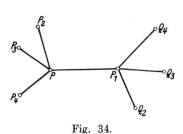

and let the number of edges of
the branches (P_1, P_1P),
(P_1, P_1Q_2), (P_1, P_1Q_3), ...
be r_1, r_2, r_3, ... So we have
$r_2 + r_3 + ... = q - 1$, since the
edges of the branches (P_1, P_1Q_i)
(i = 2, 3, ...) yield all the
edges of (P, PP_1), with the
single exception of PP_1. The

Fig. 34.

numbers r_2, r_3, ... are therefore, taken together, < q. If now it were
true that r_1 < q, then the largest of the numbers r_1, r_2, r_3, ... would
be < q, and then q would not be the minimal height. Therefore $r_1 \geq q$.
Now (by Theorem 15) every edge belongs to one of the two branches
(P, PP_1) and (P_1, P_1P), and (Theorem 16) PP_1 is the only edge which
belongs to both branches; therefore $q + r_1 = \alpha_1 + 1$ and consequently,
since $r_1 \geq q$, we have $2q \leq \alpha_1 + 1$ .///

Now we can prove the following

> Theorem 18   If P is one of the vertices of a tree which have
>
> minimal height q and U is a vertex distinct from P, then the
>
> height h of U is always $\geq \dfrac{\alpha_1 + 1}{2}$ (α_1 is the number of edges of
>
> the tree). Equality holds here if and only if $q = \dfrac{\alpha_1 + 1}{2}$ and
>
> U is the only vertex P_1 adjacent to P, for which the branch
>
> (P, PP_1) has q edges.[12]

Proof Let $P_1, P_2, \ldots, P_\mu, \ldots$; $q, p_1, p_2, \ldots, p_\mu, \ldots$ have the same
meaning as in the proof of Theorem 17 and let $q = p_1$. Furthermore let
$PP_\mu \ldots ZU$ be the path which joins P with U and let r be the number of
edges (length) of this path. We consider the branch (U, UZ); let the
number of its edges be ρ (see Fig. 35). By Theorem 15 every edge
belongs to one of the two branches (P, PP_μ) and (U, UZ). Therefore

Fig. 35.

these branches have k common edges,
and $p_\mu + \rho - k = \alpha_1$ and $\rho = \alpha_1 - p_\mu + k$.
Since the r edges of the path $PP_\mu \ldots ZU$
are common edges of these branches, $k \geq r$
and therefore $\rho \geq \alpha_1 - p_\mu + r$. This

inequality remains true if r is replaced by 1 ($\leq r$) and p_μ by $\dfrac{\alpha_1 + 1}{2}$;
since q is the height of P, on the other hand in consequence of Theorem

17 we have $p_\mu \leq q \leq \dfrac{\alpha_1 + 1}{2}$. So the inequality

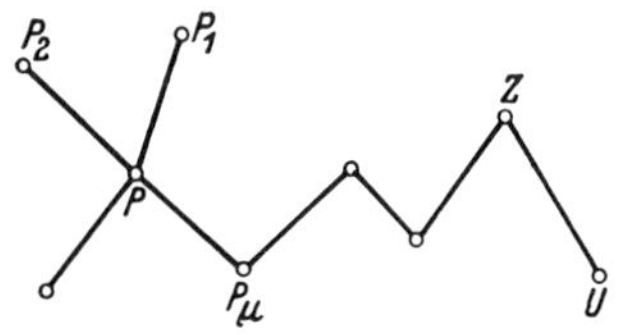

$\rho \geq \alpha_1 - \dfrac{\alpha_1 + 1}{2} + 1 = \dfrac{\alpha_1 + 1}{2}$ is obtained. It follows for the height h of U

that $h \geq \dfrac{\alpha_1 + 1}{2}$.

We would like now, in order to prove the second part of the theorem, to investigate when the "equals" sign holds in this inequality

$h \geq \dfrac{\alpha_1 + 1}{2}$. Of course, this is possible only if in the inequalities

$$r \geq 1 \text{ and } p_\mu \leq q \leq \dfrac{\alpha_1 + 1}{2}$$

equality holds in all three places; then $p_\mu = q = \dfrac{\alpha_1 + 1}{2}$ $(=p_1)$ and

$r = 1$. The latter means that $U = P_\mu$; so $U = P_\mu = P_1$, since otherwise we

would obtain the absurd result $\alpha_1 \geq p_1 + p_\mu = \dfrac{\alpha_1 + 1}{2} + \dfrac{\alpha_1 + 1}{2} = \alpha_1 + 1$.

The latter remark shows also that, in this case, among the vertices P_ν

the vertex P_1 is the only one for which the number p_ν of edges of the

branch (P, PP_ν) has the (maximal) value $p_1 = q$.

Now we must still show that conversely, in the case of

$q = \dfrac{\alpha_1 + 1}{2}$, the vertex $U = P_1$ just like P has the height $q = \dfrac{\alpha_1 + 1}{2}$. In

the opposite case, as we have just proved, the height of P_1 would be

greater than $\dfrac{\alpha_1 + 1}{2}$; so the number of edges of a branch (P_1, P_1X) would

be greater than $\dfrac{\alpha_1 + 1}{2}$. This is impossible in the case that $X \neq P$, for

then this branch would be contained, by Theorem 14, in the branch

(P, PP_1) with $q = \dfrac{\alpha_1 + 1}{2}$ edges.

But this is not possible for $X = P$, since the branch (P_1, P_1P) has

exactly $\alpha_1 - (\dfrac{\alpha_1 + 1}{2} - 1) = \dfrac{\alpha_1 + 1}{2}$ edges.///

§4. Centroid and centroid axis

Since, according to Theorem 17, a vertex U, whose height is

$> \dfrac{\alpha_1 + 1}{2}$, cannot be of minimal height, we can give the result expressed in Theorem 18 the following formulation:

> Theorem 19   In every tree with α_1 edges there is either a
>
> single vertex M or exactly two vertices, M_1 and M_2, of minimal
>
> height q, according to whether $q < \dfrac{\alpha_1 + 1}{2}$ (and therefore
>
> $q \leq \dfrac{\alpha_1}{2}$) or $q = \dfrac{\alpha_1 + 1}{2}$.   In the second case M_1 and M_2 are
>
> endpoints of the same edge.

In the first case M is called the centroid point of the tree, and
in the second case (which, of course, can happen only for odd α_1) M_1
and M_2 are called the two centroid points and the edge M_1M_2 the centroid
axis of the tree. The centroid of a tree is the set consisting of its
centroid points. So the centroid of a tree contains either one or two
vertices of the tree. These concepts were introduced by Jordan [1, pp.
186-187] in essentially the same way as described here. The tree of
Fig. 36 has one centroid point M, and the tree

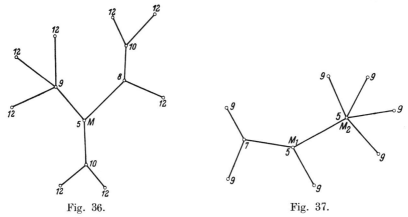

Fig. 36.   Fig. 37.

of Fig. 37 the two centroid points, M_1 and M_2. The numbers in these
figures give the height of the vertex in question. The characteristic
property of a centroid expressed in Theorem 19 can be expressed in a
somewhat modified form. We denote the numbers of edges of the branches
coming from any vertex P of the tree B by h_1, h_2, ... , h_ν . Then the

number of edges of B is $\sum_i h_i = \alpha_1$.

In the first case let one of the numbers h_i for a vertex P, say h_1, be greater than the sum of the remaining ones, and this then is the greatest and therefore the height of P. Then it follows that

$$2h_1 > h_1 + h_2 + \ldots + h_\nu = \alpha_1,$$

and therefore $h_1 > \dfrac{\alpha_1}{2}$.   This is possible only if $h_1 > \dfrac{\alpha_1 + 1}{2}$ or $h_1 = \dfrac{\alpha_1 + 1}{2}$.   If $h_1 > \dfrac{\alpha_1 + 1}{2}$, then by Theorem 17 the vertex P cannot have minimal height q and therefore cannot be in the centroid. So in this case

$$h_1 - (h_2 + h_3 + \ldots + h_\nu) = 2h_1 - \alpha_1 > 1 .$$

But if $h_1 = \dfrac{\alpha_1 + 1}{2}$ then by Theorem 18 (second part) P is of minimal height and indeed by Theorem 19 is one of two centroid points. In this case $(h_1 = \dfrac{\alpha_1 + 1}{2})$ the difference is

$$h_1 - (h_2 + h_3 + \ldots + h_\nu) = 2h_1 - \alpha_1 = 1 .$$

We consider now the second case, namely, that none of the numbers h_1, h_2, ... , h_ν is greater than the sum of the remaining ones;[13] then the following holds for the height h_1 of P:

$$h_1 \leq h_2 + h_3 + \ldots + h_\nu,$$

and therefore $h_1 \leq \dfrac{\alpha_1}{2} < \dfrac{\alpha_1 + 1}{2}$.

By Theorem 18 (first part) P is of minimal height and is therefore either the only centroid point or one of two centroid points. But the latter is impossible, since (Theorem 19) two centroid points have height $\dfrac{\alpha_1 + 1}{2}$.

Both of the results obtained can be reversed. For two centroid points (Theorem 19) we have:

$$q = h_1 = \frac{\alpha_1 + 1}{2} \text{ , and therefore}$$

$$h_1 > h_2 + h_3 + \ldots + h_\nu,$$

and $h_1 - (h_2 + h_3 + \ldots + h_\nu) = 1$.

But for a single centroid point we have (Theorem 19)

$$q = h_1 \leq \frac{\alpha_1}{2} \text{ and therefore}$$

$$h_1 \leq h_2 + h_3 + \ldots + h_\nu \text{ .}$$

The results obtained can now be summarized as follows:

Theorem 20   Let h_1, h_2, ... , h_ν be the numbers of edges of the branches coming from any definite vertex of a tree B and let $h_1 \geq h_i$ ($i = 2,3, \ldots, \nu$).

We set[14] $h_1 - (h_2 + h_3 + \ldots + h_\nu) = S$.

1) If B has a single centroid point, then this is the only vertex for which $S \leq 0$.

2) If B has two centroid points, then these are the only vertices for which $S = 1$.

3) For all other vertices $S > 1$.

Sylvester [2] and Cayley [3] introduced the concepts of a single centroid point and two centroid points by the property expressed in this theorem. [They called a single centroid point the "centre" and one of two centroid points a "bicentre of magnitude."] This definition is justified only by a theorem, and for a single centroid point (or two centroid points) no definition by direct construction is known, of the type of Jordan's definition, through which we introduced centers of linear dimension above. The effective finding of the centroid for a given tree is made easier, however, by the following theorem:

Theorem 21 If P is not in the centroid, then an edge PQ goes from P to a vertex Q, which is of smaller height than P.[15]

Theorems 18 and 19 state that for the height h_p of P the inequality $h_p > \dfrac{\alpha_1 + 1}{2}$ holds. Now among the branches (P, PX) let the one which has the maximum number of edges be (P, PQ); then h_p is the number of edges of (P, PQ); among the branches coming from Q let (Q, QR) have the maximal number of edges h_Q. Two cases are possible. 1) R $\neq$ P; then (Q, QR) is (Theorem 14) a proper subset of (P, PQ), and therefore $h_Q < h_p$. 2) R = P; then (as we have already seen above at the end of the proof of Theorem 17) $h_p = \alpha_1 + 1 - h_Q$. This is substituted into the inequality $h_p > \dfrac{\alpha_1 + 1}{2}$, and the inequality $h_Q < \dfrac{\alpha_1 + 1}{2}$ is obtained; so $h_Q < h_p$. So in both cases Q has the desired property.

If Q is also not in the centroid, then an edge QR likewise goes to a vertex R of still smaller height, etc. Finally on this path PQR ... the centroid point or one of the centroid points must be reached.

It should be noted that in addition to centers and axes of linear dimension and points and axes of centroids, Jordan [1, §5] introduced a third and fourth kind of centers and axes. We shall not go into the details of this here.

§5. Connection with certain number determinations
Applications to chemistry

We shall touch briefly upon a few more concepts and problems here, to which the preceding material of this chapter owes its origin.

By stating the group of a graph G the symmetry properties of G are determined, so that in the case of a finite graph the order of this group can[16] be designated as the degree of symmetry of G.

If a vertex P of a graph G (not necessarily a tree) is designated as the root, then the graph is referred to as a rooted graph. Two rooted graphs G_p and H_Q are called isomorphic (of the same structure) if

an invertible transformation exists for them, in which P and Q are

mapped on each other (so then G and H are also isomorphic to each

other). We consider now those transformations of G into themselves,

which map the root P into itself; these form a group (a subgroup of the

group of G), which we designate as the group of the rooted graph G_p.

Its order, if it is finite, is called the degree of symmetry of the

rooted graph G_p.

We are led to these concepts if we ask the following general

question (in which we restrict ourselves to finite graphs): How many

graphs (rooted graphs, respectively) are there of distinct structure

(nonisomorphic) and with a given number of edges? This problem is

closely connected with the following question: How can the degree of

symmetry of a given graph or rooted graph be determined? Jordan [1]

posed these problems; he also indicated a reduction process by which the

problem is finally reduced to the case of trees. In the treatment of

this simplest case the concepts of centers and axes can be used to

advantage. This is based on the following remark, which follows

immediately from the definition: In every transformation of a tree into

itself the center (the axis) of linear dimension as well as the centroid

(mass axis) corresponds to itself. So if it is desired, for example, to

determine the degree of symmetry of a tree B with center C, then it

should be noted that in every transformation of B into itself the

branches coming from the vertex C are permuted among themselves. The

question concerning the number of transformations of B into itself is

reduced to the same question with respect to these branches, whereby,

however, a branch is to be considered as a rooted tree with root C.

In this way we get a reduction of the number of edges under

consideration, whereby (for a given graph or for a given number of of organic chemistry standing at that time at the center of interest; this problem is the determination of the number of isomeric compounds.

We would like to illustrate this connection of trees with chemistry, and in fact with an example concerning paraffins, which have been studied the most in this connection.

The composition of paraffins is given by the formula $C_n H_{2n+2}$. We let correspond to the n carbon atoms one vertex each $C^{(i)}$ (i = 1,2, ..., n) and to the 2n+2 hydrogen atoms one vertex each $H^{(j)}$ (j = 1,2, ... 2n+2). The constitution of the paraffin is determined by giving a connected graph G with vertices $C^{(i)}$ and $H^{(j)}$, where every C-vertex has degree 4 and every H-vertex degree 1 (this agrees with the fact that a carbon atom has valence 4 and a hydrogen atom valence 1). This graph has $\alpha_0 = 3n + 2$ vertices; double the number of its edges is

$$2\,\alpha_1 = 4n + 1 \cdot (2n + 2),$$

and therefore we have that $\alpha_1 = 3n + 1$, so that the graph has connectivity

$$\mu = \alpha_1 - \alpha_0 + 1 = 0.$$

By Theorem IV.14, G is a tree. The H-vertices are the endpoints of the tree. If all end edges are removed, we get the tree $\bar{G}$, which is called a kenogram by Cayley [2, p. 202]. G is determined uniquely by $\bar{G}$ in its structure; if k_i (≤ 4) is the number of edges going to the vertex $C^{(i)}$ from $\bar{G}$ (i = 1,2, ..., n) then we get G from $\bar{G}$ by adding $4-k_i$ end edges to each vertex $C^{(i)}$. The determination of the number of different paraffins $C_n H_{2n+2}$ with given n is therefore equivalent to the following graph problem: determining the number of trees which have n vertices and in which at most 4 edges go to each vertex.

[We note again that all number questions mentioned here are based on the assumption that two trees (or even general graphs) are considered from $\bar{G}$ (i = 1,2, ..., n) then we get G from $\bar{G}$ by adding $4-k_i$ end edges to each vertex $C^{(i)}$. The determination of the number of different paraffins $C_n H_{2n+2}$ with given n is therefore equivalent to the following graph problem: determining the number of trees which have n vertices and in which at most 4 edges go to each vertex.

[We note again that all number questions mentioned here are based on the assumption that two trees (or even general graphs) are considered different if and only if they are of different structure (nonisomorphic). With some definitions other definitions of equality are justified. Even though it cannot be closely connected with the preceding material, we introduce an interesting result -- since it is solved[17] by a remarkably simple formula -- mentioned here without proof of the formula.

If two trees with the same vertices are considered distinct if and only if a pair of vertices exists which is joined by an edge in one of the trees but not in the other, then there are exactly n^{n-2} distinct trees with vertices P_1, P_2, ... , P_n $(n \geq 2)$[18].]

Notes on Chapter V

[1] The concept of <u>distance</u> in this sense was introduced by Cayley [3, p. 429].

[2] According to Cayley [4, p. 599, see also 3, p. 429] Sylvester is the discoverer of centers; also in Dehn and Heegaard [1, p. 176] they are named after Sylvester. Yet Sylvester [1, p. 24 and 4, p. 640] stated in 1873 and then again in 1882 that these centers were discovered by Jordan; without knowing Jordan's work Sylvester only rediscovered them but was the first to apply these concepts to finding the form and number of those trees which belong to certain chemical bonds (see §5). -- Cayley used the designations <u>center</u> and <u>bicenter</u> <u>of</u> <u>distance</u>.

[3] Cf. also de Polignac. [1, pp. 39-40].

[4] If it is assumed of W only that it is the longest of those paths of B which contain a certain vertex of W, then the theorem -- as simple examples show -- is no longer true. In this sense the definition in Sainte-Lague [4, p. 9] must be corrected. Also incorrect is the assertion there that the two longest paths which end in A together generate the longest path which runs through A; the end edges going to A of these two paths can indeed be identical.

[5] From this proof it can be seen also that in the case of two central points a longest path going to C must contain the axis.

[6] We shall understand from now on that for $\ell = 1$ the path $A\ R_1\ R_2\ \ldots\ R_{\ell-1}\ C$ will mean the edge AC.

[7] In Sainte-Lague [2, p. 5; 4, pp. 5-6], where the diameter is designated as the length (longueur) of the graph, this minimum is called for arbitrary graphs (not just for trees) the width (largeur) of the graph.

[8] In Dehn and Heegaard [1, p. 176] instead of this

number $\frac{d+1}{2}$ the number which is 1 smaller is erroneously stated.

[9] H.W.E. Jung: Über den kleinsten Kreis, der eine ebene Figur einschließt , Journal für r.u.a. Mathematik, 137, 1910, pp. 310-313.

[10] The designation (branche) was introduced by Jordan [1, p. 186]; Cayley [3, p. 429] says main branch.

[11] Jordan [1, p. 187].

[12] Jordan [1, p. 186-187].

[13] Sylvester [2] called such vertices subequal.

[14] Sylvester [2] designated this difference S, in case it is positive, as the superiority of the vertex in question.

[15] See Cayley [3, p. 600]

[16] This designation (ordre de symétrie) stems from Jordan [1, p. 186].

[17] For the above mentioned Jordan-Cayley number questions there are no explicit formulas known, only a recursive determination process.

[18] This theorem stems from Cayley [9]. Prüfer gave an elegant proof [1]. Cf. also O. Dziobek: Eine Formel der Substitutions-theorie, Sitzungsberichte der Berliner Mathematischen Gesellschaft, 17, 1917, pp. 64-67. The theorem can also be formulated as follows: the complete graph with n vertices has n^{n-2} spanning forests.

Chapter VI

Infinite Graphs

§1. Graphs of finite degree

With infinite graphs, graphs of finite degree play a distinguished
role: they form an intermediate link between finite graphs and infinite
graphs of infinite degree.

> Theorem 1   The set of vertices as well as the set of edges of
> a connected graph G of finite degree is finite or countably
> infinite.

Proof Let P be an arbitrary vertex of G. There is only a finite set of
vertices, which are at distance 1 from P, and likewise the sets of
vertices which are at distance 2,3,4, ... are finite.

The set of those vertices which are at a finite distance from P is
finite or countably infinite since it is a sum of finitely or countably
infinitely many finite sets. But since G was assumed to be connected,
this is the set of all vertices of G. Then, since G is of finite
degree, the set of its edges is also finite or countably infinite. (As
this proof shows, the theorem remains true if, instead of assuming that
G is of finite degree, it is assumed only that at most countably many
edges go to each vertex.) The theorem can then be formulated as
follows:

> Theorem 2   Every graph of finite degree consists of
> components, each of which has at most countably many vertices
> and edges.[1]

This theorem is important for the entire theory of infinite graphs

106

and for its application to set theory. As we shall see, with the help of this theorem the proofs of some theorems concerning arbitrarily large sets can be immediately reduced to the case where the sets involved are countable.

A remarkable property of graphs of finite degree is expressed by the following theorem.

> Theorem 3   Every infinite connected graph G of finite degree has a singly infinite path, where the initial vertex P_0 of this path can be arbitrarily specified.[2]

Proof   As an infinite graph of finite degree, G has infinitely many vertices. The set of those paths of G which begin at P_0 is therefore infinite, since G is connected, and consequently for every vertex P of G there is a path which ends in P $(P \neq P_0)$. Each of these paths begins with an edge which goes to P_0. There is only a finite number of such edges, and therefore there must be one -- say P_0P_1 -- among them, with which infinitely many paths begin. All paths beginning with P_0P_1 have as a second edge one of the finitely many edges which go to P_1 and therefore there must be an edge P_1P_2 $(P_2 \neq P_0)$ with the property that infinitely many paths begin with P_0P_1, P_1P_2. Continuing in this way there arises a never ending sequence P_0P_1, P_1P_2, P_2P_3, ... which in fact generates a singly infinite path. In this proof the Axiom of Choice was used.///

Neither of the two conditions of the theorem just proved can be dropped. If the graph is not of finite degree, it does not have to have a singly infinite path. This is illustrated by the example of the graph which has vertices P_i (i = 0, 1, 2, ... ad inf.) and edges P_0P_i (i = 1, 2, 3, ... ad inf.). The same is the case if the graph is not connected. This is illustrated, for example, by the graph which is the sum of a 2-cycle, a triangle, a 4-sided polygon, ... ad inf. This example also shows that a graph can contain arbitrarily long paths without having to contain an infinite path.

Another example of a graph that can contain arbitrarily long paths without having an infinite path is the connected graph arising from the last-mentioned graph by adding a new vertex and then joining every old vertex to the new one. This new graph cannot, of course, be of finite degree.

Theorem 3 is used in the proof of:

> **Theorem 4**   For every acyclic connected infinite graph G of finite degree there is a one-to-one correspondence between the set Π of its vertices and the set K of its edges, such that every vertex is an endpoint of the edge corresponding to it.

Proof   If P_0 is an arbitrary vertex of G and Π' is the set of vertices of G distinct from P_0, then Theorem IV.11 says that there exists a one-to-one correspondence Z' between K and Π', which assigns to each edge one of its endpoints. Now by Theorem 3, let $P_0P_1P_2 \ldots$ be a singly infinite path of G. Since P_0 is not assigned to an edge by Z', then P_1 must be assigned to the edge P_0P_1 by Z', and therefore the vertex P_2 must be assigned to the edge P_1P_2, $\ldots$; the vertex P_{n+1} must be assigned to the edge P_nP_{n+1}; etc. Now the correspondence Z' can be altered by assigning the edge P_nP_{n+1} ($n = 0, 1, 2, \ldots$ ad inf.) to P_n instead of to P_{n+1} and leaving Z' otherwise unaltered. Thus we get a one-to-one correspondence Z, which is clearly the one we wished to have between Π and K.///

The edge k of G can be assigned a direction in such a way that the endpoint of edge k assigned to k by Z is always the initial point (and the other endpoint the terminal point). Thus the theorem proved above can be expressed as follows:

> **Theorem 5**   Every acyclic connected infinite graph of finite

degree can be transformed into a directed graph in such a way
that every vertex is the initial point of one and only one
edge.

Theorems 4 and 5 can be extended to disconnected graphs if and only
if every component of the graph is infinite.

§2. The Infinity Lemma

The same basic idea which motivated us in the proof of Theorem 3
leads us now to the proof of the important:

Theorem 6   (Infinity Lemma) Let π_1, π_2, π_3, ... be a
countably infinite sequence of finite, nonempty, pairwise
disjoint sets of points. Let the points contained in these
sets form the vertices of a graph. If G has the property that
every point of π_{n+1} (n = 1, 2, 3,..., ad inf.) is joined with a
point of π_n by an edge of G, then G has a singly infinite path
$P_1 P_2 P_3 ...$, where P_n (n = 1, 2, 3, ... ad inf.) is a point
of π_n. [3]

Proof   In the proof of this theorem a (finite) path of G shall be
designated as an S-path, provided its vertices belong in their proper
order, respectively, to π_1, π_2, ... , π_k. There are infinitely many
S-paths in G, since, with the exception of the points of π_1, each vertex
of G is the second endpoint of an S-path. Every S-path begins with an
edge which joins a point P_1 of π_1 with a point X_2 of π_2. Since such
edges are finite in number, it must be that one of these edges, say
$P_1 P_2$, is in infinitely many S-paths. All these S-paths contain as their
second edge one of the finitely many edges $P_2 X_3$, where X_3 belongs
to π_3, and so there must be a point P_3 in π_3 with the property that
infinitely many S-paths, which begin with $P_1 P_2$, also contain $P_2 P_3$.
Continuing in this way a point P_4 of π_4 is defined, P_5 of π_5, etc. The

process cannot end and leads to an infinite path $P_1 P_2 P_3$ ... with the desired property.///

The Infinity Lemma, just now proved,[4] can be used not only in graph theory -- where we shall later (Chap. VIII, XII, XIII) deal with some examples -- but in the most diverse mathematical disciplines, since it often furnishes a useful method of carrying over certain results from the finite to the infinite. We give now three examples.[5]

The first example is concerned with kinship relations, which give rise to an old and well known application of graphs in the form of family trees. We show that if the hypothesis that mankind will never become extinct is assumed, then there exists a person today who is the ancestor of an infinite sequence of descendants.[6]

Let E_1 be the set of all people living today, E_2 the children of the elements of E_1; E_3 the set of the children of the elements of E_2; etc. According to the above hypothesis -- because of the finiteness of human life -- none of the sets E_1, E_2, E_3, ... is empty. Since a person can have only finitely many children, it follows from the finiteness of E_1 that all the sets E_i are finite. To each element which is contained in one of the sets E_i, we let correspond a point, where the point set E_i corresponds to the set π_i ($i = 1, 2, 3, ...$).[7] We choose the points of these sets π_i as vertices of a graph G. A vertex A of π_{n+1} is joined with a vertex B of π_n by an edge of G if the person corresponding to the vertex A is a child of the person corresponding to the vertex B; other edges are not introduced. The graph G so defined and the sets π_i fulfill the conditions of the Infinity Lemma. Using the lemma, we get an infinite sequence a_1, a_2, a_3, ... with the property that a_i is an element of E_i and a_{i+1} is a child of a_i. So a_1 is a person living today with the desired property.///

In a similar way it can be shown that the existence of an infinite male line can be concluded from the infinite lifespan of humanity.

Some applications of the Infinity Lemma are analogous to
applications of the Heine-Borel covering theorem. Because of this it
seems interesting to remark that, from a certain standpoint, the
Infinity Lemma can be thought of as the proper foundation of this
covering theorem. We shall reduce the following theorem of de la Vallee
Poussin,[8] which clearly contains the Heine-Borel Theorem as a special
case, to the Infinity Lemma:

> Let E be a closed subset of the interval (0,1) and let I be a
> family of intervals which have the property that every point of
> E is contained in one of these intervals. Then there is a
> natural number n such that if (0,1) is divided into 2^n equal
> subintervals, those subintervals which contain a point of E are
> contained in an interval of the family I.

Proof   If the theorem were not true then there would be for every n at
least one interval $(\frac{m}{2^n} , \frac{m+1}{2^n})$, where m = 0, 1, 2, ... , or 2^n - 1,
which contains a point of E and which is not contained in any interval
of I. We designate the set of these intervals by E_n. To every element
of the sets E_i we let a point correspond, whereby the set E_i corresponds
to the point set π_i (i = 1, 2, 3, ...). We choose the points of these
sets π_i as vertices of a graph G. A vertex A of π_{n+1} is joined with a
vertex B of π_n by an edge of G if the interval which corresponds to the
point A comes from the interval which corresponds to the point B by
bisection of that interval; other edges are not introduced. The graph G
so defined and the sets π_i fulfill the conditions of the Infinity
Lemma. This lemma is used, and we get the following result. There
exists an infinite sequence a_1, a_2, a_3, ... of intervals, which all

1) come from the preceding one by bisection

2) contain a point of E

111

and   3)   are not contained in any interval of I.
But then the common point α of the intervals a_1, a_2, a_3, ... is also not
contained in any interval of I. This is impossible since, because E is
closed, α must belong to E. (This proof makes use of the Nested
Intervals Theorem but not the Bolzano-Weierstrass theorem and holds also
for the plane, 3 dimensional space, etc.)

The third application of the Infinity Lemma relates to the
following so-called Baudet conjecture, which was proved by van der
Waerden.[9]

α) Let k and m be two arbitrary natural numbers. Then there is a
natural number N (depending on k and m) with the property that, no
matter how the set {1, 2, ..., N} is split into k pairwise disjoint
subsets, at least one of these subsets contains an arithmetic sequence
with m terms.

We give here no proof of this theorem, but we show that it is
equivalent to the following theorem:

β) If k and m are arbitrary natural numbers and the set of all
natural numbers is divided completely arbitrarily into k pairwise
disjoint subsets, then at least one of these subsets contains an
arithmetic sequence with m terms.

It is clear that (β) follows from (α). The converse of this
assertion can be obtained with the help of the Infinity Lemma in the
following way.

We consider the set E_n of those partitions of the set
Z_n = {1, 2, ..., n} into k disjoint subsets such that none of the k
subsets contains an arithmetic sequence with m terms; E_n is, of course,
finite. We assume that theorem (α) is false; then none of the sets E_n
is empty. We let a point correspond to each element of E_n, and in this
way there is a set π_n corresponding to E_n (n=1,2,...ad inf.). A point

from π_{n+1} is joined now with a point from π_n by an edge if and only if the corresponding elements A of π_{n+1} and B of π_n are in the following relation: the partition B of Z_n comes from the partition A of Z_{n+1} by removing the number n+1. The graph so defined and the sets π_i fulfill the conditions of the Infinity Lemma. By using the lemma, we get an infinite sequence A_1, A_2, A_3,... with the property that for every n

      1) A_n is an element of E_n

and   2)   two numbers which belong to the same subset of the

              partition A_{n-1} also belong to the same subset of the

              partition A_n (and the same for A_{n-2}, A_{n-3}, ...)

Two natural numbers are put into the same class if and only if these two numbers belong to the same subset in a partition A_n (and therefore in every partition in which both numbers occur), and in this way a partition of the set of all natural numbers into k disjoint subsets is obtained. By Theorem (β) one of these subsets contains an arithmetic sequence with m terms; if N is the largest number of this sequence, then, of course, this sequence must belong to a subset of the partition A_N of E_N, and this contradicts the definition of the sets E_n. (It is easily seen that this proof of the equivalence of (α) and (β) remains valid if instead of arithmetic sequences other classes of finite sets of numbers are considered, geometric sequences, for example, etc.)

§3. The sharpened Equivalence Theorem.

The material to be considered now represents an important application of graph theory to abstract set theory. We prove

    <u>Theorem 7</u> Let the set π of vertices of the graph G come from

two disjoint sets π_1 and π_2 and let the set K of edges of G
come from two disjoint sets K_1 and K_2 in such a way that

1) every edge of G joins a π_1-vertex with a π_2-vertex;

2) every π_1-vertex is the endpoint of one and only one
K_1-edge and every π_2-vertex is the endpoint of at most
one K_1-edge

3) every π_2-vertex is the endpoint of one and only one
K_2-edge and every π_1-vertex is the endpoint of at most
one K_2-edge.

Then G has a factor of first degree.

Proof Each vertex is of degree either 1 or 2; so Theorem I.27 can be
used. Of the four possibilities which this theorem allows for the
components of G we can exclude the first one by showing that no
component of G can be a (finite) path.

Suppose the opposite were true. Suppose that the path
$W = P_1 P_2 \ldots P_n$ of G were a component of G. By symmetry we can assume
without loss of generality that P_1 belongs to π_1. Then $P_1 P_2$ belongs to
K_1, since, if it belonged to K_2, another second edge, namely an edge
from K_1, would have to end in P_1, which would, of course, have to belong
to the same component W. On the other hand, P_1 is an endpoint of W.
Likewise $P_2 P_3$ belongs to K_2 (since according to (1) P_2 is
a π_2-vertex), $P_3 P_4$ belongs again to K_1, $P_4 P_5$ to K_2, etc. The last
edge $P_{n-1} P_n$ of W belongs therefore to K_1 or to K_2 depending on whether n
is even or odd. In the first case P_n is a π_2-vertex, and in the second
a π_1-vertex. In both cases there is a second edge of G which ends in P_n
(in the first case there is also one which belongs to K_2, and in the
second case one which belongs to K_1). This second edge must belong to
the same component W of G as $P_{n-1} P_n$, while, on the other hand, no second
edge in W goes to P_n. This contradiction shows that in fact no

component of G can be a path.

So the possibilities for the components of our graph G consist only of possibilities 2), 3), and 4) of Theorem I.27. In Case 2 (a cycle) we can add that the cycle contains an even number of edges, since the vertices belong alternately to π_1 and π_2. So Theorem I.12 can be applied to all components G_α of G: every G_α has a factor of first degree. The sum of all these factors is naturally a factor of first degree of G. Theorem 7 is hereby proved.///

Theorem 7 is the graph theoretical formulation of a very general theorem of abstract set theory. In order to make the transition to set theoretical terminology possible, we must always be careful about the following. Let there be two arbitrary disjoint sets π_1 and π_2, whose elements we shall designate as points. The necessary and sufficient condition that π_1 and π_2 be equivalent is the existence of a graph with the following property: every edge of the graph joins a π_1-point with a π_2-point and every point of $\pi_1 + \pi_2$ is the endpoint of one and only one edge of G. Every graph of this type determines a one-to-one correspondence between the sets π_1 and π_2 and vice versa.

The graph G, formed by the set K_1 of edges of Theorem 7, therefore determines a one-to-one correspondence for π_1 and a subset π_2' of π_2, and the graph G_2 formed by the edges of K_2 determines a one-to-one correspondence for π_2 and a subset π_1' of π_1. (Here π_2' consists of those points of π_2 which are endpoints of a K_1-edge, and π_1' consists of those points of π_1 which are endpoints of a K_2-edge.) On the other hand, a factor of first degree of G determines a one-to-one correspondence for π_1 and π_2. Theorem 7 can be expressed in set theoretical terminology as follows:

If there is a one-to-one correspondence Z_1 for the set π_1 and a subset π_2' of the set π_2 and at the same time a one-to-one correspondence Z_2 for Π_2 and a subset Π_1' of Π_1,

        α) then there is a one-to-one correspondence Z

        for Π_1 and Π_2 ;

and   β) Z has the property that two elements correspond to each

        other through Z if and only if they correspond to each

        other either through Z_1 or Z_2.

The assertion (α) is the well-known Cantor-Bernstein equivalence theorem, to which the added statement (β) gives a substantial sharpening.

In the proof given here the terminology of graph theory could be avoided. The result would be the simplest proof of the Equivalence Theorem, that given by Julius König.[10] It cannot be asserted that this proof was substantially simplified by the graph theoretical formulation. It was only made clearer. Later (Chap. XIII) we shall deal graph theoretically with other theorems of general set theory, where avoiding graph theoretical terminology would substantially complicate the proofs. As preparation for this material it seemed expedient at this point - where it is a matter of very simple graphs (namely graphs which contain only vertices of first and second degree) -- to show in detail the applicability of graph theory to general set theory. The added statement (β) was proved by J. König without being explicitly expressed. It was explicitly expressed in an essentially unmodified form by St. Banach[11], who gave a generalization and -- in collaboration with A. Tarski -- noteworthy applications of this generalization.[12]

Notes on Chapter VI

[1]König [5, p. 460].

[2]König [8, p. 122].

[3]König [7, §3 and 8, §1].

[4]This proof uses the Axiom of Choice just like the proof of Theorem 3 given above. With most applications of the Infinity Lemma this axiom can nevertheless be avoided, into the details of which we shall not go.

[5]The first two are contained in the work of König [8, §4 and §2].

[6]This says more than the assertion of the existence of a person living today who will have infinitely many descendants. The latter is obvious.

[7]A person can belong to several of the sets E_i; then we let correspond to him distinct vertices according to whether he is included in one or the other generation E_i; the π_i's are therefore pairwise disjoint.

[8]Intégrales de Lebesgue. Functions d'ensemble. Classes de Baire, Paris 1916, p. 14.

[9]Beweis einer Baudetschen Vermutung, Nieuf Archiev voor Wiskunde (2), 15, 1927, pp. 212-216.

[10]Sur la théorie des ensembles, Comptes Rendus, Paris, 146, 1906, p. 110.

[11]Un théorème sur les transformations biunivoques, Fundamenta Mathematicae, 6, 1924, p. 236, théorème 1. --With respect to this reference W. Sierpinski (Leçons sur les nombres transfinis, Paris 1928, p. 90) designated this addendum β) as "Théorème de M. Banach". This designation does not seem quite suitable, since, as already said, this addendum was already proved in the work of J. König and the Banach proof

can be seen essentially as a translation of this proof into the
"language of substitution theory," just as we have translated the same
proof here into the "language of graph theory" (cf. the remark in
D. König [7, p. 130] where in the 8th line from the bottom "theorem 1"
should be substituted for "theorem 2"). A comparison of these two
translations, at least as far as ease of understandability is concerned,
will doubtless turn out in favor of graph theory. Certainly Sierpiński
(loc. cit.) gave the Banach proof a more lucid presentation.

[12]Banach et Tarski, Sur la décomposition de points en parties
respectivement congruentes, Fundamenta Mathematicae, 6, 1924,
particularly p. 251.

Chapter VII

Basis problems for directed graphs

§1. The vertex basis

The properties with which we are now concerned depend on whether or

not for two vertices P and Q of a directed graph an edge $\overrightarrow{P\,Q}$ is present

in the graph. We are able to and wish to restrict ourselves to such

graphs in this and the next chapter, in which for every two vertices P

and Q at most one edge $\overrightarrow{P\,Q}$ is contained and at most one edge $\overrightarrow{Q\,P}$.

Here the concept of (directed) path plays a role. This concept was

introduced in Chap. II, §4. The same chain of reasoning which led us to

Theorem I.5 gives us

Theorem 1 If a (directed) path goes from A to B and a

(directed) path goes from B to C (where C ≠ A) then there is a

(directed) path from A to C.

Let π be the set of vertices of an arbitrary directed graph G. We

designate by π_A that subset of π which contains the vertex A of G and

also those vertices P of G for which a (directed) path in G from A to P

exists. If there is no vertex B in G such that π_A is a proper subset

of π_B, we call π_A a fundamental set of G and A a source of π_A. By

specifying the source the fundamental set is, of course, determined.

But a fundamental set $\pi_A = \pi_B$ can have different sources, A and B. If,

for example, G is a (directed) cycle, then the set of all its vertices

is a fundamental set and every vertex is one of its sources. We now

prove

Theorem 2 Every vertex A of a finite directed graph belongs to

a fundamental set of G.

119

<u>Proof</u> This is clear if π_A is a fundamental set; otherwise there is a
vertex B which does not belong to π_A and for which π_A is contained
in π_B. Now again either π_B is a fundamental set which contains A, or
there is a vertex C not contained in π_B for which π_B, and therefore
also π_A, is contained in π_C, etc. Because of the finite number of
vertices this process must end, and we get a fundamental set π_z which
contains A.///

For infinite graphs the theorem just proved is not valid; this is shown
by the example of the continuously directed singly infinite
path $\overleftarrow{P_1 P_2 P_3} \ldots$. This graph does not have any fundamental set at all.

    <u>Theorem 3</u>   A proper subset of a fundamental set is not a
    fundamental set.

<u>Proof</u> We assume that the fundamental set π_A of G contains a fundamental
set π_B of G as a proper subset (and so B $\neq$ A). From the definition of
the fundamental set π_B it follows then that A is contained in π_B and
that therefore a (directed) path goes from B to A. So if there is a
(directed) path in G which goes from A to P then by Theorem 1 there is
also a (directed) path from B to P. So π_A is a subset of π_B and
hence $\pi_A = \pi_B$; so π_B is not a proper subset of π_A. ///

    A (finite or infinite) subset B of π is designated as a vertex
basis for G, provided that it has the following two properties:

    1) If P is an arbitrary vertex of G which is not contained in B,
    then there is a (directed) path in G going from a vertex of B to P.

    2) No (directed) path of G joins two distinct vertices of B.

    <u>Theorem 4</u>   Every finite directed graph has a vertex basis.

<u>Proof</u> We determine for every fundamental set of G one of its sources.
The set of these sources is a vertex basis. The first defining basis
property follows from Theorem 2. As for the second defining basis
property, let Q_1 and Q_2 be sources of two distinct fundamental

sets π_{Q_1} and π_{Q_2} of G, and we assume that a (directed) path of G goes from Q_1 to Q_2. Then, by Theorem 1, π_{Q_2} is a subset of π_{Q_1}. But since $\pi_{Q_1} \neq \pi_{Q_2}$, this contradicts Theorem 3.///

Theorem 4 is not true for infinite graphs. For example, the continuously directed singly infinite path $\overleftarrow{P_1 P_2 P_3} \ldots$ has no vertex basis (on the other hand, the continuously directed singly infinite path $\overrightarrow{P_1 P_2 P_3} \ldots$ has a unique vertex basis, which consists of the single vertex P_1).

Theorem 5   If the vertex P is contained in a vertex basis B, then π_P is a fundamental set.

Proof   If this were not true, then there would be a vertex P' not contained in π_P, such that π_P would be a subset of $\pi_{P'}$, and therefore there would be a (directed) path from P' to P. But if this is the case, then P' does not belong to B, and there must therefore be a (directed) path from a vertex Q of B to P'. Here $Q \neq P$, since otherwise P' would belong to π_P. The (directed) paths $\overrightarrow{Q...P'}$ and $\overrightarrow{P'...P}$ give rise by Theorem 1 to a (directed) path $\overrightarrow{Q...P}$, and this contradicts the second defining property of a basis for B.///

Now we prove, for both finite and infinite graphs,

Theorem 6   Every vertex basis B consists of one source each of all fundamental sets.

Proof   By Theorem 5 every vertex of B is a source of a fundamental set. Also two distinct vertices P_1 and P_2 of B are never sources of the same fundamental set since, as a consequence of the second defining property of a basis, no (directed) path goes from P_1 to P_2. It remains now only to show that every fundamental set π_Q has a source contained in B. If Q is not itself such a source of π_Q, then there is (by the first defining property of a basis) such a vertex P in B, from which a (directed) path leads to Q. But then (Theorem 1) $\pi_Q = \pi_P$, and therefore (by definition of source) P is also a source of π_Q and is contained in B, as the theorem requires.///

From this theorem we get

Theorem 7   Two vertex bases of the same directed graph are equivalent.

Proof By Theorem 6 their cardinalities agree with the cardinality of the set of all fundamental sets.///

The preceding developments are particularly simple if the graph contains with every edge $\overrightarrow{P\,Q}$ also an edge $\overrightarrow{Q\,P}$. In this case a vertex basis also always exists for infinite graphs, because the fundamental sets are identical to the sets of vertices of the individual components.[1] Here every vertex of a fundamental set is a source of this fundamental set. Furthermore in this case two fundamental sets are disjoint from each other (which is not in general the case), so that every vertex basis is obtained (Theorem 6) by choosing a vertex from every component; if there is an infinite number of components, then the Axiom of Choice must be applied. The problem of vertex bases dealt with here can be modified by replacing the word "path" by "edge" in both places of the definition of vertex basis given above. For these vertex bases of the second kind Theorem 7 is not true[2] (even for finite graphs) as can easily be shown with very simple examples. We shall try to find a minimal vertex basis of the second kind. In other words we pose the following problem; let as few vertices as possible of a given graph be chosen (call the set of them S) such that for every vertex $P \notin S$ there is an edge going from a vertex of S to P (in the direction towards P), but no two vertices of S are joined by an edge. If the graph contains with every edge $\overrightarrow{P\,Q}$ also an edge $\overleftarrow{P\,Q}$ and if every pair of edges $(\overrightarrow{P\,Q}, \overleftarrow{P\,Q})$ is replaced by an undirected edge PQ, then we have a problem for undirected graphs.[3] To show an application of this problem, we shall define a special undirected graph as follows. We let a vertex correspond to each of the 64 squares of a chess board and we join two vertices by an edge if and only if the straight line joining the midpoints of these squares is parallel to one of the edges or to one of the diagonals of the board (and therefore if and only if a move of the queen between the two squares is possible.) It is easily seen that for

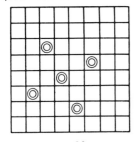

Fig. 38.

this graph (where the degrees of the vertices range from 21 to 27) our task amounts to the problem of the five queens on the chess board. This problem requires that the least number of queens be set up on the board in such a way that every unoccupied square -- and only these squares -- can be attacked by one of the queens[4] (Fig. 38).

§2.   The edge basis. The condensation of a graph.

In analogy to the concept of vertex basis introduced above we present now the following definition. A subset B of the set of edges of a directed graph G (and the subgraph of G formed by these edges) is designated as an edge basis for G if it has the following two properties:

1)   If $\overrightarrow{P\,Q}$ is an arbitrary edge (which is not contained in B) then a (directed) path consisting of edges of B goes from P to Q.

2)   If $\overrightarrow{P\,Q}$ is an arbitrary edge of B, then no (directed) path from P to Q can be formed by the other edges in B.

The theory of edge bases to be considered now -- so far as it relates to finite graphs -- is essentially the graph theoretical interpretation of those investigations, which P. Hertz [1] carried out for certain problems of logic (axiomatic theory). In his introduction Hertz already referred to this "geometrical" interpretation, particularly for the proof of Theorem 9, which follows below. We return

to the application to logic in the next chapter (§1).

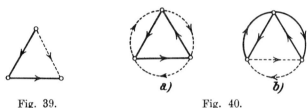

Fig. 39.   Fig. 40.

For the graphs of Figures 39-42 the solid lines form an edge

basis. For Fig. 39 this is the only edge basis. The graphs of Fig. 40,

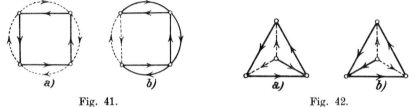

Fig. 41.   Fig. 42.

41, and 42, however, have several edge bases. Figures 40a and 40b (and

likewise 41a and 41b) show that two edge bases of the same finite graph

do not have to have the same number of edges.

We give now some definitions. We shall designate the edge $\overrightarrow{P\,Q}$ as a

simple edge if the graph does not contain an edge $\overleftarrow{P\,Q}$. If there is an

edge $\overrightarrow{P\,Q}$ as well as an edge $\overleftarrow{P\,Q}$ then both are called double edges. A

directed graph is called transitive[5] if it always contains along with

the edges $\overrightarrow{P\,Q}$ and $\overrightarrow{Q\,R}$, where R ≠ P, an edge $\overrightarrow{P\,R}$. If in a transitive

graph a (directed) path goes from P to Q (≠ P) then the graph contains

an edge $\overrightarrow{P\,Q}$. Every directed graph can by adding new edges be made into

a transitive graph. A directed graph is called a net if for every two

of its vertices P and Q, there are both an edge $\overrightarrow{P\,Q}$ and $\overrightarrow{Q\,P}$ [6] . Every

net is connected and transitive; all its edges are double edges. A

subgraph of a directed graph G is called a maximal net of G, if it is a

net and it is not contained in any subgraph of G, which is a net, as a proper subset.

Theorem 8   If the double edges of the transitive graph G form

the subgraph N of G, then the components N_α of $N = \Sigma N_\alpha$ agree

with the maximal nets of G.

Proof We show first that N_α is a net. Along with the edge $\overrightarrow{R\,S}$ the

edge $\overleftarrow{R\,S}$ of N is also contained in N_α. So N_α contains only double

edges. Since N_α is connected there is for every two vertices P and Q

of N_α a path $PA_1A_2\ldots A_\nu Q$ of N_α, and therefore, since the edges are

double, a directed path $\overrightarrow{PA_1A_2\ldots A_\nu Q}$ of N_α. Because of the transitivity

of G there is therefore in G an edge $\overrightarrow{P\,Q}$. Likewise there is also an

edge $\overleftarrow{P\,Q}$ in G. So $\overrightarrow{P\,Q}$ and $\overleftarrow{P\,Q}$ are contained in N as double edges and,

of course, in the same component N_α of N, which contains P and Q.

Secondly we must show that N_α is a maximal net of G. We suppose

that N_α is contained as a proper subset in a net M of G. If the sets of

vertices of M and of N_α agreed with each other, then we would have

$M = N_\alpha$. So there is in M a vertex P which is not contained in N_α. If

now Q ($\neq$ P) is a vertex of N_α and therefore also of M, then $\overrightarrow{P\,Q}$ and

$\overleftarrow{P\,Q}$ are contained in the net M. As a double edge $\overrightarrow{P\,Q}$ belongs to N and

to N_α, since Q is a vertex of N_α. So P would have to belong to N_α.

Thirdly it must be shown that any maximal net M of G is identical to

an N_α. M is a connected subgraph of N and so by Theorem I.26 is

contained in a component N_α of N. Now M as maximal net cannot be a

proper subset of the net N_α, and so $M = N_\alpha$. ///

With regard to the uniqueness of the edge basis we have

Theorem 9   A transitive graph G, which has only simple edges,

cannot have two distinct edge bases.

Proof If there are two distinct bases, B and B', then there must be an

edge $\overrightarrow{P\,Q}$ in B, which is not contained in B'. From the first defining

property for a basis B' it follows that there exists a (directed) path
$\overrightarrow{P R_1 R_2 \ldots R_\mu Q}$ in G consisting of edges of B' and containing at least
two edges. If we replace every such edge $\overrightarrow{X Y}$ of this path, which is not
contained in B, by a (directed) path $\overrightarrow{X \ldots Y}$ consisting of edges of B,
there results a sequence of edges from B of the type
$\overrightarrow{P S_1}, \overrightarrow{S_1 S_2}, \ldots, \overrightarrow{S_{\nu-1} S_\nu}, \overrightarrow{S_\nu Q}$, where all $\nu + 1$ edges need not be
different, but where again at least two distinct edges are involved,
$\nu \geq 1$, so that this sequence cannot be reduced to the one edge $\overrightarrow{P Q}$. Now
a (directed) path can always be formed from the edges of a continuously
directed sequence of edges, which connect the endpoints of this sequence
of edges (see the proof of Theorem I.3). From the second defining
property for the basis B it follows that $\overrightarrow{P Q}$ is contained in this
sequence, so that either P or Q is identical to one of the points
$S_1, S_2, \ldots, S_\nu$. Say $P = S_k$, where $k \neq 1$. Now certain edges of the
sequence

$$\overrightarrow{S_1 S_2}, \overrightarrow{S_2 S_3}, \ldots, \overrightarrow{S_{k-2} S_{k-1}}, \overrightarrow{S_{k-1} P}$$

form a (directed) path leading from S_1 to P. Because of the
transitivity of G there is therefore in G an edge $\overrightarrow{S_1 P}$, and, since G
also contains an edge $\overleftarrow{S_1 P}$, this is not a simple edge.///

For the graph of Fig. 43 the solid edges give its only edge basis.
Both hypotheses of the theorem just proved are essential: if there are
also double edges Fig. 40 or 41 shows that distinct edge bases can

Fig. 43.

exist; but if the graph is not transitive, then the
graphs of Figs. 41 and 42 give an example.

Let the simple edges of the transitive graph G
form the subgraph M and let the double edges form the
subgraph N and let it be partitioned into its components, $N = \sum N_\alpha$,
where (Theorem 8) the N_α are the maximal nets of G. We now define the
condensation of the transitive graph G in the following way. We let to

126

every vertex P_i of G, which does not belong to N, correspond a vertex Q_i

and to every maximal net N_α of G (if N_α is not a component of G) a

vertex R_α. We introduce three kinds of edges.

1) If neither P_i nor P_j belongs to N, we let an edge Q_iQ_j

correspond to the edge P_iP_j of G;

2) if P_i does not belong to N, but P_j does belong to N and, say,

to N_α, then to all such edges $\overrightarrow{P_i P_j}$ ($\overleftarrow{P_i P_j}$, respectively) of G we let

correspond one edge $\overrightarrow{Q_i R_\alpha}$ ($\overleftarrow{Q_i R_\alpha}$, respectively);

3) if P_i and P_j both belong to N (and, say, P_i belongs to N_α and

P_j belongs to N_β, where $N_\alpha \neq N_\beta$, then to all such edges $\overrightarrow{P_i P_j}$ we let

correspond one edge $\overrightarrow{R_\alpha R_\beta}$ (in case $N_\alpha = N_\beta$ we let no edge at all

correspond to the edge $\overrightarrow{P_i P_j}$).

The edges thus introduced form the condensation of G. (Graphically

expressed, this definition says: the condensation of G arises from G by

"shrinking" every maximal net N_α to a point R_α, whereby the double

edges, that is, the edges of N, disappear; but if several edges $\overrightarrow{P Q}$ come

out of the process, only one edge $\overrightarrow{P Q}$ is retained.)

So to every edge of the condensation of G corresponds at least one

simple edge of G and to distinct edges of the condensation of G

correspond distinct edges of G. And vice versa: to every simple edge

of G corresponds an edge of the condensation; but the same edge of the

condensation can correspond to distinct edges $\overleftarrow{P_i P_j}$ and $\overleftarrow{P'_i P'_j}$ if and

only if P_i and P'_i are identical or belong to the same maximal net of G

and at the same time P_j and P_j' are either identical or belong to the

same maximal net. Now we prove three theorems concerning the

condensation of a transitive graph.

Theorem 10   If G is a transitive graph, then every edge of its

127

condensation is simple.

Proof We suppose that both $\overrightarrow{P\,Q}$ and $\overleftarrow{P\,Q}$ are in the condensation. To these edges should correspond in G the simple edges $\overrightarrow{P_1\,Q_1}$ and $\overleftarrow{P_2\,Q_2}$, respectively. First, let $P_1 \neq P_2$ and $Q_1 \neq Q_2$; then P_1 and P_2, since the same vertex P in the condensation corresponds to them, are contained in the same maximal net N_α of G, so that G contains an edge $\overrightarrow{P_2\,P_1}$. Likewise Q_1 and Q_2 are contained in a maximal net N_β (where $N_\alpha \neq N_\beta$, since otherwise $\overrightarrow{P_1\,Q_1}$ would not be a simple edge of G); so there is an edge $\overrightarrow{Q_1\,Q_2}$ in G. Now the three edges $\overrightarrow{Q_1\,Q_2}$, $\overrightarrow{Q_2\,P_2}$, $\overrightarrow{P_2\,P_1}$ form a (directed) path going from Q_1 to P_1. By transitivity there is an edge $\overrightarrow{Q_1\,P_1}$ in G. This leads to the impossible result that $\overrightarrow{P_1\,Q_1}$ is a double edge of G. Secondly, if $P_1 = P_2$ and $Q_1 \neq Q_2$, then we get an edge $\overrightarrow{Q_1\,Q_2}$ of G, which gives, along with the edge $\overrightarrow{Q_2\,P_2} = \overrightarrow{Q_2\,P_1}$, a (directed) path from Q_1 to P_1 and therefore an edge $\overrightarrow{Q_1\,P_1}$ of G, which again contradicts the fact that the edge $\overrightarrow{P_1\,Q_1}$ is simple. In the same way we can deal with the third case: $P_1 \neq P_2$, $Q_1 = Q_2$. Finally we have the fourth case: $P_1 = P_2$, $Q_1 = Q_2$ is impossible since, because $\overrightarrow{P_2\,Q_2} = \overrightarrow{P_1\,Q_1}$, $\overleftarrow{P_2\,Q_2}$ would be a double edge.///

Theorem 11 If G is a transitive graph, then its condensation is also transitive.

Proof Let $\overrightarrow{P'\,Q'}$ and $\overrightarrow{Q'\,R'}$ be two edges of the condensation of G, to which the edges $\overrightarrow{P\,Q_1}$ and $\overrightarrow{Q_2\,R}$ in G correspond. We must show that the condensation of G contains an edge $\overrightarrow{P'\,R'}$. We have $P' \neq R'$, since otherwise, in contradiction to Theorem 10, $\overrightarrow{P'\,Q'}$ would not be simple; therefore we also have $P \neq R$. If $Q_1 = Q_2$, then it follows from the transitivity of G that G has an edge $\overrightarrow{P\,R}$. But if $Q_1 \neq Q_2$, then Q_1 and Q_2 belong to one and the same maximal net of G, since the same point Q' in the condensation corresponds to them, so that an edge $\overrightarrow{Q_1\,Q_2}$ exists in

G. The edges $\overrightarrow{P\,Q_1}$, $\overrightarrow{Q_1\,Q_2}$, $\overrightarrow{Q_2\,R}$ together form a (directed) path from P to R ($\neq$ P). Because of the transitivity of G there is therefore also an edge $\overrightarrow{P\,R}$ in G in this case. And $\overrightarrow{P\,R}$ is a simple edge of G, since otherwise it would belong to a maximal net of G and then to the vertices P and R of G there could not correspond distinct vertices P' and R' of the condensation. But an edge $\overrightarrow{P'\,R'}$ in the condensation of G must correspond to the simple edge $\overrightarrow{P\,R}$ of G.///

From Theorems 10, 11, and 9 the following theorem follows immediately.

Theorem 12   The condensation of a transitive graph cannot have two distinct edge bases.

§3.   Reduction of the edge basis problem

We wish now to reduce the problem of the determination of an edge basis for an arbitrary transitive graph to the case in which the graph either is a net or has only simple edges. This will be accomplished if we prove the following theorem.

Theorem 13   Let G be an arbitrary transitive graph and $\{N_\alpha\}$ the set of its maximal nets. Let B* be an edge basis for the condensation of G, B_α an edge basis for N_α, and $\hat{B}$ the sum of the (pairwise disjoint) sets B_α. If each edge from B* is replaced by an edge of G corresponding to it, then the set B' results from B*. Then the sum B of the (disjoint) sets $\hat{B}$ and B' is an edge basis for G.

Proof   First, we prove for B the first defining property of a basis. Let $\overrightarrow{P_1\,P_2}$ be an arbitrary edge of G, which is not contained in B. If $\overrightarrow{P_1\,P_2}$ is double and therefore contained in a maximal net N_α, then there is a (directed) path formed out of edges from B_α (and therefore B) going from P_1 to P_2. But if the edge $\overrightarrow{P_1\,P_2}$ is simple, then a distinct edge $\overrightarrow{Q_1 Q_2}$ of the condensation of G corresponds to it. There is a

(directed) path $\overrightarrow{A_1 A_2 \ldots A_\nu}$ of the condensation consisting of edges

from B*, which goes from $A_1 = Q_1$ to $A_\nu = Q_2$ (if $\overrightarrow{Q_1 Q_2}$ belongs to B* then

the edge $\overrightarrow{Q_1 Q_2} = \overrightarrow{A_1 A_2}$ is itself the (directed) path). To the edges

$\overrightarrow{A_1 A_2}, \overrightarrow{A_2 A_3}, \ldots, \overrightarrow{A_{\nu-1} A_\nu}$ of this (directed) path the edges

$\overrightarrow{A_1'' A_2'}, \overrightarrow{A_2'' A_3'}, \ldots, \overrightarrow{A_{\nu-1}'' A_\nu'}$ in B' should correspond, where

A_1'' (A_ν' resp.) belongs to the same maximal net N_1 (N_ν, resp.)

of G as $P_1 = A_1'$ ($P_2 = A_\nu''$ resp.). Also, in general, (and not only

for $\rho = 1$ or ν) A_ρ' and A_ρ'' belong, for $\rho = 1, 2, \ldots, \nu$, to the same

maximal net N_ρ, since the same vertex A_ρ corresponds to them in the

condensation of G. So if A_ρ' and A_ρ'' are distinct, then a (directed) path

formed out of edges from B_ρ goes from A_ρ' to A_ρ''. Certain edges of these

(directed) paths, together with the edges

$A_\rho'' A_{\rho+1}'$ ($\rho = 1, 2, \ldots, \nu-1$), generate a (directed) path consisting of

edges of B going from $A_1' = P_1$ to $A_\nu'' = P_2$.

We now prove for B the second defining basis property.

Let $\overrightarrow{P_1 P_2}$ be an arbitrary edge of B.

First, let $\overrightarrow{P_1 P_2}$ be double and let it belong to the maximal net

N_α of G, and therefore to B_α. We show that no (directed) path of G

consisting of edges of B, which does not contain $\overrightarrow{P_1 P_2}$, goes from P_1 to

P_2. We assume that there is such a (directed) path $\overrightarrow{A_1 A_2 \ldots A_\nu}$ where

$A_1 = P_1$ and $A_\nu = P_2$. Because of the second defining basis property

for B_α not all of the edges of this path can belong to N_α (and so to

B_α) nor to the totality $\sum N_\alpha$ of double edges of G, since a (directed)

path (as a connected subgraph) would have to (Theorem I.26) belong completely to one component N_α. So this (directed) path contains a

simple edge $\overrightarrow{A_\rho A_{\rho+1}}$. Now since $\overrightarrow{P_1 P_2}$ is double, $\overrightarrow{P_2 P_1}$ is contained in

G, so that (see Fig. 44) $\overrightarrow{A_{\rho+1} A_{\rho+2} \dots P_2 P_1 A_2 \dots A_\rho}$ is a (directed)

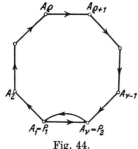

Fig. 44.

path of G. Because of the transitivity of G,

$\overrightarrow{A_{\rho+1} A_\rho}$ is contained in G, which contradicts

the fact that $\overrightarrow{A_\rho A_{\rho+1}}$ is simple.

Secondly, let $\overrightarrow{P_1 P_2}$ be simple. We assume

again that there is a (directed) path

$\overrightarrow{A_1 A_2 \dots A_\nu}$ $(A_1 = P_1, A_\nu = P_2)$ consisting of edges from B and not

containing $\overrightarrow{P_1 P_2}$. Not all the edges of this (directed) path belong to

the same maximal net, since then $\overrightarrow{A_1 A_\nu} = \overrightarrow{P_1 P_2}$ would also have to belong

to this maximal net and could not be simple. So, as in the first case,

it follows that this (directed) path contains at least one simple edge

$\overrightarrow{A_\rho A_{\rho+1}}$. To the simple edges of this (directed) path correspond edges

of the condensation of G, which form a (directed) path $\overrightarrow{X \dots Y}$

consisting of edges of B*. By the definition of B the edge $\overrightarrow{X Y}$

corresponding to the simple B-edge $\overrightarrow{P_1 P_2}$ in the condensation of G must

belong to B*, and this contradicts the second defining property of a

basis for B*.///

Theorem 13, just proved, can be supplemented by

Theorem 14 If G is an arbitrary transitive graph, then each of

its edge bases is generated in the way described in Theorem 13.

Proof Let B be an arbitrary edge basis for G. We show first: 1) The set B_α of the edges of B contained in the maximal net N_α of G forms an edge basis for N_α.

Let $\overrightarrow{P_1 P_2}$ be an arbitrary edge of N_α not contained in B (and therefore not in B_α). Because of the first defining basis property for B there is a (directed) path $\overrightarrow{A_1 A_2 \ldots A_\nu}$ going from $A_1 = P_1$ to $A_\nu = P_2$ and consisting of edges from B. If the edges of this (directed) path were not all contained in N_α, then (as we saw above) an edge $\overrightarrow{A_\rho A_{\rho+1}}$ of this path would be a simple edge of G. This, however, is impossible. Since $\overrightarrow{P_1 P_2}$ is a double edge, $\overrightarrow{A_{\rho+1} A_{\rho+2} \ldots P_2 P_1 A_2 A_3 \ldots A_\rho}$ is a (directed) path of G (see Fig. 44); so because of the transitivity of G an edge $\overrightarrow{A_{\rho+1} A_\rho}$ would also be contained in G. So $\overrightarrow{A_1 A_2 \ldots A_\nu}$ contains only edges from B_α, by which the first defining basis property for B_α is proved. The second defining basis property for B_α follows from the second defining basis property for B, so that assertion (1) is proved.

We now prove

2) The set B_r of those edges of the condensation of G (each counted once) which correspond to the simple edges of B is an[7] edge basis for the condensation of G.

As for the first defining basis property for B_r, let $\overrightarrow{Q_1 Q_2}$ be an arbitrary edge of the condensation of G not contained in B_r; let the edge $\overrightarrow{P_1 P_2}$ in G correspond to it. There is a (directed) path of G consisting of edges from B which goes from P_1 to P_2. It is clear that those edges of the condensation of G which correspond to the simple edges of this (directed) path form a (directed) path of the condensation of G going from Q_1 to Q_2.

Now we prove the second defining basis property for B_r. If B_r did not have this property there would be in B_r an edge $\overline{Q_1\,Q_2}^>$ and a (directed) path $\overline{A_1\,A_2\,\dots\,A_\nu}^>$ of the condensation of G, going from $A_1 = Q_1$ to $A_\nu = Q_2$ and consisting of edges from B_r and not containing $\overline{Q_1\,Q_2}^>$. The (simple) edges $\overline{A_1''\,A_2'}^>$, $\overline{A_2''\,A_3'}^>$, $\dots$, $\overline{A_{\nu-1}''\,A_\nu'}^>$ of G would correspond to the edges of this (directed) path in B, which are all distinct from the edge $\overline{P_1\,P_2}^>$, which corresponds to the edge $\overline{Q_1\,Q_2}^>$ of B_r in B. Here A_1'' belongs to the same maximal net N_1 as the vertex $P_1 = A_1'$ corresponding to the vertex $A_1 = Q_1$ of the condensation of G. Likewise A_ν' belongs to the same maximal net N_ν as the vertex $P_2 = A_\nu''$ corresponding to the vertex $A_\nu = Q_2$ of the condensation of G. Also in general A_ρ' and A_ρ'' ($\rho = 1, 2, \dots, \nu$) belong to the same maximal net N_ρ, since the same vertex A_ρ corresponds to them in the condensation of G. According to (1) a (directed) path of N_ρ, consisting of double (and therefore distinct from $\overline{P_1\,P_2}^>$) edges from B, goes from A_ρ' to A_ρ'' if A_ρ' and A_ρ'' are distinct. Certain edges of these (directed) paths, together with the edges $A_\rho''\ A_\rho'$ which likewise belong to B, generate a (directed) path, consisting of edges from B, which goes from $A_1' = P_1$ to $A_\nu'' = P_2$ and does not contain $\overline{P_1\,P_2}^>$. But this contradicts the second defining basis property for B. In order to be able to deduce Theorem 14 from theorems (1) and (2), just proved, we need to prove the following:

3)   If $\overrightarrow{P_1' \ P_2'}$ and $\overrightarrow{P_1'' \ P_2''}$ are two edges of the edge basis B of G, then the

same edge $\overrightarrow{P_1 \ P_2}$ in the condensation of G cannot correspond to them.

If this were the case, then P_1' and P_1'' would either have to be

identical or belong to the same maximal net N_α of G, and P_2' and P_2'' would

also either have to be identical or belong to the same maximal net

N_β ($\neq N_\alpha$). Since by (1) every two vertices (in both directions) in a

maximal net can be joined by such a (directed) path of the net which has

only edges from B, there would be a (directed) path consisting of edges

from B which contains the edge $\overrightarrow{P_1' \ P_2'}$ but otherwise has only double

edges, and therefore does not contain $\overrightarrow{P_1'' \ P_2''}$, going from P_1'' to P_2'' .

This, however, would contradict the second defining basis property for

B.

By this we have proved (3) and thus proved Theorem 14.///

We wish to emphasize that in the above material relating to

Theorems 8-14, a separate treatment of finite and infinite graphs could

have been avoided. But this distinction is essential if we now wish to

consider the question of the existence of an edge basis.

§4.   Existence of an edge basis.

Minimal edge bases.

Most importantly, we have

Theorem 15   Every finite directed graph has an edge basis.

Proof   If the graph G is not itself an edge basis for G, then it has an

edge $\overrightarrow{P_1 \ Q_1}$ with the property that there is a (directed) path which does

not contain $\overrightarrow{P_1 \ Q_1}$ going from P_1 to Q_1. Now either the graph G_1,

obtained from G by removing $\overrightarrow{P_1 \ Q_1}$, is an edge basis for G or there is in

G_1 a (directed) path from P_2 to Q_2 which joins the endpoints of an

edge $\overrightarrow{P_2 Q_2}$ of G_1 with each other and which does not contain

$\overrightarrow{P_2 Q_2}$, etc. Because there is a finite number of edges the process must

end with a subgraph G_n. Then G_n is an edge basis. It is clear that the

second defining basis property holds, and the fact that the first

property is not lost by removal of the edges $\overrightarrow{P_1 Q_1}$, $\overrightarrow{P_2 Q_2}$, ... follows

immediately from Theorem 1.///

The following theorem can be contrasted with the one just proved.

Theorem 16   There exist infinite transitive graphs containing

nothing but simple edges, which have no edge basis.

Proof   The simplest example of this can be given in the following way.

Let the vertices of G be Q and P_1, P_2, P_3, ... ad inf.; let the edges of

G be $\overrightarrow{P_i P_j}$ ($i < j$: i,j = 1, 2, 3, ..., ad inf.) and

$\overrightarrow{P_i Q}$ (i = 1, 2, 3, ... ad inf.). It is clear that this graph has no

double edges and is transitive. We suppose that G has an edge basis

B. First, every edge of the form $\overrightarrow{P_i P_{i+1}}$ is contained in B, because if

this were not the case, then there would have to be a (directed) path

consisting of edges from B going from P_i to P_{i+1}; but since every

(directed) path of G beginning at P_i is either of the form

$\overrightarrow{P_i P_{i_1} P_{i_2} \ldots P_{i_\nu}}$ or $\overrightarrow{P_i P_{i_1} P_{i_2} \ldots P_{i_\nu} Q}$, where

$i < i_1 < i_2 < \ldots < i_\nu$, this path can be reduced to the single

edge $\overrightarrow{P_i P_{i+1}}$, which then is contained in B. Since there are edges of G

that end in Q, B must contain an edge $\overrightarrow{P_\rho Q}$. Now either B also contains

the edge $\overrightarrow{P_{\rho+1} Q}$ or a (directed) path $\overrightarrow{P_{\rho+1} \ldots Q}$ consisting of edges from B goes from $P_{\rho+1}$ to Q. Along with the edge $\overrightarrow{P_\rho\ P_{\rho+1}}$ there results

in both cases a (directed) path consisting of edges from B and not

containing $\overrightarrow{P_\rho\ Q}$ going from P_ρ to Q. But since $\overrightarrow{P_\rho\ Q}$ is contained in B,

this contradicts the second defining basis property for B.///

There is, however, a certain class of infinite graphs for which the

existence of an edge basis can easily be proved. As we have seen in §1,

a graph with nothing but double edges always has a vertex basis. In the

same way:

Theorem 17   If a directed graph G has only double edges, it has

an edge basis. In particular, every net has an edge basis.

Proof   In order to show this, we shall let correspond to every vertex P

of G a vertex P' and to every pair of edges $\overrightarrow{P\ Q}$, $\overleftarrow{P\ Q}$ of G an edge P' Q'.

These edges P' Q' form an undirected graph G'. Let S' be a frame of G'

(see Theorem IV.27) and let B be the set of those edges of G which

correspond to the edges of S' (so that with $\overrightarrow{P\ Q}$ in B we always have

that $\overleftarrow{P\ Q}$ also belongs to B). We prove that B is an edge basis for G.

Let $\overrightarrow{P\ Q}$ be an arbitrary edge of G not contained in B. By Theorem IV.21

and IV.22 there is a path consisting of edges of S' going from P' to Q';

to this path corresponds a path P ... Q in G consisting of nothing but

B-edges. Since all edges are double, there therefore results a

(directed) path consisting of B-edges from G, which goes from P to Q.

Now as for the second defining basis property let $\overrightarrow{P\ Q}$ be an arbitrary

B-edge. If a (directed) path $\overrightarrow{P\ A_1\ A_2\ ...\ A_\nu\ Q}$ consisting of B-edges and

not containing $\overrightarrow{P\ Q}$, and therefore a cycle $P\ A_1\ A_2\ ...\ A_\nu\ Q\ P$ of G

consisting of (at least three) B-edges, were to exist, then there would

be a cycle $P'\ A_1'\ A_2'\ ...\ A_\nu'\ Q'\ P'$ consisting of edges from S'

corresponding to this in G'. But a frame cannot contain a cycle.///

If, in particular, G is a net, and G' is therefore a complete graph, then the edges A'X' of G', where A' is an arbitrarily chosen fixed vertex of G' and X' runs through all the remaining vertices of G', would form a frame S' of G'. The edge basis B of G corresponding to this frame consists of all the edges $\overrightarrow{A\,X}$ and $\overleftarrow{A\,X}$ of G, where A is a fixed vertex of G.

If the net G is finite and contains n vertices, then this edge basis consists of $2(n-1)$ edges (for $n = 3$ see Fig. 40b). A finite net, however -- if $n > 2$ -- always has an edge basis, which consists of fewer edges.

> Theorem 18 A finite net N with n vertices has an edge basis consisting of n edges but none which has fewer than n edges.

Proof If P_1, P_2, ..., P_n are the vertices of N in arbitrary order, then the edges of the (directed) Hamiltonian cycle $Z = \overrightarrow{P_1 P_2 \ldots P_n P_1}$ of N form an edge basis for N consisting of n edges. If $\overrightarrow{P_i P_j}$ is an arbitrary edge of N not contained in Z, then either $\overrightarrow{P_i P_{i+1} P_{i+2} \ldots P_j}$ or $\overrightarrow{P_i P_{i+1} \ldots P_n P_1 P_2 \ldots P_j}$ is a (directed) path going from P_i to P_j, which contains only edges from Z, depending on whether $i < j$ or $i > j$.

As for the second defining basis property, there cannot be a (directed) path consisting of edges from Z which does not contain $\overrightarrow{P_i P_{i+1}}$ and which goes from P_i to P_{i+1}, since $\overrightarrow{P_i P_{i+1}}$ is the only edge of Z whose initial vertex is P_i (in case $i = n$, set 1 for $i + 1$ here).

On the other hand let B be an arbitrary edge basis for N. It is clear that every one of the n vertices of N must be the initial vertex of an edge of B. So therefore B contains at least n edges.///

The following converse holds.

Theorem 19   If N is a finite net with n vertices, then the
edges of any edge basis B for N with a minimal number of edges
form a continuously directed Hamiltonian cycle of N.

Proof   Since each of the n vertices of N is an initial vertex as well as
a terminal vertex of an edge of B and B (by Theorem 18) has n edges,
then every vertex is the initial vertex and also the terminal vertex of
only one edge of B, so that every vertex is the endpoint of exactly two
edges of B.   Therefore by Theorem I.28 the graph B (considered as an
undirected graph) is a sum of cycles.   If there were two distinct
cycles, one containing P and the other Q, then the edge $\overrightarrow{P\,Q}$ of N could
not belong to B (by Theorem I.21) and therefore there would have to be a
(directed) path going from P to Q formed by B-edges, but this is
impossible (again by Theorem I.21).   Therefore the edges of B form a
unique cycle which contains all vertices of N.   If it were not
continuously directed, then one of its vertices would either be the
initial vertex of two B-edges or the terminal vertex of two B-edges.///

Theorem 19, which is thus proved, also gives the result that the
net with n vertices has (n-1)! minimal edge bases.

Theorem 18 can be extended to infinite nets:

Theorem 20   If n is the cardinality of the set of vertices of
an infinite net N, then every edge basis for N is also of
cardinality n.

Proof   Let b be the cardinality of an edge basis B for N.   The
cardinality of the set of edges of N is $2n^2$, and therefore by the Well-
ordering Principle is equal to n; hence $b \leq n$.   Since on the other hand
every vertex is the initial vertex of a B-edge, $b \geq n$.   From the fact
that $b \leq n$ and $b \geq n$ it follows as a consequence of the equivalence
theorem that b = n.///

With respect to the determination of a minimal edge basis for an arbitrary transitive graph G, our results can be summarized in the following way. By Theorem 14 every edge basis is obtained by determining an edge basis B_α for every maximal net N_α of G (Theorem 17 states that this is possible), by combining these B_α and then adding the edges of that set B' of edges, which corresponds to the edge basis B_r of the condensation of G (if such a set exists). If a minimal edge basis is desired, then, since B_r (in case it exists) is uniquely determined (Theorem 12, cf. Theorem 11) we need only take care that the B_α are minimally chosen. We have just seen (Theorems 17-20) how this can happen.

Finally we would like to point out something which was also remarked by Hertz [1, p. 248]. Graph theoretically this remark of Hertz can be formulated in the following way. A graph can have a smaller edge basis than a subgraph of itself. The simplest example (given by Hertz) of this is given by the graph G of Fig. 45, which has six vertices and eleven edges. The five edges shown as dotted lines form, as one can

Fig. 45.

easily convince oneself, an edge basis for G. But if we consider the subgraph G' of G which is formed by the six solid edges, then G' has as its only basis the set of all six of its edges. By adding new (ideal) vertices and edges it is possible that the minimal number of edges of edge bases can be reduced. The question can be asked: how can a given graph be added to, so that the resulting graph has as small an edge basis as possible?[8] This problem was also posed by Hertz [1, §3] and its solution worked on.

Notes on Chapter VII

[1] The introduction of fundamental sets means in a certain sense the carrying over of the concept of "component" to the theory of directed graphs.

[2] As the example of the continuously directed triangle shows, a finite graph does not need to have a vertex basis of the second kind.

[3] For the graphs of a particular class of graphs -- which we shall designate as bipartite graphs -- a similar problem will be treated in Chapter XIV. (cf. Theorem XIV. 13).

[4] See, for example, Ahrens [2, Vol. II, p. 285]; to distinguish two similar problems, this problem is designated there as "Nebenproblem mit Angriffsverbot" ("subsidiary problem with attack prohibition"). There is also to be found there, for all three problems, a complete list of solutions, which was prepared by K. von Szily (1901). Also the well-known problem of the eight queens, which contrasts with this minimal problem in a certain sense dually as a maximal problem, can be thought of as a special case (for the graph G defined above) of a general graph problem. This general problem runs as follows: let as many vertices as possible of a given graph be distinguished in such a way that for none of its edges both endpoints are distinguished.

[5] In Hertz the corresponding designation is closed (abgeschlossen).

[6] This for directed graphs is the analogous thing to a complete graph.

[7] According to Theorem 12 it is also the only edge basis for the condensation of G.

[8] For the problem of the vertex basis the answer to this question is trivial.

Chapter VIII

Various applications of directed graphs

(Logic -- Theory of Games -- Group theory)

§1. Axiomatic theory

The concepts and results of §1 of the preceding chapter can be
applied to formal logic (axiomatic theory). Let $A = \{A_1, A_2, \ldots\}$
be a set of statements, to each of which we assign a vertex A_i; we
introduce an edge $\overrightarrow{A_i A_j}$ if and only if the statement A_j follows from the
statement A_i, which we designate by $A_i \rightarrow A_j$. (What we mean here by
"follow" we do not need to explain any more precisely; the only property
which we postulate for the binary relation designated by the symbol $\rightarrow$ is
transitivity: if $A \rightarrow B$ and $B \rightarrow C$, and if $C \neq A$,[1] then $A \rightarrow C$. If
Theorem VII.4 is applied to the directed graph so defined, then we have
the following: If A is finite, then A has a subset B with the property
that every statement in A which is not contained in B follows from a
statement in B, but no statement in B follows from another statement in
B.   Theorem VII.7 states that for a finite as well as an infinite system
A every such system B has the same number (or cardinality) of
statements. For infinite systems $A = \{A_1, A_2, \ldots\}$ no such system B
needs to exist, if, for example, A_i means the following statement: "the
natural number x is divisible by 2^i (i = 1, 2, ... ad inf.) Here we
have $A_i \rightarrow A_j$, if i > j.

[The meaning of these results becomes clearer, if we consider the
general case where not only implications of the form "A → B", but also
such of the form "A_1 & A_2 & A_3 & ...) → B" are considered. In this

141

general case -- which, as it seems, cannot be interpreted graph
theoretically -- a subsystem B of the statement system $A = \{A_1, A_2, \ldots\}$
is designated as a <u>basis</u> or as an <u>independent axiom system</u> of A, in case
1) every statement in A which is not contained in B follows from the
statements in B and 2) no statement in B follows from the other
statements in B. The existence of an independent axiom system for a
finite system A (cf. Theorem VII.4) can also be proved in the same
way. But here -- in contrast to Theorem VII.7 -- two independent axiom
systems for the same system A can have a different number of axioms.
For example, consider a system $A = \{A_1, A_2, A_3\}$ with three statements.
The three statements A_1, A_2, A_3 shall mean, respectively, "x is
divisible by 10," "x is divisible by 5," and "x is divisible by 2."
Then A_1 forms for itself, and also A_2 and A_3 together, an independent
axiom system for A.]

The material of §§2-4 in Chapter 7 admits of a similar
interpretation. We consider again a system $A = \{A_1, A_2, \ldots\}$ of
statements and the collection K of certain statements of the form
$A_i \rightarrow A_j$, which we shall designate by K_{ij}. We investigate (independently
of the meaning of the A_i's which can be considered as "logical
variables") the logical dependence or independence of these statements
k_{ij} (and not, as before, of the statements A_i), where now not only
statements of the form $k_1 \rightarrow k_2$ come into consideration, but also such
statements of the form $(k_1 \ \& \ k_2 \ \& \ k_3 \ \& \ \ldots) \rightarrow k$. The meaning of a
statement of this form we define as follows. If K' is a subset of K,
then we say that a statement $k = A_\alpha \rightarrow A_\beta$ of K, not contained in K',
follows from K', if a finite sequence of statements of the form

$$A_\alpha \rightarrow A_{\alpha_1}, \ A_{\alpha_1} \rightarrow A_{\alpha_2}, \ A_{\alpha_2} \rightarrow A_{\alpha_3}, \ \ldots, \ A_{\alpha_{n-1}} \rightarrow A_{\alpha_n}, \ A_{\alpha_n} \rightarrow A_\beta$$

is contained in K' (and so if k can be derived by syllogisms from the statements in K'). The subsystem K' of K is called an independent axiom system for K, if every statement of K not contained in K' follows from K', but no statement of K' follows from the set of the remaining statements of K'.

Hertz [1, p. 250] set for himself the task "of getting an overview of the multiplicity of possibilities of independent axiom systems, and, in particular, of investigating the case in which there is only one [independent] axiom system and the question of how the ones with the least possible theorems can be gotten."

If the elements A_i of A are interpreted as vertices and the elements $A_i \rightarrow A_j$ as edges $\overrightarrow{A_i A_j}$ then a directed graph G corresponds to the system K and it is clear that an edge basis for G corresponds to an independent axiom system for K. The transitivity of G means that K contains, along with statements $A_1 \rightarrow A_2$ and $A_2 \rightarrow A_3$ where $A_3 \neq A_1^2$, the statement $A_1 \rightarrow A_3$. This can be assumed, so that we can restrict ourselves to transitive graphs. The results of §§2-4 of Chapter 7 can be directly translated into logic, and we thereby get Hertz's results. In particular, we get a process to determine the smallest independent system of axioms for a "discipline" of the type K consisting of finitely many statements.

§2. Binary relations

We wish now to consider a still more general application of directed graphs to logic. Let D be an arbitrary finite or infinite set, which we shall think of as fixed. We say that a binary relation R is given for D if for every ordered pair (A,B) of elements A, B of D it is established whether the relation A R B is true or not. If the elements of D are interpreted as vertices and an edge $\overrightarrow{A B}$ is introduced if and only if A R B is true, then a directed graph is uniquely assigned by this rule to the relation R. (It must be assumed here that for every element a of D an element b of D exists, for which either a R b or b R a is true.) Conversely, every directed graph determines in this way a relation with respect to its set D of vertices. So it is clear that the theory of binary relations can be identified with the theory of directed graphs (in case the graph -- as we assume in this chapter -- contains no more than one edge $\overrightarrow{A\ B}$ for any pair of vertices A, B). Ch. S. Peirce developed a theory for binary relations, and this theory was further developed by E. Schröder. We shall be satisfied here to explain the basic concepts of this theory[3] in a graph theoretical interpretation.

Since we think of D as fixed, all graphs G to be considered are subgraphs of the graph (net) Γ which is obtained if any two points A, B of D are joined by an edge $\overrightarrow{A\ B}$ as well as by an edge $\overleftarrow{A\ B}$. The negation $\overline{G}$ of G is formed by those edges of Γ, which are not contained in G. The identical product $(G_1\ G_2)$ of two graphs G_1, G_2 consists of those edges which are contained in both G_1 and G_2. The identical sum $(G_1 + G_2)$ of G_1 and G_2 consists of those edges which are contained[4] either in G_1 or in G_2 (or in both). The conversion $\overset{u}{G}$ of G is formed by those edges $\overrightarrow{A\ B}$ of Γ, for which $\overleftarrow{A\ B}$ is contained in G. Characteristic of the Peirce - Schröder theory are the concepts of relative multiplication and addition. The relative product $(G_1; G_2)$ consists of those edges $\overrightarrow{A\ B}$ of Γ for which a vertex C of Γ exists with the property that $\overrightarrow{A\ C}$ is an edge of G_1 and $\overrightarrow{C\ B}$ is an edge of G_2. The relative sum $(G_1 \not+ G_2)$ consists of those edges $\overrightarrow{A\ B}$ of Γ, which are such that for every vertex C of Γ either an edge $\overrightarrow{A\ C}$ in G_1 exists or an edge $\overrightarrow{C\ B}$ in G_2 (the word "or" here is meant in the inclusive sense).

The Peirce - Schröder theory is an algebra of these six basic operations which is concerned with equations $G_1 = G_2$ and "inequations " $G_1 \not\subseteq G_2$ ("G_1 is a subgraph of G_2") and with their solutions. Our graph theoretical interpretation is perhaps of little use for this algebra. These operations, however, particularly the relative ones, can perhaps be used elsewhere in graph theory. We give here several simple examples.

    1.   The transitivity of a graph can be characterized by whether or not the relative product (G; G) is a subgraph of G:

(G; G) $\nleq$ G.

    2.   The equation $G\overset{u}{G} = 0$ (where 0 means the null graph) says that every edge of G is simple.

    3.   The equation $G + \overset{u}{G} = \Gamma$ says that, for every two elements A and B of D, either $\overset{\longrightarrow}{A\ B}$ or $\overset{\longleftarrow}{A\ B}$ (or both) are edges of G.

Those graphs for which these three relationships

$$(G;\ G) \nleq G,\ G\overset{u}{G} = 0,\ G + \overset{u}{G} = \Gamma$$

are simultaneously true (see, for example, the graph of Fig. 43) can be designated as ordered graphs[5]. If the binary relation corresponding to such a graph is designated by <, then these three properties say the following:

    1.)   If A < B and B < C then A < C.

    2.)   if A < B, then it is not true that B < A.

    3.)   Either A < B or B < A (or both).

By Cantor's definition binary relations with these three properties are the ones by which the set D is simply ordered. In the case that D is finite it is easily seen by Theorem VII.9 that the only edge basis of an ordered graph G is a (directed) path which joins[6] all the vertices of G with one another in the order of their simple ordering. If of the properties 1), 2), and 3) only 3) is required and not 1) and 2), then the Redei Theorem (II.10) states that if D is finite, there is a simple ordering consistent with this relation; in this general case, however, this simple ordering is not uniquely determined by the relation.

§3. Solitaire games

The application of graph theory to a type of game will now be
illustrated with two examples. The first example concerns the so-called
jug problems. The oldest of these problems goes as follows:

A person has three jugs; the first, which holds eight liters, is
full of wine, the second and third, which hold 5 liters and three
liters, respectively, are empty. By a series of pourings -- without
using any other measuring devices -- it should be brought about that the
first jug contains four liters of wine, and so does the second jug.[7]

Any state of the three jugs can be indicated by an ordered triple
(x, y, z) of numbers, where x, y, and z tell how many liters of wine are
contained in the first, second, and third jug, respectively.

The numbers x, y and z are, of course, whole numbers satisfying the
relations

$$x + y + z = 8$$
$$0 \leq x \leq 8, \quad 0 \leq y \leq 5, \quad 0 \leq z \leq 3.$$

Also it is clear that in every state that can be brought about by
these pourings, there must be either an empty or a full jug, so that one
of the following six equations must also hold:

$$x = 0, \ x = 8, \ y = 0, \ y = 5, \ z = 0, \ z = 3.$$

It is easy to see that all these conditions are satisfied only by the
following 16 ordered triples:

(8,0,0) (7,1,0) (7,0,1) (6,2,0) (6,0,2) (5,3,0) (5,0,3) (4,4,0)
(4,1,3) (3,5,0) (3,2,3) (2,5,1) (2,3,3) (1,5,2) (1,4,3) (0,5,3).

We wish now to interpret these ordered triples as vertices of a
directed graph. A vertex (x,y,z) is joined to a vertex (x',y',z') by an
edge directed towards (x',y',z') if and only if the state (x',y',z') can
be reached from the state (x,y,z) by a single pouring. In this way a
directed graph is defined (with 16 vertices and 58 edges). It is shown

146

in Fig. 46, where for two edges which join the same pair of vertices in opposite directions, only one edge is indicated but it has two arrows. Our problem requires us now to find a (directed) path of the graph, which goes from (8,0,0) to (4,4,0). With the help of the illustrated graph the shortest path of this kind consisting of seven edges (pourings), can easily be found:

    (8,0,0) (3,5,0) (3,2,3) (6,2,0) (6,0,2) (1,5,2) (1,4,3) (4,4,0).

If the only solutions allowed are ones in which no state can be repeated then our problem has 16 solutions (see Brunel [1]). The other solutions

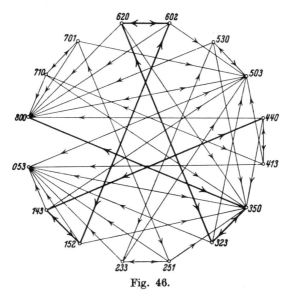

Fig. 46.

can be read off Fig. 46. The figure also shows that from any one of the 16 states any one of the other 15 states can be reached by appropriate pourings. (In other words this means: the set of all vertices of our graph is a fundamental set of the graph and every vertex is a source of this fundamental set.)

Our second example is concerned with the problems of ferry crossings. Instead of a pouring we have a ferry crossing. Since -- in contrast to pourings -- the ferry crossings are reversible, with the

147

appearance of the edge $\overrightarrow{P\ Q}$ in our graph we shall always have $\overleftarrow{P\ Q}$ in our graph also. If these two edges are always replaced by the single undirected edge PQ, then it is seen that we have an undirected graph which concerns us here: a path must be found which joins two given vertices.

The oldest and best known of the problems belonging to this type requires a ferryman to ferry a wolf, a goat, and a head of cabbage across a river in a boat which can hold only one of the three besides the ferryman, and the goat may not be left alone with either the wolf or

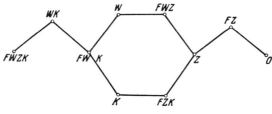

Fig. 47.

the cabbage in the absence of the ferryman. This problem corresponds to the graph of Fig. 47 where, for example, "WC" means that the wolf and cabbage are together on this side of the river while the goat and the ferryman (and therefore also the boat) are on the other side; O is the symbol for all four being on the other side. It is really not difficult to find a path from FWGC to O. A graphical picture of the whole matter is given by Fig. 47.

The solitaire game called the Boss Puzzle[8] (see, for example, B. Ahrens [2, Vol. 1, Chap. 8]; also called the 15 game; see, for example, B. Ahrens [2, Vol. 1, Chap. 19]) and many other games for one person can be interpreted graph theoretically in a similar way, namely, all those games which deal with bringing about a certain state (position), or one of several given states, from a given state by a (finite) sequence of steps, such that the steps can be chosen[9] freely from a given set of steps (which make up the rules of the game).

§4. Games for two people

In certain other games not every step can be freely chosen, but
every second step depends on the decision of a second player. Graph
theory can also be used, and indeed to a greater degree, for the theory
of such games for two people (to which chess belongs as a most popular
game). Games of understanding, like chess, can be characterized in the
following way. The game is played by two players alternating turns; the
first play of the first player (white) changes the beginning position P_0
to a second position P_1, from which follows the first play of the second
player (black), which changes the position P_1 to the position P_2, etc.
The rules of play determine, on the one hand, the set of positions
(which can be replaced by an equivalent set); on the other hand, the
rules of play determine whether or not one player or the other can reach
a position P_β from a position P_α, i.e. whether a move from P_α to P_β
follows the rules or not. Which of the permissible plays is chosen
depends only on the decision of the player whose turn it is (and not on
manual skill or on chance). If a player reaches a certain position,
from which no play on the part of the other player is allowed according
to the rules of the game (a losing position for the second player) then
he has won the game.

If this is not attained by either of the players, i.e. if the game
is infinite, then we say that the game is undecided, or it ends in a
draw.

Let W and B be the set of positions in which it is White's (or
Black's) turn. We consider two positions distinct if in one it is one
player's turn and in the other it is the other player's turn. It is not
assumed that W and B are finite. To every chess-like game we can now
assign a definite directed graph in the following way. To the elements

of W and B we let a point P_α (Q_β , resp.) correspond and introduce an edge $\overrightarrow{P_\alpha Q_\beta}$ if and only if a legal move from the position corresponding to P_α leads to the position corresponding to Q_β. So there results for chess, checkers, etc. a definite directed graph.

All chess-like games are therefore -- according to the choice of the graph -- contained in the following "most general game."

Let G be an arbitrary directed graph with vertices P_1, P_2, ..., Q_1, Q_2, ..., in which every edge has the form $P_\alpha Q_\beta$ (i.e. $\overrightarrow{P_\alpha Q_\beta}$ or $\overleftarrow{P_\alpha Q_\beta}$). At the beginning of the game a single game piece is at a given vertex (starting point). The game consists of the two players' changing the piece alternately along an edge -- in the direction proper to the directed edge -- to another vertex. When a player reaches a vertex Z, out of which there is no directed edge $\overrightarrow{Z X}$, then he has won the game. If this is not reached by either player, then the game is undecided (a draw).

A game is represented, for example, by the solid edges of the graph of Fig. 49, which the person who made the first move has won in 3 moves (P$_1$ is the starting point).

For chess (and for much simpler games as well) the construction of the corresponding graph would, of course, be from a practical point of view scarcely possible. For general theoretical investigations concerning games this graph theoretical representation can be applied to advantage. We think especially of investigations which relate to the concept of a <u>winning position</u>. What is to be understood by this designation for the solitaire games treated above (§3) is clear: a (directed) path in the corresponding graph goes from A to Z, where the vertex Z corresponds to the position which is the goal

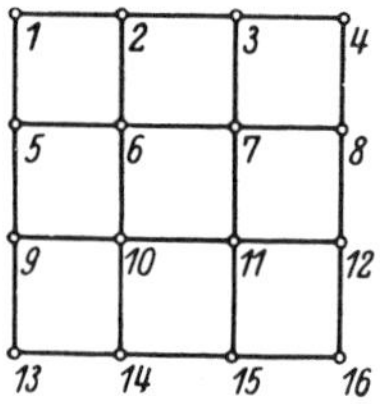

Fig. 48.

to be reached (in the first example of §3 this is (4,4,0); in the second

0), and so A corresponds to a winning position.

The meaning of this designation is not so clear for the game of

chess, even if -- from the "theory" of "end plays" and "chess problems"

-- the concept of the winning position is known to the practical chess

player (or seems to be). That this concept, i.e. the meaning of such

statements as "the player A can be assured of victory in position P",

must and can be given strictly mathematically -- in which every

subjective and psychological impulse is excluded -- was recognized by

Zermelo in his Cambridge address.[10]

In the interpretation[11] of this statement we can make use of the

graph theoretical interpretation introduced above, by which we wish to

identify the positions and moves with vertices and edges of the directed

graph G, which represents the game. We consider the totality T of the

sequences of edges, which begin at P_{α_1}, of the type

$$F = \overrightarrow{P_{\alpha_1} Q_{\beta_1}}, \overrightarrow{Q_{\beta_1} P_{\alpha_2}}, \overrightarrow{P_{\alpha_2} Q_{\beta_2}}, \ldots, \overrightarrow{Q_{\beta_{n-1}} P_{\alpha_n}}, \overrightarrow{P_{\alpha_n} Q_{\beta_n}}.$$

Now P_{α_1} is called a winning position for White if the set T has a subset

R with the following three properties:

1) There is an element in R, which contains a single edge

   $$\overrightarrow{P_{\alpha_1} Q_\beta}.$$

2) If F is an element of R which ends in Q_{β_n} and $\overrightarrow{Q_{\beta_n} P_{\alpha_{n+1}}}$ is an

   arbitrary edge of G, then there is an edge $\overrightarrow{P_{\alpha_{n+1}} Q_{\beta_{n+1}}}$ of G

   with the property that the sequence of edges which results from

   F by adding the edges $\overrightarrow{Q_{\beta_n} P_{\alpha_{n+1}}}$ and $\overrightarrow{P_{\alpha_{n+1}} Q_{\beta_{n+1}}}$ also belongs to

   R.

3) If $\overrightarrow{P_{\alpha_1} Q_{\beta_1}}$, $\overrightarrow{Q_{\beta_1} P_{\alpha_2}}$, $\overrightarrow{P_{\alpha_2} Q_{\beta_2}}$, ... is a finite or infinite

sequence of edges of G whose segments consisting of the first

1, 3, 5, 7, ... edges belong to R and it is continued as far as

possible to the right, then this sequence of edges ends with an

edge $\overrightarrow{P_{\alpha_n} Q_{\beta_n}}$ (and therefore the sequence cannot be infinite).

By specification of such a subset R a strategy is given White in

position P_{α_1} for playing, namely the following: he should play so that

the sequences of edges corresponding to the first 1, 3, 5, ... moves are

all elements of R. By (1) and (2) White can always follow this strategy

and (3) guarantees that he will win if he follows this strategy.

It could be assumed that if P_1 is a winning position for White a

number r always exists, dependent only on P_1 , such that White can be

assured of winning when he is at most r moves from P_1. The following

example shows that this is, in general, not the case. Let the vertices

of the graph be:

P_1^1 ; P_2^1 , P_2^2 ; P_3^1 , P_3^2 , P_3^3; ...; P_α^1, P_α^2, ..., P_α^α; ... ad inf.

Q_1^1; Q_2^1, Q_2^2; Q_3^1, Q_3^2, Q_3^3; ...; Q_α^1, Q_α^2, ..., Q_α^α, ... ad inf.

and let the edges be:

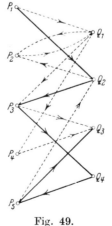

Fig. 49.

152

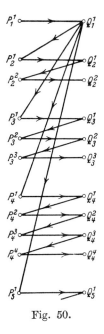

Fig. 50.

$$\overset{\longrightarrow}{P_\alpha^\beta \, Q_\alpha^\beta} \quad (\alpha = 1, 2, 3, \ldots \text{ ad inf.}; \ \beta = 1, 2, \ldots, \alpha),$$

$$\overset{\longrightarrow}{Q_\alpha^\beta \, P_\alpha^{\beta+1}} \quad (\alpha = 2, 3, 4, \ldots, \text{ ad inf.}; \ \beta = 1, 2, 3, \ldots, \alpha - 1) \text{ and}$$

$$\overset{\longrightarrow}{Q_1^1 \, P_\alpha^1} \quad (\alpha = 2, 3, 4, \ldots, \text{ ad inf.}).$$ See Fig. 50, which is to be thought of as continuing down indefinitely. Now the starting position P_1^1 is a winning position, for White begins with the move $\overset{\longrightarrow}{P_1^1 \, Q_1^1}$, whereupon Black must move to one of the points P_2^1, P_3^1, P_4^1, ... If he chooses any one of these moves, say $\overset{\longrightarrow}{Q_1^1 \, P_\rho^1}$, then the continuation of the game is uniquely determined, (namely: $\overset{\longrightarrow}{P_\rho^1 \, Q_\rho^1}$, $\overset{\longrightarrow}{Q_\rho^1 \, P_\rho^2}$, $\overset{\longrightarrow}{P_\rho^2 \, Q_\rho^2}$, $\ldots \overset{\longrightarrow}{P_\rho^\rho \, Q_\rho^\rho}$ and leads to White's winning in $\rho + 1$ moves (and not sooner), namely to the losing position Q_ρ^ρ for Black.) But now the number ρ can be chosen arbitrarily large by Black so that he can postpone losing arbitrarily long.

Consequently White cannot predict in position P_1^1 the largest number of moves necessary for his winning -- although his winning (in finitely many moves) is certain.

This phenomenon cannot happen in chess or in the rest of the well-known games. This is due to the fact that in all these games only finitely many moves are allowed in any position, i.e., that the graph is of finite degree.

> Theorem   If, in a game, only finitely many moves are permitted in any position, then there is for every winning position P for White a number N, dependent on P, with the property that White, playing from this position P, can be assured of winning in at most N moves.[12]

Proof   The proof of this theorem can be reduced to our Infinity Lemma (Theorem VI.6) in the following way. Since P is a winning position, a set R corresponds to it with the three properties given above. Let E_n be the set of the R-elements consisting of 2n - 1 edges (n = 1, 2, 3, ...). By our finiteness assumption E_1, E_2, E_3, ... are all finite sets. In the proof we are, of course, allowed to assume that White plays according to the strategy given by the set R and that therefore the first 2n - 1 moves counted from P for n = 1, 2, 3, ... give a sequence of edges contained in R. We suppose that the theorem is false. This says that arbitrarily "long" sequences of edges are contained in R and that therefore none of the sets E_1, E_2, E_3, ... is empty. We let a point correspond to the elements of E_i, by which the set E_i corresponds to the point set π_i (i = 1, 2, 3, ...). We choose the points of these sets as vertices of a graph U. A vertex A of $\pi_{\nu+1}$ is joined with a vertex B of π_ν by an edge of U, if the sequence of

U so defined and the sets π_i fulfill the conditions of the Infinity

Lemma. If the lemma is applied, there results an infinite sequence a_1,

a_2, a_3, ... of R-elements, where a_i consists of $2i - 1$ edges and is a

segment of a_{i+1}. If these elements a_1, a_2, a_3, ... are "blended" in,

there results an infinite game, whose segments consisting of the first

1, 3, 5, 7, ... moves belong to R. But this is impossible, since such a

game -- according to property (3) of R -- must be finite.

The example given earlier (Fig. 50, where only one vertex Q_1^1 is of

infinite degree) shows that the assumption in this proof that the graph

of the game is of finite degree is essential. The finiteness of the

graph, however, did not have to be assumed, so that the theorem proved

remains true for such games in which infinitely many positions occur, if

at the same time only finitely many moves are allowed from any position

by the rules of the game.[13] For the case in which there are only

finitely many positions, the theorem can be further sharpened. If t is

the (finite) number of distinct positions, then this number t can always

be chosen for the upper bound N -- independent of the winning position P

-- so that t is a universal upper bound for the necessary number of

moves for any winning position.[14] We omit the proof of this theorem

here.

In this connection we state the following theorem, without proof,

from which the theorem just stated can easily be derived.

If a player is in a winning position, then he can win from this

position in such a way that, however his opponent may play, no position

appears twice.[15]

In all this work graph theory offers a natural means to picture

clearly relationships which are often quite complicated.

§5. Cayley Group - diagrams

Cayley [5, 6, and 8] has given a graphical method for representing an abstract group of finite order by a "colored" directed graph.[16] Let

$$\Gamma = \{S_1, S_2, \ldots, S_{n-1}, S_n = 1\}$$

be an arbitrary group of finite order n. To every element S_i we let correspond a vertex P_i (i = 1, 2, ..., n) and join every pair of vertices P_i, P_j (i, j = 1, 2, ..., n; i ≠ j) by an edge $\overrightarrow{P_i P_j}$ and by an edge $\overleftarrow{P_i P_j}$. There results a directed graph (net) G determined by Γ (according to its structure) with n vertices and n(n-1) edges. Furthermore let there be 1, 2, ..., α, ..., n-1 distinct colors (indices). We give the edge $\overrightarrow{P_i P_j}$ the color α, so that $S_\alpha S_i = S_j$, and therefore $S_\alpha = S_j S_i^{-1}$.[17] Then the n-1 edges, which go out from P_i, get, for each i, the n-1 distinct colors, and likewise the n-1 edges, which go to P_i. This follows from the fact that $\{S_1 S_\alpha, S_2 S_\alpha, \ldots, S_n S_\alpha\}$ and $\{S_\alpha S_1, S_\alpha S_2, \ldots, S_\alpha S_n\}$ represent a permutation of $\{S_1, S_2, \ldots, S_n\}$.

Also we get the following more general result:

If α_1, α_2 ..., α_r is an arbitrary sequence of the colors 1, 2, ..., n-1 -- with or without repetitions -- there is for each i one and only one continuously directed sequence of edges with the initial vertex P_i:

$$\overrightarrow{P_i P_{i_1}}, \quad \overrightarrow{P_{i_1} P_{i_2}}, \quad \ldots, \quad \overrightarrow{P_{i_{r-1}} P_{i_r}}, \tag{1}$$

such that these edges have in order the colors α_1, α_2, ..., α_r and so such that

$$S_{\alpha_1} S_i = S_{i_1}, \quad S_{\alpha_2} S_{i_1} = S_{i_2}, \quad \ldots, \quad S_{\alpha_r} S_{i_{r-1}} = S_{i_r}.$$

From these equations we get by successive substitutions

$$S_{\alpha_r} S_{\alpha_{r-1}} \cdots S_{\alpha_2} S_{\alpha_1} S_i = S_{i_r} \tag{2}$$

This sequence of edges is closed (i.e. $P_{i_r} = P_i$ and so $S_{i_r} = S_i$) if and only if

$$S_\alpha \equiv S_{\alpha_r} S_{\alpha_{r-1}} \cdots S_{\alpha_2} S_{\alpha_1} = 1. \tag{3}$$

Since S_α, and therefore also the existence of this relation, is independent of i, we have the result that if a sequence of edges starting out from a vertex and specified by the sequence of colors $\alpha_1, \alpha_2, \ldots, \alpha_r$ returns to this vertex, then this holds for every vertex. So with the help of this graph G it can be decided immediately whether a relation of the type (3) exists or not for the group involved.

But if the sequence (1) of edges is not closed: $P_{i_r} \neq P_i$, $S_{i_r} \neq S_i$, then $S_\alpha \neq 1$ and then (2) says, i.e. the equation $S_\alpha S_i = S_{i_r}$, that $\overrightarrow{P_i P_{i_r}}$ has the color α. By this we have the result: If a sequence of edges characterized by the sequence of colors $\alpha_1, \alpha_2, \ldots, \alpha_r$ goes from P_i to Q, then the color of the edge $\overrightarrow{P_i Q}$ is independent of i.

If α is an arbitrary color of the $n-1$ colors, then every vertex is the endpoint of two edges of the color α. By Theorem I.28 the edges of color α form cycles; these are, of course, continuously directed, and therefore directed cycles. All these (directed) cycles contain the same number r_α of edges; and, in fact, r_α is equal to the order ρ_α of S_α. If we set $\alpha_1 = \alpha_2 = \ldots = \alpha_r = \alpha$, then equation (3) says: $S_\alpha^{r_\alpha} = 1$; so r_α is a multiple of ρ_α, and, in fact, $r_\alpha = \rho_\alpha$; for if $\rho_\alpha < r_\alpha$, then a proper subset of the set of edges of the (directed) cycle under consideration would give rise to a closed sequence of edges. If, for example, $\rho_\alpha = 2$ is the order of S_α, then $S_\alpha^2 = 1$, and so $\overrightarrow{P_i P_j}$ has the color α along with $\overleftarrow{P_i P_j}$, since in this case $S_\alpha S_i = S_j$ and $S_\alpha S_j = S_i$

157

are equivalent and the edges of color α form a 2-cycle $P_i \, P_j \, P_i$.

All the properties of the abstract group Γ can be recognized by the colored graph G -- the so-called Cayley color-group -- so that it determines this group uniquely and indeed even if the designation of the vertices is not given by P_1, P_2, ..., P_n, i.e. their correspondence to the elements of Γ.

As an example we consider the group of order 6, which can be specified as the permutation group on 3 elements in the following way:

$$S_1 = \begin{pmatrix} abc \\ bca \end{pmatrix}, \; S_2 = \begin{pmatrix} abc \\ cab \end{pmatrix}, \; S_3 = \begin{pmatrix} abc \\ acb \end{pmatrix}, \; S_4 = \begin{pmatrix} abc \\ bac \end{pmatrix}, \; S_5 = \begin{pmatrix} abc \\ cba \end{pmatrix}, \; S_6 = 1;$$

or as substitution group:

$$x' = \frac{1}{1-x} \; , \; \frac{x-1}{x} \; , \; \frac{1}{x} \; , \; \frac{x}{x-1} \; , \; 1-x, \; x.$$

Fig. 51 shows the Cayley diagram of this group. The colors 1, 2, 3, 4, and 5 added to the edges of this group correspond to the elements S_1, S_2, S_3, S_4, and S_5 of the group. Corresponding to the relations

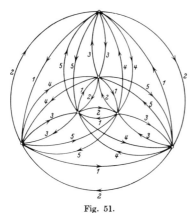

$$S_1^3 = S_2^3 = S_3^2 = S_4^2 = S_5^2 = 1$$

the edges of colors 1 and 2 form triangles; those of colors 3, 4,

Fig. 51.

and 5 form 2-cycles. It can also be seen from the figure, for example, that $S_1 \, S_3 \, S_1 \, S_3 = 1$, since -- beginning at any vertex -- the sequence of edges of color 1, 3, 1, 3 is closed. The diagram can be simplified by removing all edges of color 2, 4, and 5. In Fig. 52 only edges of color 1 and 3 are drawn; this simpler diagram determines our group, since it is generated by S_1 and S_3

$$(S_2 = S_1^2, \; S_4 = S_1 \, S_3, \; S_5 = S_3 \, S_1).$$

We borrow these figures from Hilton's book [1, pp. 84-90], where other examples and figures for Cayley color-groups are also to be found.

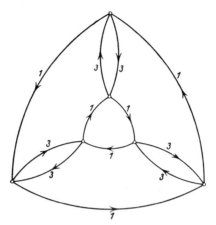

Fig. 52.

Notes on Chapter VIII

[1]As one is accustomed to establish the concept "follows" and to establish the meaning of the symbol → in logic, this restriction C≠A can be done away with, since in this generally usual interpretation A → A is always true. This can be justified by adding to each vertex A of the graph a loop $\overrightarrow{AA}$. It is clear that this would not affect the results of §1 of Chapter VII. (The same holds for §§2-4 of Chapter VII.)

[2]Compare the preceding footnote.

[3]A short description of this theory is to be found in E. Schröder, Note über die Algebra der binären Relative, Mathematische Annalen, 46, 1895, pp. 144-158.

[4]If G_1 and G_2 are disjoint, then this is the same summation which we introduced in Chapter I.

[5]Hertz [1, §2, Def. 45] uses for this the designation "Zug" (or, to be exact, for the corresponding logical structure).

[6]See Hertz [1, §2, Lehrsatz 49]; cf. Fig. 43.

[7]The idea of interpreting this question graph theoretically stems from Brunel [1].

[8]The Boss-puzzle can be connected with graph theory in another respect. The problem of this game can be thought of as a question about the graph of Fig. 48. If this graph is replaced by an arbitrary graph

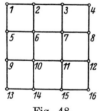 (finite and connected, for example), we get here a general problem of graph theory, which is treated in detail in Kowalewski [1, chapters II, III and 2 pp. 61-81], where the question as to whether the graph has a Hamiltonian cycle plays a

Fig. 48.

decisive rôle. In a similar way the solitaire game leads also to a general graph problem; cf. Lucas [1, Vol. 4, pp. 222-223, problems V, VI].

[9]The solution of any mathematical problem actually falls under this game definition; the rules of the game are furnished by logic.

[10]Über eine Anwendung der Mengenlehre auf die Theorie des Schachspiels, Proceedings of the fifth International Congress of Mathematicians (Cambridge, 1912), vol. 2, Cambridge, 1913, pp. 501-504. In this reference the sense of the following statement is explained: "der Gewinn läßt sich in höchstens r Zügen erzwingen" ("Winning can be forced in at most r moves"), while in the definition that follows here there is no question of any higher bound r.

[11]König [8, §5].

[12]König [8, §5; cf. also the addendum.] Mr. J. von Neumann brought to my attention the connection between this theorem and the Infinity Lemma.

[13]Such a game would come about, for example, if the finite number of chess pieces is retained, but chess is played on an infinite board and only moves of such "length" are allowed which are possible on the usual board of 64 squares. The given proof holds for this "infinite" game of chess.

[14]This theorem is to be found in the above mentioned paper of Zermelo's. A proof by Zermelo is to be found in König's paper [8, Zusatz].

[15]This theorem is also expressed in Zermelo's lecture. L. Kalmár gave a proof, Zur Theorie der abstrakten Spiele, Acta Litterarum ac Scientiarum (Sectio Scientiarum Mathematicarum), Szeged, 4, 1928, pp. 65-85. This work by Kalmár contains far-reaching generalizations of the investigations mentioned here. -- From the standpoint of Brouwer's intuitionism M. Euwe has concerned himself with similar questions (Mengentheoretische Betrachtungen über das Schachspiel, Proceedings of the Akad. Wet. Amsterdam, 32, 1929, pp. 633-644).

[16]Cf. also Maschke [1] and Baker [1].

[17]A loop could be added to every vertex and an nth color assigned to all the loops.

Chapter IX

(Directed) Cycles and Stars and the corresponding linear forms

§1. Linear Forms

Let G be an arbitrary finite or infinite directed graph. We let correspond to every edge k_α of G a variable x_α (independent of the others), which we regard as the representative of the edge k_α. If we assign [1] a number c_α to every edge k_α of G, then a valuation $\{c_\alpha\}$ of G is thus given and G is transformed into a graph with valuation. If only finitely many of the c_α are different from 0 (for finite graphs this is, of course, always the case), then the valuation is designated as a finite valuation. A certain linear form

$$\sum_{(\alpha)} c_\alpha x_\alpha$$

is defined by every finite valuation of G and, conversely, every linear form of the variables x_α corresponding to the edges determines a finite valuation of G.

If G denotes an infinite graph then it is appropriate to consider here the finite sum $\sum c_\alpha x_\alpha$ formally as an infinite sum in which all x_α occur, but only finitely many of them with coefficients c_α different from zero. By a sum of infinitely many zeros it is understood that we mean 0. Accordingly we shall say, for example, that in the linear form $3x_1 + 2x_5$ the coefficient c_2 is equal to 0, so that in the specification of a linear form a value c_α is ascribed to the x_α of every edge k_α (the cardinality of the edges can be arbitrarily large).

If in a finite or infinite valuation the values different from zero

are all equal to 1 or -1, then a directed subgraph G' of G is defined by
this valuation (and, in case it is finite, also by its corresponding
linear form):

G' consists of those edges k_α of G for which c_α is not zero
and k_α has the same direction as in G or the opposite direction
according to whether $c_\alpha = 1$ or $c_\alpha = -1$.

Conversely, a valuation (with values 0, 1, -1) is determined by an
arbitrarily directed subgraph G' of G, and therefore, if the subgraph is
finite, a linear form with coefficients 1, -1 (and 0).

As is well known, the linear form L with variables x_α is said to be
composed from the linear forms L_1, L_2, ..., L_ν provided that for certain
constants[2] a_1, a_2, ..., a_ν the equation

$$L = a_1 L_1 + a_2 L_2 + ... + a_\nu L_\nu$$

is identically true in the x_α. The words linear and homogeneous are
omitted in what follows, as all relations are assumed to be linear and
homogeneous. The linear forms L_i are called independent of one another
if an equation

$$\sum_i a_i L_i = 0$$

is true (identically in the x_α) only if all $a_i = 0$. It is often
expedient to regard the sums $\sum_i a_i L_i$ formally as infinite sums in which
only finitely many of the coefficients a_i are different from zero.

§2. Cycle forms

A particular role is played by linear forms which belong[3] to
(directed) cycles of graphs and which we designate as <u>cycle forms</u>. For
example, the linear form

$$x_1 - x_2 - x_3 + x_4 + x_5$$

corresponds to the cycle described in the clockwise sense in Fig. 53.

The question to which we now turn is: Which linear forms can be
composed from cycle forms? The answer is contained in the following

theorem.

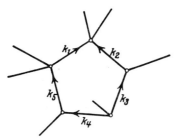

Fig. 53.

__Theorem 1__ A linear form $L = \sum c_\alpha x_\alpha$ can be composed from cycle

forms of a graph if and only if the following condition is

satisfied. If k_1, k_2, ... are edges (to which the variables

x_1, x_2, ... correspond) ending at vertex P, then for every

vertex P the equation[4]

$$\sum \varepsilon_i c_i = 0$$

holds, where ε_i = 1 or -1, depending on whether k_i is in the

direction toward P or in the opposite direction.[5]

__Proof__ The necessity of this condition is easy to see in the simplest

case, where L is itself a cycle form. If P is not a vertex of this

(directed) cycle, then for this P the equation $\sum \varepsilon_i c_i = 0$ must be

considered to be satisfied. So let P be a vertex of this (directed)

cycle ... $\overrightarrow{R\ P\ Q}$ ... Depending on the directions of the edges PR and PQ

there are four possibilities for Fig. 54. We have written the values of

the corresponding ε_i's over the edges and the values of the

corresponding c_i's under the edges. In the four cases $\sum \varepsilon_i c_i$ is

165

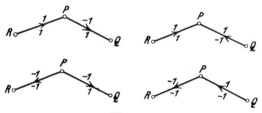

Fig. 54.

reduced to two terms for the vertex P and gives rise to the values

$$1 \cdot 1 + (-1) \cdot 1, \qquad\qquad\qquad 1 \cdot 1 + 1 \cdot (-1),$$

$$(-1) \cdot (-1) + (-1) \cdot 1, \qquad\qquad (-1) \cdot (-1) + 1 \cdot (-1),$$

which are all equal to 0. But if $\sum \epsilon_i \, c_i$ has for the linear forms

L_1, L_2, ..., L_ν, with respect to the vertex P, the values

S_1, S_2, ..., S_ν, then it has for the form $\sum_{\rho=1}^{\nu} a_\rho L_\rho$, again with respect

to P, the value $S = \sum_{\rho=1}^{\nu} a_\rho S_\rho$. If all the S_ρ's are 0, then S is also

0. But since we have just proved $\sum \epsilon_i \, c_i = 0$ for cycle forms, this

equation (with respect to each vertex) also holds for every linear form

which can be composed from cycle forms. With this the necessity of the

condition is proved in general.

We now wish to show that the condition is also sufficient, and we

may assume that L does not vanish identically. Those edges to which a

coefficient c_α of L different from 0 corresponds form a nonempty finite

subgraph G' of the original graph G. If G' does not contain a cycle

then it is -- by Theorem IV. 7 -- the sum of trees, and so has -- by

Theorem IV. 4 -- an endpoint P. With respect to this endpoint the

condition $\sum \varepsilon_i c_i = 0$ is reduced to c = 0, where c is the coefficient of

L corresponding to the only edge of G' ending in P. But now c = 0

contradicts the definition of G' and consequently, if G' contains no

cycle, the conditions $\sum \varepsilon_i c_i = 0$ cannot be satisfied for every

vertex. We may therefore assume that G' has a (directed) cycle Z_1, to

which the linear form L_1 corresponds. Let k_1 be an edge of Z_1 to which

the coefficient c_1 in L corresponds. Now we consider the form

$L - \varepsilon_1 c_1 L_1$, where $\varepsilon_1 = \pm 1$ depending on whether k_1 has the same

direction in G and Z_1 or not. The variable x_1 corresponding to the edge

k_1 is not contained in this form (its coefficient is $c_1 - \varepsilon_1 c_1 \varepsilon_1 = 0$).

Since Z_1 is a subgraph of G', those edges of G to which a coefficient

different from 0 in $L - \varepsilon_1 c_1 L_1$ corresponds form a subgraph G" of G'

and, in fact, a proper subgraph, since k_1 is not contained in G". If G"

also contains a (directed) cycle Z_2, then we can likewise "remove" an

edge of this cycle, i.e. change to a form

$L - \varepsilon_1 c_1 L_1 - \varepsilon_2 c_2 L_2$, such that those edges to which a coefficient

different from 0 corresponds form a proper subgraph G''' of G", etc.

Because of the finite number of edges of G' this process must end, and

we finally reach the form

$$N = L - \varepsilon_1 c_1 L_1 - \varepsilon_2 c_2 L_2 - \ldots - \varepsilon_\nu c_\nu L_\nu,$$

which has the property that those edges to which a coefficient in N

different from 0 corresponds form an acyclic subgraph $G^{(\nu+1)}$ of G.

We show that N vanishes identically, i.e. that $G^{(\nu+1)}$ is the null

graph. As cycle forms the forms $L_1, L_2, \ldots, L_\nu$ satisfy the condition

in question (namely, $\sum \varepsilon_i c_i = 0$ for all vertices), and this was assumed

for L. Then the form N composed from the forms $L, L_1, L_2, \ldots, L_\nu$ must

also satisfy this condition (as we have seen in the proof of the

necessity of the condition). But consequently, since the graph $G^{(\nu+1)}$

corresponding to N contains no cycle, N must be identically 0. From the
fact that N = 0 we get that

$$L = \varepsilon_1 \, c_1 \, L_1 + \varepsilon_2 \, c_2 \, L_2 + \ldots + \varepsilon_\nu \, c_\nu \, L_\nu,$$

and so L is, in fact, composed of cycle forms.///

Theorem 1 is thus completely proved. The proof given shows that
the theorem also holds true in the case where only whole numbers are
allowed as coefficients of the linear forms and the compositions.

§3. Basis of cycle forms

By Theorem 1 we have characterized the system T of those linear
forms which can be composed from cycle forms of a graph. We raise the
question: isn't there also a proper subsystem T_z' of the system T_z of
all cycle forms of a graph, from whose elements the elements of T can
all be composed? A "smallest possible" such subsystem T_z' should be
determined; more precisely, let there be no element of T_z' which can be
composed from all the other elements of T_z'. In other words, we are
looking for a basis for the system T, where B is called a basis for a
system if every element of the system can be composed from elements of
B, while no element of B can be composed from the other elements of B.
Naturally T and T_z have the same bases. With the help of the concept of
frame a basis can now be specified for T_z. Let S be a frame of the
graph G. If a direction of rotation is assigned in an arbitrary but
definite way to every cycle of the fundamental system of cycles (see
Chap. IV §3) belonging to S, a fundamental system F_S of (directed)
cycles belonging to S is obtained. Now Kirchhoff's [1, §1] Theorem
holds:

Theorem 2 The system B of linear forms which belong to the
(directed) cycles of any fundamental system F_S of (directed)

cycles forms a basis for the totality of cycle forms.

Proof First, no form in B can be composed from the other forms in B, since -- by Theorem IV. 29 -- every form in B contains a variable which is not contained in any other form in B.

Secondly, we wish to show that if Z is any (directed) cycle and L is the linear form belonging to it, L can be composed[6] from the linear forms in B. Let k_1, k_2, ..., k_ν be those edges of Z which do not belong to S, and let Z_i be (see Theorem IV. 28) the only (directed) cycle of the fundamental system F_S which contains the edge k_i but otherwise contains only edges of S ($i = 1, 2, ..., \nu$). If the linear form corresponding to the (directed) cycle Z_i is denoted by L_i, and $\varepsilon_i = \pm 1$ depending on whether k_i has the same direction in Z and in Z_i or not, we prove that

$$L = \varepsilon_1 L_1 + \varepsilon_2 L_2 + \ldots + \varepsilon_\nu L_\nu. \tag{1}$$

By the definition of ε none of the variables x_1, x_2, ..., x_ν (which correspond to the edges k_1, k_2, ..., k_ν) occurs in

$$N = \varepsilon_1 L_1 + \varepsilon_2 L_2 + \ldots + \varepsilon_\nu L_\nu - L$$

and therefore N contains only the variables which correspond to edges of S. Since a frame contains no cycle, there is therefore no cycle with the property that the variables belonging to its edges are all contained in N. But now, since N is composed of cycle forms, it satisfies the conditions of Theorem 1. As we have already seen in the proof of this theorem, N must therefore vanish identically, and so (1) is proved.///

With regard to the proof just completed, let it be particularly emphasized that we have not only shown that every cycle form can be composed from the linear forms which correspond to the (directed) cycles of a fundamental system, but it has also turned out that this is true if only whole number values for the coefficients of the composition are allowed, and, in fact, only 0, 1, and -1.

We mention also here that the converse of Theorem 2 is not true,[7] as can be seen by the example of the graph of Fig. 55. As the edges of this graph may be directed, the linear forms corresponding to the four

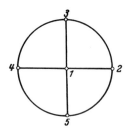

Fig. 55.

(directed) cycles

2 3 4 5 2, 1 3 4 1, 1 4 5 1, 1 5 2 1

form a basis for the totality of cycle forms, as can easily be seen. But since all three edges of 1 4 5 1 are contained in another (directed) cycle of this basis, these four (directed) cycles, by Theorem IV. 29, do not form a fundamental system.

It is known from algebra that two bases of a finite system of linear forms have the same number of elements. The following theorem therefore follows from Theorems 2 and IV. 30.

Theorem 3   If the graph is finite and has connectivity μ, then every basis of the totality of cycle forms consists of μ forms.

This theorem can be extended to infinite graphs.

Theorem 4   All bases of the totality of cycle forms are equivalent to one another and have -- if there are infinitely many cycles -- the same cardinality as the set of cycles of the graph.

Proof   If all edges which do not belong to any cycle are removed, there results the subgraph G' of the graph G. Then the theorem says the same for G' as for G. On the other hand the theorem holds if G is finite (Theorem 3). We can therefore assume in the proof without loss of generality that G has infinitely many edges and that every edge of G belongs to a cycle of G. But then, since every cycle contains only finitely many edges, the cardinality m of the cycles is also infinite, as is the cardinality α_1 of the edges. Let B be any one of the bases in question. Its cardinality b is also infinite; otherwise at least one of the variables, say x_ρ, would be missing from every element of B, and so the cycle form corresponding to a cycle which contains the edge k_ρ (corresponding to the variable x_ρ) cannot be composed from elements of B.

We now consider some simple inequalities. First every element of B contains only finitely many variables and therefore at most $\aleph_0 b = b$ variables occur in all the elements of B (this equation holds as a consequence of Zermelo's Well-ordering Principle). As we have just seen, all variables must occur; so $\alpha_1 \leq b$. Second, $b \leq m$. Third the cardinality of the cycles which have ν edges is at most α_1^ν, and so the cardinality m of all cycles is at most

$$\alpha_1^2 + \alpha_1^3 + \alpha_1^4 + \ldots = \alpha_1 + \alpha_1 + \alpha_1 + \ldots = \aleph_0 \alpha_1 = \alpha_1$$

(here the Well-ordering Principle is used again). Therefore $m \leq \alpha_1$. The triple inequality so obtained

$$\alpha_1 \leq b \leq m \leq \alpha_1$$

gives by the Equivalence Theorem the following result:

$$\alpha_1 = b = m.$$

So our theorem is proved.///

The cardinality m of the cycles plays the same role here as connectivity plays for finite graphs (see Theorem 3). Out of this also comes the result that if μ denotes the cardinality of the edges which

are <u>not</u> contained in a frame of G, then in the case of an infinite

graph μ has the same value $\mu = b = m$ for every frame.

§4. Star forms

In a certain way we can contrast the cycles of a graph as dual[8]

structures with those subgraphs which consist of all edges which meet

at the same vertex P. The subgraph of G formed by those edges of G

ending in vertex P_α (both those directed towards P_α and those directed

away from P_α), is called the (directed) <u>star</u>[9] B_α. As a directed graph

(and in this chapter, unless we speak explicitly of an undirected star,

stars will be considered directed) the star is always directed, so that

all its edges are directed towards the "center" P_α. Like every directed

subgraph, all these B_α's can be defined by valuations of G. If we wish

to assign linear forms (star forms) to the stars and treat them in a way

similar to the way this has been done for cycles, we must assume that

all stars are finite, i.e. the graph is of finite degree. We will

always assume this when we are dealing with star forms.

A star form can never also be a cycle form. A

subgraph G' can be a cycle as well as (undirected)

star, if it is a 2-cycle. In order to be able to

get from a 2-cycle first a directed 2-cycle, and

then a directed star, one of the two edges (see Fig.

56) must have its direction changed.

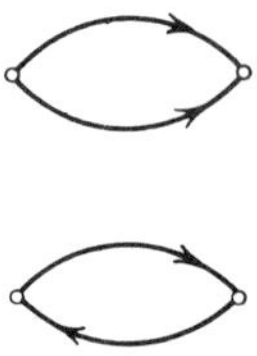

We raise the question: which linear forms can

be composed from star forms? The answer is

Fig. 56.   analogous to Theorem 1; here, however, the

corresponding condition is sufficient if and only if the graph is

172

finite.

We prove

Theorem 5   A linear form $L = \sum c_\alpha x_\alpha$ can be composed from star

forms if the graph is finite, if and only if the following

conditions are satisfied. If k_1, k_2, ..., k_ν are the edges of

the (directed) cycle Z (to which the variables x_1, x_2, ..., x_ν

correspond), then the equation $\sum_{i=1}^{\nu} \varepsilon_i c_i = 0$ holds for every

(directed) cycle Z, where $\varepsilon_i = \pm 1$, depending on whether k_i in

Z has the same direction as in the original graph G (of finite

degree) or the opposite one.

Proof   As for the necessity of the condition we assume that $L = \sum c_\alpha x_\alpha$

can be composed from star forms B_α:

$$L = \sum b_\alpha B_\alpha.$$

Every edge occurs in exactly two stars of the graph, and indeed with

opposite directions. So for any edge $k_i = k_{rs} = P_r P_s$:

$$n_i c_i = b_s - b_r,$$

where $n = \pm 1$, depending on whether k_i is directed toward P_s or P_r. So

for the edges of a cycle

$$Z = \overrightarrow{P_1 P_2 \ldots P_\nu P_1}$$

(since the n_i's agree now with the ε_i's of the theorem):

$$\sum_{i=1}^{\nu} \varepsilon_i c_i = (b_2 - b_1) + (b_3 - b_2) + \ldots + (b_\nu - b_{\nu-1}) + (b_1 - b_\nu) = 0.$$

In order to show that the condition proved necessary for finite graphs

is also sufficient, we prove first the following theorem, which is also

of interest in other connections and which does not need the hypothesis

that the graph is of finite degree.

Theorem 6   If a valuation $\{c_\alpha\}$ of G satisfies the

condition $\sum_{i=1}^{\nu} \varepsilon_i c_i = 0$ for every (directed) cycle Z of G,

where the ε_i's are defined as in Theorem 5, then a number

b_γ can be assigned to every vertex P_γ of G, such that for every edge $k_\alpha = \overrightarrow{P_r P_s}$ the difference $b_s - b_r$ is equal to c_α.

Proof   First, let G contain no cycle; then the condition is without content, and the assertion must be proved for every valuation $\{c_\alpha\}$. Let G_1 be any component of G and let P_0 be any vertex of G_1. We assign the (rational whole) number b_0 to the vertex P_0 completely arbitrarily. The number b_α, which is assigned to any other vertex P_α of G_1, we define as follows. We join P_0 with P_α by the unique path (see Theorem IV.2):

$$P^{(0)}\ P^{(1)}\ P^{(2)}\ \ldots\ P^{(k)}\ \ldots\ P^{(\nu-1)}\ P^{(\nu)} \qquad (P^{(0)} = P_0,\ P^{(\nu)} = P_\alpha).$$

Let $c^{(k)}$ be the value, which is assigned to the edge $P^{(k-1)}\ P^{(k)}$ by the valuation $\{c_\alpha\}$ ($k = 1, 2, \ldots, \nu$); we then set
$b_\alpha = b_0 + n^{(1)}\ c^{(1)} + n^{(2)}\ c^{(2)} + \ldots + n^{(\nu)}c^{(\nu)}$, where $n^{(k)}$ is set $= \pm 1$, depending on whether $P^{(k-1)}\ P^{(k)}$ is directed toward $P^{(k)}$ or $P^{(k-1)}$ ($k = 1, 2, \ldots, \nu$). Now let $P_r\ P_s$ be any edge of G_1; then this edge is, by Theorem IV. 6, either the last edge of the path

$$W = P^{(0)}\ P^{(1)}\ P^{(2)}\ldots\ P^{(\nu-1)}\ P^{(\nu)}$$
$$(P^{(0)} = P_0,\ P^{(\nu)} = P_s),$$

which joins P_0 with P_s, or the last edge of the path which joins P_0 with P_r. We assume the first case (therefore $P^{(\nu-1)} = P_r$). By the definition of the b_α's, if $c^{(1)}, c^{(2)}, \ldots, c^{(\nu)}$ are the values c_α which by the valuation $\{c_\alpha\}$ correspond to the edges $P^{(k-1)}\ P^{(k)}$ ($k = 1, 2, \ldots, \nu$) and $n^{(k)} = \pm 1$, depending on whether $P^{(k-1)}\ P^{(k)}$ in G is directed toward $P^{(k)}$ or toward $P^{(k-1)}$:

$$b_s = b_0 + n^{(1)}\ c^{(1)} + n^{(2)}\ c^{(2)} + \ldots + n^{(\nu)}c^{(\nu)},$$

$$b_r = b_o + \eta^{(1)} c^{(1)} + \eta^{(2)} c^{(2)} + \ldots + \eta^{(\nu-1)} c^{(\nu-1)},$$

and therefore

$$b_s - b_r = \eta^{(\nu)} c^{(\nu)}.$$

Here $c^{(\nu)} = c_\alpha$ is the value assigned to the edge $k_\alpha = P_r P_s$ and, if $P_r P_s$ in G is directed from P_r to P_s, $\eta^{(\nu)} = 1$. For the directed edge

$\overrightarrow{P_r P_s}$ of G we have

$$c_\alpha = b_s - b_r.$$

We get the same equation in the second case. This determination of the numbers b_α can be carried out for the vertices of every component of G, in the course of which the choice of the vertices P_0 requires, in general, the Axiom of Choice. Since an edge never joins two vertices belonging to different components, our theorem is proved for acyclic graphs.

Second, let G be an arbitrary graph and let S be a frame of G. The valuation $\{c_\alpha\}$ is supposed to satisfy the condition $\sum \varepsilon_i c_i = 0$ for every (directed) cycle of G. Since S contains no cycle, a number b_α can be assigned to the vertices of S in such a way that the condition $c_\alpha = b_s - b_r$ is satisfied for every edge $\overrightarrow{P_r P_s}$ of S. By Theorem IV. 21 a number b_α is assigned in this way to every vertex of G. We prove that this correspondence also satisfies the condition $c_\alpha = b_s - b_r$ for those edges $k_\alpha = \overrightarrow{P_r P_s}$ of G which do not belong to the frame S. By definition of the frame for every edge $P_r P_s$ which does not belong to S there is a (directed) cycle

$$Z = \overrightarrow{P^{(o)} P^{(1)} P^{(2)} \ldots P^{(\rho)} \ldots P^{(\nu)} P^{(o)}}$$
$$(P^{(o)} = P_s, \quad P^{(\nu)} = P_r),$$

whose edges, with the exception of $P_r P_s = P^{(\nu)} P^{(o)}$, all belong to S.

Let the numbers $b^{(\rho)}$ be assigned to the vertices $P^{(\rho)}$ of this (directed) cycle and let the edges $P^{(\beta-1)} P^{(\beta)}$ have the values $c^{(\rho)}$ in the valuation $\{c_\alpha\}$. We know already that for the edges $P^{(\rho-1)} P^{(\rho)}$ of S, the relation $c^{(\rho)} = b^{(\rho)} - b^{(\rho-1)}$ holds, if the edge is directed towards $P^{(\rho)}$. In general, therefore, we have:

$$\varepsilon^{(\rho)} c^{(\rho)} = b^{(\rho)} - b^{(\rho-1)} \tag{1}$$

where $\varepsilon^{(\rho)} = \pm 1$, depending on whether $P^{(\rho-1)} P^{(\rho)}$ in G is directed towards $P^{(\rho)}$ or $P^{(\rho-1)}$. Now we apply the assumed condition $\sum \varepsilon_i c_i = 0$ to the (directed) cycle Z. We get:

$$\varepsilon^{(1)} c^{(1)} + \varepsilon^{(2)} c^{(2)} + \ldots + \varepsilon^{(\nu)} c^{(\nu)} + c_\alpha = 0.$$

If (1) is used here for $\rho = 1, 2, \ldots, \nu$, we get

$$\begin{aligned}
c_\alpha &= - (\varepsilon^{(1)} c^{(1)} + \varepsilon^{(2)} c^{(2)} + \ldots + \varepsilon^{(\nu)} c^{(\nu)}) \\
&= - [(b^{(1)} - b^{(0)}) + (b^{(2)} - b^{(1)}) + \ldots + (b^{(\nu)} - b^{(\nu-1)})] \\
&= - (b^{(\nu)} - b^{(0)}) = - (b_r - b_s) = b_s - b_r ./ / /
\end{aligned}$$

With this Theorem 6 is completely proved and for both finite and infinite graphs. The theorem holds, as the proof shows, if the numbers b_γ are taken only from the integral domain to which the numbers c_α belong. So the b_γ's can be chosen as whole numbers if the c_α's are whole numbers. With the help of Theorem 6 the second part of Theorem 5 can be easily proved. We assume that the linear form $L = \sum c_\alpha x_\alpha$ satisfies the conditions $\sum \varepsilon_i c_i = 0$ for every (directed) cycle of the finite graph G. We assign a number b_α to the vertices P_α corresponding to Theorem 6 in such a way that for every edge $k_\alpha = \overrightarrow{P_r P_s}$ the difference $b_s - b_r$ is equal to c_α. The linear form assigned to the (directed) star with center P_ρ is designated by B_ρ; so

$$L = \sum b_\rho B_\rho ,$$

since in this sum the variable x_α is multiplied by the coefficient

$$b_s \cdot 1 + b_r \cdot (-1) = b_s - b_r = c_\alpha.$$

L is therefore composed from star forms.///

The proof of Theorem 5, just completed, shows that the theorem also holds if only whole numbers are allowed as coefficients of the linear forms and of the compositions. As for the finiteness of the graph, we add the following remark to the last part of the proof of Theorem 5. In contrast to Theorem 6 the finiteness of the graph had to be assumed here, since otherwise $\sum b_\rho B_\rho$ would, in general, be an infinite sum and, if only linear combinations in which only finitely many nonvanishing coefficients occur were introduced for linear forms, the equation $L = \sum b_\rho B_\rho$ would be meaningless. Given any infinite graph it cannot be concluded from the condition $\sum \varepsilon_i c_i = 0$, for every (directed) cycle, that the linear form $\sum c_\alpha x_\alpha$ can be composed from star forms. The continuously directed doubly infinite path $\ldots P_{-2} P_{-1} P_0 P_1 P_2 \ldots$ (which contains no cycle, and so for which the condition $\sum \varepsilon_i c_i = 0$ is without content) furnishes an example of this. The variable x_i ($i = 0, \pm 1, \pm 2, \ldots$) corresponds to the edge $\overrightarrow{P_i P_{i+1}}$ and so we can immediately convince ourselves that the linear form x_1, for example, cannot be composed from (finitely many) star forms $B_i = x_{i+1} - x_i$.

We would like to add a second remark to Theorem 5, and in doing so we must assume again that the graph is finite. In order to be able to compose a linear form from star forms, it suffices to postulate the relations $\sum \varepsilon_i c_i = 0$ only for the (directed) cycles of a fundamental system. It easily follows from Theorem 2 (second part of the proof) that these relations hold for all (directed) cycles. A corresponding remark also holds for Theorem 6, and even in the case of infinite valuations.

§5. Basis of star forms

We would now like to consider connections which exist among star forms of a graph of finite degree, just as we have already considered them for cycle forms. But here finite and infinite graphs (of finite degree) must be treated separately.

We prove first the following theorem of Kirchhoff [1, §5]:

Theorem 7 If B_i $(i = 1, 2, \ldots, \alpha_o)$ are the star forms of a finite connected graph G, then an equation

$$\sum_{i=1}^{\alpha_o} b_i \, B_i = 0$$

is identically true in the variables x_ρ if and only if all b_i are equal to one another.

Proof If this is the case, $b_i = b$ and $\sum_{i=1}^{\alpha_o} b \, B_i = b \sum_{i=1}^{\alpha_o} B_i = 0$, since any variable x_j which corresponds to the edge $k_j = \overrightarrow{P_r P_s}$ is contained only in B_r and in B_s, and in B_r its coefficient is -1, but in B_s its coefficient is 1.

On the other hand, let

$$\sum_{i=1}^{\alpha_o} b_i \, B_i = 0$$

be any equation which is identically true in the x_j's. If every $b_i = 0$, then the theorem is true. We may therefore assume that $b_1 \neq 0$. If the equation

$$b_1 \sum_{i=1}^{\alpha_o} B_i = \sum_{i=1}^{\alpha_o} b_1 \, B_i = 0$$

is subtracted from this equation, we get the equation

$$\sum_{i=1}^{\alpha_0} (b_i - b_1) B_i = 0 \tag{1}$$

in which B_1 does not occur. We show that (for every i) $b_i = b_1$. We suppose that this is not the case, and let π_1 be the totality of those vertices P_i for which $b_i \neq b_1$. Let the set of all other vertices to which P_1 belongs be designated by π_2. By Theorem I.25 there is an edge $k_j = P_r P_s$, which joins vertex P_s from π_1 with a vertex P_r from π_2. The coefficient of x_j is in (1):

$$\pm [(b_s - b_1) - (b_r - b_1)] = \pm (b_s - b_1),$$

since $b_r = b_1$. This must vanish, so that, in contradiction to the definition of π_1, $b_s = b_1$. Therefore π_1 must be empty, so that, for each i, $b_i = b_1$.///

From the theorem just proved it follows -- and the theorem was stated in this form by Kirchhoff -- that among the star forms B_1, B_2, ... B_{α_0} exactly $\alpha_0 - 1$ of them are independent of one another. If any one of them is omitted, then the remaining ones form a basis.

Theorem 7 can be supplemented for infinite graphs in the following way:

Theorem 8 The star forms of an infinite connected graph (of finite degree) are independent of one another.

Proof An infinite graph of finite degree has infinitely many vertices (stars). Therefore, for every linear form $\sum b_i B_i$ there is some B_α which does not occur in this sum ($b_\alpha = 0$). Now let $\sum b_i B_i = 0$ identically in the x, where not all b_i vanish. Just as above it can be concluded that an edge $k_j = P_r P_s$ exists, where $b_s \neq 0$ and $b_r = 0$. But then the coefficient $\pm (b_s - b_r) = \pm b_s$ of x_j cannot vanish in $\sum b_i B_i$, and the theorem is proved.///

179

We consider now a sum $\sum b_i B_i$ for a disconnected graph $G = \sum G^{(\rho)}$ of finite degree. If those terms $b_i B_i$ for which the vertices P_i belong to the same component $G^{(\rho)}$ of G are collected together to form a sum $\sum^{(\rho)}$, then we have

$$\sum b_i B_i = \sum^{(1)} + \sum^{(2)} + \ldots + \sum^{(\nu)}.$$

Since no two $G^{(\rho)}$ have a common edge, no variable x_j can appear in two different sums $\sum^{(\rho)}$. If, therefore, $\sum b_i B_i$ vanishes identically in the x, then the same is true for all sums $\sum^{(\rho)}$. In other words, every identical equation $\sum b_i B_i = 0$ can be composed from identical equations $\sum b_i B_i = 0$ such that all P_i belong to the same component $G^{(\rho)}$. So if we wish to count up the relations independent of one another among the B_i, then it must be done, instead of for G, for all $G^{(\rho)}$. These are then independent of one another, even when thought of as relations for G. These remarks allow us to draw the following conclusions from Theorems 7 and 8:

> Theorem 9 If B_i are the star forms of a finite graph,
> then $\sum b_i B_i$ is identically 0 if and only if the b_i which
> correspond to the vertices P_i of the same component $G^{(\rho)}$ are
> equal to one another for every ρ.

> Theorem 10 If the number or cardinality of the finite
> components of a graph G of finite degree is ν, then there are
> among the star forms exactly ν relations independent of one
> another (which can all be written with coefficients 0 and 1).

> Theorem 11 If in every finite component of a graph of finite
> degree one vertex is chosen, then those star forms which do not
> belong to chosen vertices form a basis for the totality of star
> forms.[10] Every basis can be obtained in this way.

We now prove the following analog of Theorem 4.

Theorem 12 The bases of the totality of star forms of a graph
of finite degree are equivalent to one another.

Proof Let α_0 be the number (cardinality) of vertices of the finite
components and ν the number (cardinality) of finite components of the
graph G. If G is finite, Theorem 11 says that every basis has $\alpha_0 - \nu$
elements. But if G, and therefore also α_0, is infinite, then any basis,
and therefore also its cardinality b, is infinite. Now by Theorem 11
$b + \nu = \alpha_0$. But since $b \geq \nu$ (b is greater than or equal to the
cardinality of all finite components), it follows that $b = \alpha_0$ as a
consequence of the Well-ordering Principle. So Theorem 12 is proved:
the cardinality of every basis is $\alpha_0 - \nu$ or α_0, depending on whether G
is finite or infinite.///

We can get another proof of Theorem 12 by making appropriate
changes in the proof given above for Theorem 4. The m of the proof is
replaced by the cardinality α_0 of the vertices (of the stars).

§6. Simultaneous consideration of cycle forms and star forms

In this section we restrict ourselves to the case where the
coefficients of the linear forms and compositions are taken from the
field of real (complex) numbers.

We would like to set cycle forms and star forms in relation to each
other. With this goal in mind we prove now the following theorem, which
is also of interest in other connections and is concerned with
undirected graphs.

Theorem 13 Let a real or complex number b_i be assigned to
every vertex P_i of a finite graph G, and a positive number p_r
to every edge k_r of G, so that for every i the following
condition is satisfied. If

$$k_i^{(1)} = P_i P_1', \quad k_i^{(2)} = P_i P_2', \quad \ldots, \quad k_i^{(\nu_i)} = P_i P_{\nu_i}'$$

designate the edges of G going to P_i;

$p_i^{(1)}, p_i^{(2)}, \ldots, p_i^{(\nu_i)}$ denote the numbers p in order

assigned to these edges, and $b_1', b_2', \ldots, b_{\nu_i}'$ denote the

numbers b in order assigned to the vertices

$P_1', P_2', \ldots, P_{\nu_i}'$ then the following relation[11] is true:

$$\sum_{\rho=1}^{\nu_i} p_i^{(\rho)} (b_\rho' - b_i) = 0.$$

Then all b_i corresponding to the vertices P_i of the same

component of G are equal to one another.

<u>Proof</u> First of all, let the numbers b be real. Since every p is
positive, this equation shows that, if one of the numbers b_ρ' is greater
than b_i, another of these numbers must be smaller than b_i. If the
theorem were not true for a component G_1 of G, then there would be by
Theorem I.25 an edge $P_1 P_2$ in G_1, so that $b_1 \neq b_2$ and therefore, say,
$b_1 < b_2$. Now b_1 occurs among the numbers b assigned to the vertices
which are joined with P_2 by an edge, and so one of them, b_3, must be
greater than b_2. Likewise, one of the numbers b, b_4, is greater than
b_3, etc., ad inf. But then there would be among finitely many numbers
b_i no greatest number. For complex b the theorem can be gotten from the
case of real b simply by splitting the b_i into real and imaginary
components.///

For infinite graphs the theorem just proved does not hold. This is
shown again by the doubly infinite path

$$\ldots P_{-2} \, P_{-1} \, P_o \, P_1 \, P_2 \, \ldots,$$

if p_r is set equal to 1 and if b_i is set equal to i. Then all the relations

$$\sum_{\rho=1}^{2} (b_{\rho}{'} - b_{i}) = [(i - 1) - i] + [(i + 1) - i] = 0$$

are satisfied and yet all b_i are distinct from one another.

Now we prove for arbitrary directed graphs of finite degree

Theorem 14 A nonvanishing linear form cannot be composed from cycle forms and at the same time from star forms.

Instead of proving this theorem, we shall prove the following more general theorem, which becomes the theorem just stated in the special case where $\omega_\alpha = 1$.

Theorem 15 Let $Z_\beta = \sum_{(\alpha)} c_\alpha^{(\beta)} x_\alpha$ be the cycle forms of a graph G of finite degree and let the ω_α be arbitrary positive numbers assigned to the edges k_α. Then if a nonvanishing linear form $L = \sum_{(\alpha)} c_\alpha x_\alpha$ can be composed from the linear forms

$$Z_\beta^{(\omega)} = \sum_{(\alpha)} c_\alpha^{(\beta)} \omega_\alpha x_\alpha,$$

then it cannot be composed from the star forms B_γ of the graph. Proof We show first that we can restrict ourselves in the proof to finite graphs. We assume that we could compose L not only from certain forms $Z_\beta^{(\omega)}$ but also from certain forms B_γ. The variables x_α which actually occur in the first or second representation of L form a finite set M; the endpoints of the edges which correspond to the elements of M therefore also form a finite set N. Let G' be that subgraph of the original graph G formed by those edges of G which have one (or both) of its endpoints from N. Since N is finite and G is of finite degree, G' is finite and the double representation of the form L is then also possible with respect to the finite graph G', since the cycle forms and star forms involved here are also cycle forms and star forms, respectively, with respect to G'. Therefore we may assume the graph G as finite.

The $Z_\beta^{(\omega)}$ come from the cycle forms Z_β by replacing all x_i by $\omega_i \, x_i$. The condition for the composability of a form L from the forms $Z_\beta^{(\omega)}$ comes therefore from the condition ($\sum \epsilon_i \, c_i = 0$, Theorem 1) for the composability of the forms Z_β by replacing the c_i there by $\dfrac{c_i}{\omega_i}$. If therefore $k_i^{(\rho)} = P_i \, P_\rho{}'$ ($\rho = 1, 2, \ldots, \nu_i$) are the edges of G going to P_i, to which the numbers $c_i^{(\rho)}$ and $\omega_i^{(\rho)}$ ($\rho = 1, 2, \ldots, \nu_i$) are assigned as c-values and ω-values and if $\epsilon_\rho = \pm 1$, depending on whether $k_i^{(\rho)}$ is directed toward P_i or not, this condition is now: for every vertex P_i it must be that

$$\sum_{\rho=1}^{\nu_i} \frac{\epsilon_i^{(\rho)} \, c_i^{(\rho)}}{\omega_i^{(\rho)}} = 0.$$

If L can also be composed from star forms B_j, and therefore $L = \sum_{(j)} b_j \, B_j$, then (see the beginning of the proof of Theorem 5)

$$\epsilon_i^{(\rho)} \, c_i^{(\rho)} = b_i - b_\rho{}'$$

where $b_\rho{}'$ is the coefficient of the star form belonging to $P_\rho{}'$.

If this is substituted, we get

$$\sum_{\rho=1}^{\nu_i} \frac{1}{\omega_i^{(\rho)}} (b_\rho{}' - b_i) = 0.$$

This is (for $p_\alpha = \dfrac{1}{\omega_\alpha}$) precisely the condition of Theorem 13; so this theorem says that all b_j, corresponding to the vertices P_j of the same component of G are equal to one another. But by Theorem 9

$$L = \sum_{(j)} b_j \, B_j$$

is identically 0. Since this contradicts our assumption, we have proved Theorem 15 and therefore also Theorem 14.///

We would like to give a second proof for Theorem 14, which does not use Theorem 13, and in which we restrict ourselves again to finite graphs G. This proof is based on the first part of Theorem 1 (necessity

of the condition), which we now express as follows:

<u>Theorem 1*</u> If $B_\beta = \sum_{(\alpha)} \lambda_\alpha^{(\beta)} x_\alpha$ are the star forms of a graph

G of finite degree, then a linear form $L = \sum_{(\alpha)} c_\alpha x_\alpha$ can be

composed of cycle forms Z_γ of G if and only if every B_β

vanishes[12] for $x_1 = c_1$, $x_2 = c_2$, ...

If this condition is satisfied, then every linear form which can be

composed from the B_β must vanish for these values. If L can be composed

from the Z_γ as well as the B_β, then L itself must vanish for these

values, and therefore $\sum_{(\alpha)} c_\alpha^2 = 0$; this implies that all $c_\alpha = 0$ for real

c_α. With this, Theorem 14 is proved for real c_α. The general case,

where we allow complex numbers, can be reduced to this case by splitting

the c_α, the b_β, and the z_γ into real and imaginary components in

$L = \sum c_\alpha x_\alpha = \sum b_\beta B_\beta = \sum z_\gamma Z_\gamma$.

This reasoning can also be generalized to give a proof of Theorem

15.

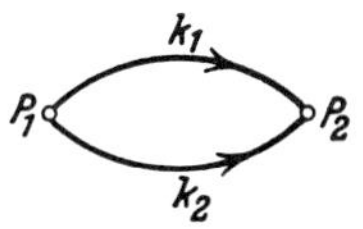

In Theorem 15 the assumption that the ω be

positive is essential. We consider the graph

which consists of two edges k_1, k_2 which join P_1

and P_2 with each other; let both edges be

directed towards P_2 (Fig. 57). We set

$\omega_1 = 1$, $\omega_2 = -1$. The graph has two (directed)

Fig. 57.

cycles and two (directed) stars, and

$$Z_1^{(\omega)} = x_1 + x_2, \quad Z_2^{(\omega)} = -x_1 - x_2,$$
$$B_1 = -x_1 - x_2, \quad B_2 = x_1 + x_2.$$

Now, the linear form

$$L = x_1 + x_2$$

can be composed from the $Z_\beta^{(\omega)}$ as well as from the B_γ; we have

$$L = Z_1^{(\omega)} = B_2.$$

As a consequence of Theorem 14 we now prove the following theorem, in which a particular property of (directed) cycles and (directed) stars, distinguishing this kind of subgraph from other subgraphs, appears.

Theorem 16   For a finite graph G every linear

form $L = \sum_i c_i x_i$ (and therefore, for example, also every x_i)

can be composed from cycle forms[13] and star forms[14].

Proof Let α_o be the number of vertices of G, and let α_1 be the number of edges. Let μ be the connectivity of G, and let ν be the number of components. Since the number of variables is α_1, it suffices to show, using a well known theorem of linear algebra, that α_1 forms independent of one another can be chosen from the totality of cycle forms and star forms. Let F' be a basis for the totality of cycle forms and let F" be a basis for the totality of star forms. (The existence of F' and F" is -- even for infinite graphs -- assured by Theorems 2 and 11.) By Theorem 3, F' consists of $\mu = \alpha_1 - \alpha_o + \nu$ forms and by Theorem 11 (second part) F" consists of $\alpha_o - \nu$ forms. We have seen above (this also follows from Theorem 14) that F' and F" do not have a common element. If F' and F" are combined, we get a system F which contains

$$\mu + (\alpha_o - \nu) = (\alpha_1 - \alpha_o + \nu) + (\alpha_o - \nu) = \alpha_1$$

forms. We show that these α_1 forms are independent of one another. We assume that $\sum_i a_i Z_i + \sum_i b_i B_i = 0$ or, in other words,

$$\sum_i a_i Z_i = \sum_i (-b_i) B_i \qquad (1)$$

with not all coefficients a_i and b_i equal to 0, where the Z_i belong to F' and the B_i belong to F". Because of the basis property of F" (and

F') neither all a_i nor all b_i can vanish, but by Theorem 14 equation (1) cannot hold, since in this case $\sum a_i Z_i$ and $\sum (-b_i) B_i$ do not vanish identically in the x_i.

The result obtained can also be expressed in such a way that F forms a basis for the totality of linear forms $\sum_i c_i x_i$ (and so does the totality K of the x_i). F and K both contain α_1 elements.

For infinite graphs Theorem 16 does not hold. The continuously directed doubly infinite path offers an example of this. As was already mentioned, x_1, for example, cannot be composed from star forms $x_{i+1} - x_i$ ($i = 0, \pm 1, \pm 2, \ldots$) (and there are no (directed) cycles here).

§7. Application in the Theory of Electricity

The preceding material owes its origin partly to questions concerning the theory of electricity, which were raised and solved by Kirchhoff in 1845. In a finite connected directed graph the edges k_1, k_2, ..., k_{α_1} can be thought of as wires in which an electric current circulates. For every (directed) edge k_i let the electrical resistance be ω_i (> 0), and let the electromotive force found in k_i (measured in the direction of k_i) be E_i. How can we determine from this information the intensities J_i of the currents (measured in the direction of k_i), traversing the edges k_i? Kirchhoff stated the following two laws, named after him:

If the edges k_{i_1}, k_{i_2}, ... form a (directed) cycle Z of G, then

$$\varepsilon_{i_1} \omega_{i_1} J_{i_1} + \varepsilon_{i_2} \omega_{i_2} J_{i_2} + \ldots = \varepsilon_{i_1} E_{i_1} + \varepsilon_{i_2} E_{i_2} + \ldots, \tag{I}$$

where $\varepsilon_\alpha = \pm 1$, depending on whether k_α has the same or opposite direction in G and in Z.

If k_{j_1}, k_{j_2}, ... are the edges of a (directed) star B of G, then

$$\varepsilon_{j_1} J_{j_1} + \varepsilon_{j_2} J_{j_2} + \ldots = 0, \tag{II}$$

where $\varepsilon_\alpha = \pm 1$, depending on whether k_α has the same or opposite direction in G and in B.

It is clear that equations (I) are consistent, since they are satisfied at the same time by $J_\alpha = \dfrac{E_\alpha}{\omega_\alpha}$. If $\omega_\alpha J_\alpha$ is set equal to x_α, the left side of (I) is changed to the cycle form corresponding to the cycle Z. So by Theorem 3 the Kirchhoff's first law furnishes $\mu = \alpha_1 - \alpha_0 + 1$ equations independent of one another; the rest of the equations (I) can be composed from them. Likewise, by Theorem 11, among the equations (II) (where the J_α themselves can now be considered as the variables x_α assigned to the edges) exactly $\alpha_0 - 1$ are independent of one another, while the $\alpha_0{}^{\text{th}}$ one can be derived (by addition) from the others. These two assertions were proved by Kirchhoff [1][15].

But now it does not at all directly follow from these two results -- as was erroneously assumed by Ahrens [1, p. 318] -- that the combined system of equations (I, II) contains exactly

$$(\alpha_1 - \alpha_0 + 1) + (\alpha_0 - 1) = \alpha_1$$

equations independent of one another, and that therefore the Kirchhoff laws suffice to calculate all intensities. Only by our Theorem 15 proved above is this assertion of Kirchhoff's guaranteed, which -- as already mentioned -- Kirchhoff assumed without proof (probably less from general physical grounds than through verification of the simplest examples). It is noteworthy that Weyl [I] was the first to recognize the necessity of proving[16] this fact. In proving this, it is essential that the resistances be positive, as we have already seen.

§8. Matrices

Now we would like to tie the material of this chapter in with the Poincaré-Veblen theory of matrices. We restrict ourselves to finite graphs for the sake of brevity.

Let P_i ($i = 1, 2, \ldots, \alpha_o$) be the vertices and k_j ($j = 1, 2, \ldots, \alpha_1$) the edges of a finite directed graph G. We define $\alpha_o \alpha_1$ numbers ε_{ij} ($i = 1, 2, \ldots, \alpha_o$; $j = 1, 2, \ldots, \alpha_1$) in the following way:

$$\varepsilon_{ij} = 1, \text{ if } P_i \text{ is the initial vertex of } k_j;$$
$$\varepsilon_{ij} = 1, \text{ if } P_i \text{ is the terminal vertex of } k_j;$$
$$\varepsilon_{ij} = 0, \text{ if } P_i \text{ is not an endpoint of } k_j.$$

The matrix

$$T = || \, \varepsilon_{ij} \, || \quad (i = 1, 2, \ldots, \alpha_o; \, j = 1, 2, \ldots, \alpha_1)$$

so defined, which has α_o rows and α_1 columns, is called the incidence matrix of G. It is the first (T_1) in the sequence of those matrices (T_q) which Poincaré introduced [2, p. 280] in his Analysis Situs for multidimensional manifolds.

The following matrix corresponds to the graph of Figure 58 in this way:

$$T = \begin{Vmatrix} -1 & 1 & 0 & 0 & 1 & 0 & 0 & 0 & 0 \\ 1 & 0 & 0 & 1 & 0 & 0 & 0 & 0 & -1 \\ 0 & -1 & -1 & 0 & 0 & -1 & 0 & 0 & 0 \\ 0 & 0 & 1 & -1 & 0 & 0 & -1 & 0 & 0 \\ 0 & 0 & 0 & 0 & 0 & 1 & 1 & -1 & 0 \\ 0 & 0 & 0 & 0 & -1 & 0 & 0 & 1 & 1 \end{Vmatrix}$$

$$T = \begin{Vmatrix} -1 & 1 & 0 & 0 & 1 & 0 & 0 & 0 & 0 \\ 1 & 0 & 0 & 1 & 0 & 0 & 0 & 0 & -1 \\ 0 & -1 & -1 & 0 & 0 & -1 & 0 & 0 & 0 \\ 0 & 0 & 1 & -1 & 0 & 0 & -1 & 0 & 0 \\ 0 & 0 & 0 & 0 & 0 & 1 & 1 & -1 & 0 \\ 0 & 0 & 0 & 0 & -1 & 0 & 0 & 1 & 1 \end{Vmatrix}$$

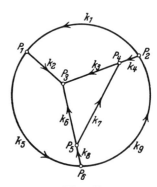

Fig. 58.

In every column there is one element equal to 1, one element equal to -1, and the rest of the elements are equal[17] to 0, so that every column has a sum of 0. To a permutation of the rows or columns there corresponds here a permutation in the numbering of the vertices and of the edges, respectively. The order of the rows is therefore not essential, and we would like to consider two matrices that differ from each other only in this ordering as not distinct (i.e., as equal). It is clear that the incidence matrices of two graphs are equal if and only if the graphs are equal. So we must be able to read all structure properties of a graph from its incidence matrix (e.g. the number of components, the connectivity, its cycles, etc.).

It can be seen immediately that the ith row of T yields the coefficients taken with opposite sign to the star form

$$\varepsilon_{i1} x_1 + \varepsilon_{i2} x_2 + \cdots + \varepsilon_{i\alpha_1} x_{\alpha_1}$$

which corresponds to the (directed) star with center P_i $(i = 1, 2, \ldots, \alpha_0)$, and therefore T is the matrix of star forms of the graph. The theory of matrices shows that Theorem 11 (or 10) implies the following:

If a finite graph contains α_o vertices and ν components, then the rank of its incidence matrix is $\alpha_o - \nu$. [18]

In order to specify a second matrix, let k_i $(i = 1, 2, \ldots, \alpha_1)$ be the edges and Z_j $(j = 1, 2, \ldots, r)$ the (directed) cycles of a finite directed graph G.

We define $\alpha_1 r$ numbers n_{ij} $(i = 1, 2, \ldots, \alpha_1; j = 1, 2, \ldots, r)$ as follows:

$n_{ij} = 1$, if k_i belongs to Z_j and has the same direction in Z_j as in G;

$n_{ij} = -1$, if k_i belongs to Z_j and has the opposite direction in Z_j as in G;

$n_{ij} = 0$, if k_i does not belong to Z_j.

The matrix

$$U = ||n_{ij}|| \qquad (i = 1, 2, \ldots, \alpha_1; j = 1, 2, \ldots, r)$$

so defined, which has α_1 rows and r columns, is called the edge-cycle-matrix[19] of G. The order of the rows and columns does not play any role here either (equal matrices).

It is seen immediately that the jth column of U yields the coefficients of the cycle form

$$n_{1j} x_1 + n_{2j} x_2 + \ldots + n_{\alpha_1 j} x_{\alpha_1}$$

which corresponds to the (directed) cycle Z_j $(j = 1, 2, \ldots, r)$, and therefore U is the matrix of the cycle forms of the graph. So Theorem 3 yields the following result:

For any finite graph of connectivity μ the rank of its edge-cycle-matrix is equal to μ.

We multiply the ρth row of T with the σth column of U and therefore the sequences:

$$\varepsilon_{\rho 1}, \varepsilon_{\rho 2}, \ldots, \varepsilon_{\rho \alpha_1} \text{ and } n_{1\sigma}, n_{2\sigma}, \ldots, n_{\alpha_1 \sigma}$$

191

and we get the sum $\sum\limits_{k=1}^{\alpha_1} \varepsilon_{\rho k}\, \eta_{k\sigma}$. From the proof we gave for the first

part of Theorem 1 (where we now put $-\varepsilon_{\rho i}$ for ε_i and $\eta_{\alpha\sigma}$ for c_α), or

likewise the first part of Theorem 5, it is evident that all these sums

(for $\rho = 1, 2, \ldots, \alpha_0$; $\sigma = 1, 2, \ldots, r$) vanish. In the terminology of

matrix calculation this implies that the product TU is the zero matrix.[20]

This result can also be formulated in the following way:

Every column of U represents a solution of that system of homogeneous

linear equations, whose matrix is T and vice versa.[21]

If certain columns of U are removed in such a way that the

remaining columns correspond to the (directed) cycles of a fundamental

system, then Theorem 1 implies that the columns of this matrix represent

a fundamental system of solutions to that system of equations. This

holds correspondingly for T (see Theorem 5).

Notes on Chapter IX

[1] It does not matter, in the material following here, whether we think of the numbers involved as integers, real numbers, or complex numbers. Up to the end of §5 any integral domain could be the source of the numbers c_α. Later, however, special properties of real (complex, resp.) numbers (ordering by magnitude) are used.

[2] Here and in the whole chapter (as long as no further restrictions are introduced) the coefficients (a_i) of the linear combinations are arbitrary elements of the quotient field of I (cf. the preceding footnote).

[3] These were introduced by Poincaré [1, p. 291]: if Z is the (directed) cycle (or, more precisely, the elementary surface, which is bordered by this cycle), to which the linear form corresponds, then Poincaré writes the congruence $Z \equiv L$. (This is merely the simplest case, q = 2, of the organization of ideas of Poincaré, which relates to manifolds of arbitrarily high dimension and which forms an essential part of Poincaré's analysis situs.) But the fundamental idea goes back to Kirchhoff [1].

[4] This sum is finite even if infinitely many edges go to P, since in this sum only finitely many of the c_i's may be distinct from zero, or can be, resp.: each c_i which does not occur in L is to be set here equal to zero.

[5] For the case where the graph is finite and the coefficients are whole numbers, cf. Veblen and Alexander [1, §4]. Veblen [2, chapter I, §§42 and 45], Chuard [1, §11], Dumas et Chuard [1], Weyl [1].

[6] In Kirchhoff [1, p. 24] this is proved by a recursion process which is useless for our purposes, since we do not restrict ourselves to finite graphs.

[7] The opposite assertion in Dehn and Heegaard [1, p. 173, lines 6-9] is therefore not right. ("Fundamentalsystem" there means what we have called basis here.)

[8] It is this duality which was always considered for polyhedra and leads to the polyèdre réciproque of Poincaré [1, §7] for multidimensional manifolds. It first becomes evident if the graphs are considered on surfaces; (cf., however, the dual graphs of Whitney [1, §4] where the duality refers to the surface of a sphere). That the duality is not complete can be partly explained by the fact that a cycle of a subgraph is also a cycle of the original graph, while this is not the case for stars.

[9] Alexander [1, §5] uses for this concept and for the corresponding multidimensional structure the word star and the symbol $S(P_\alpha)$.

[10] And indeed every star form can be composed from the elements of this basis with the help of the coefficients 0, 1, and -1.

[11] This equation can also be written in the following way:

$$b_i = \sum_{\rho=1}^{\nu_i} p_i^{(\rho)} b'_\rho \bigg/ \sum_{\rho=1}^{\nu_i} p_i^{(\rho)}$$

and can then be interpreted in a known way as a statement about centers of gravity. In the simplest case, where all p's are equal to 1, this equation states that b_i is the arithmetic mean of b_1', b_2', ...

[12] An analogous (dual) formulation can also be given for Theorem 5.

[13] Here the cycle forms can be replaced by the forms $z_\beta^{(\omega)}$ ($\omega_i > 0$) of Theorem 15.

[14] And indeed the coefficients can be taken from the same integral domain, to which the coefficients c_i of L belong.

[15] As far as the former is concerned, the number μ was not defined in Kirchhoff by the formula $\alpha_1 - \alpha_0 + 1$, but as the number of wires "at least as many as which must be removed in order to destroy all closed figures." That this number is equal to $\alpha_1 - \alpha_0 + 1$ comes up in Kirchhoff only subsequently and because of the theorem which he assumed as evident without proof (and had also already stated two years earlier, in his first work) that the systems of equations (I) and (II) are sufficient to calculate the α_1 intensities. Without this theorem, to which we shall soon revert, Ahrens [1] showed, 50 years later, that among the equations (I), $\alpha_1 - \alpha_0 + 1$ of them are independent of one another; scarcely anything had to be added to Kirchhoff's ideas for this purpose.

[16] In Weyl [1] the problem is intepreted in α_1- dimensional Euclidean space, by his regarding the equation $\sum_{(\alpha)} \lambda_\alpha c_\alpha = 0$ as the condition for orthogonality of two vectors, (λ) and (c).

[17] This suffices to show that every subdeterminant of the matrix T can assume only one of the values 0, 1, -1. See Poincaré [2, §6] and also Chuard [1, §6].

[18] Veblen [2, p. 27]; cf. for $\nu = 1$ also Chuard [1, p. 196]. The essence of this theorem (see Theorem 7) stems from Kirchhoff.

[19] It is closely connected with the matrix T_2 of Poincaré [2]; only those (directed) cycles are considered in Poincaré, as well as in Veblen [2], which border a "constituting elementary surface" of the surface of a polyhedron; the graph is interpreted there as the system of edges of a general polyhedron surface.

[20] Veblen [2, §68]

[21] The matrix of a fundamental system of whole number solutions in connection with Poincaré is studied in Dumas and Chuard [1] from the number theoretical standpoint.

Chapter X

Composition of Cycles and Stars.

§1. Composition of graphs

The material of this chapter is connected very closely with the
material of the preceding chapter. For good reasons, however, it is a
good idea to present the work here independently of the results of
Chapter IX. We shall show this intimate connection later by using
number theoretical congruences modulo 2.

Here we are dealing with undirected graphs. Along with addition of
graphs a second, likewise commutative and associative operation on
graphs plays a rule in graph theory. In this new operation the graphs
need not be disjoint, but only finitely many graphs can be dealt with by
this operation. The graphs may be infinite. We shall see immediately
that in the case where the graphs are pairwise disjoint, the new
operation reduces to addition of graphs. So the + symbol can also be
used. (The word "sum" and the $\sum$ symbol will be used only in the case
where the graphs being added are pairwise disjoint.)

Let $\{G_1, G_2, \ldots, G_r\}$ be an arbitrary finite set of arbitrary
graphs. We define a composition $G_1 + G_2 + \ldots + G_r$ of these graphs as
the graph which consists of those edges contained in an odd number of
component parts $G_1, G_2, \ldots, G_r$ (and therefore in at least one of the
component parts). For $r = 2$ our definition states that $G_1 + G_2$ consists
of those edges contained[1] in one and only one of the graphs G_1 and G_2.
Therefore, $G + G$ is always the null graph. From the definition it
follows that $(G_1 + G_2) + G_3 = G_1 + G_2 + G_3$.

196

In order to see this, we need only consider the $2^3 = 8$

possibilities of whether or not an arbitrary edge belongs to the graphs

G_1, G_2, and G_3. We can in the same way show that

$$G_1 + (G_2 + G_3) = G_1 + G_2 + G_3$$

and so $G_1 + (G_2 + G_3) = (G_1 + G_2) + G_3$, so that composition satisfies

not only the commutative law but the associative law as well. The

composition of compositions is also always a composition of certain

original graphs. The following theorem expresses an essential property

of composition.

Theorem 1   Both expressions $G_1 = G_2 + G_3 + \ldots + G_r$ and

$G_2 = G_1 + G_3 + \ldots + G_r$ are equivalent to each other, and each

implies that $G_1 + G_2 + \ldots + G_r$ is the null graph.

Each of these three expressions asserts only that any edge is

contained in any even number of graphs G_1, G_2, $\ldots$, G_r. It follows from

$G_1 = G_2 + G_3$ that $G_2 = G_1 + G_3$ and $G_3 = G_1 + G_2$.

The graphs of a system S of graphs are called independent of one

another with respect to composition if there is no finite subsystem of

the system S with the property that its elements yield the null graph in

composition, and therefore if (Theorem 1) there is no graph of the

system S that is the composition of certain other graphs of the

system. A subsystem B of S is called a composition basis for S if the

graphs of B are independent with respect to composition and every graph

of S is the composition of certain graphs of B. It can be immediately

seen that every graph of S can be composed of graphs of B in only one

way.

§2.   Composition of cycles

We now take up composition of cycles (we are dealing always here

with the case of composition of finite graphs.) For this material the
following theorem is of significance:

Theorem 2   The composition of Euler graphs is itself an Euler
graph.

Since we always deal with a finite number of component parts, it is
sufficient to prove the theorem for two component parts G_1, G_2. If P is
any vertex of the composition $G = G_1 + G_2$, then let $2n_1$ and $2n_2$ be the
number of edges of G_1 and G_2, respectively, which go to P, and
furthermore let n be the number of common edges of G_1 and G_2 that go to
P. Then there are

$$(2n_1 - n) + (2n_2 - n) = 2 (n_1 + n_2 - n)$$

edges of G which go to P. This is an even number.///

We would like to give a second, more complicated, and yet perhaps
more instructive proof of the theorem just proved, which relates
directly to an arbitrary number of component parts. Let G_1, G_2, ..., G_n
be the component parts and k_1, k_2, ..., k_ν be those edges which are
contained in one of these component parts and go to an arbitrarily
chosen vertex P of the composition $G = G_1 + G_2 + ... + G_n$. Let the edge
k_i be in exactly ρ_i distinct component parts. Since G_j is an Euler
graph, the number of those of the edges k_1, k_2, ..., k_ν which are
contained in G_j is always even; we designate it by $2\sigma_j$. We consider now
the set of those pairs (k_i, G_j) (i = 1, 2, ..., ν; j = 1, 2, ..., n) for
which k_i is contained in G_j. The number of these pairs is, on the one
hand $\sum \rho_i$, and on the other hand $\sum 2 \sigma_j$. So these two sums are equal,
so that $\sum \rho_i = \rho_1 + \rho_2 + ... + \rho_\nu$ is even. But this implies that among
the numbers ρ_1, ρ_2, ..., ρ_ν an even number of these ρ's are odd; so
there is in G an even number of edges that go to P.///

The theorem just proved says that the Euler subgraphs of a graph
form a group with respect to composition -- and, in fact, an abelian

group. The identity element of the group is the null graph and the
inverse element of any element is this element itself.

In cycles two edges go to each vertex. Theorem 2 can be applied to
cycles: every composition of cycles is an Euler graph. The converse of
this is also true. This follows immediately from Theorem II.6: using
the notation of that theorem, for every Euler graph G,

$G = K_1 + K_2 + \ldots + K_n$, where the component part cycles
K_1, K_2, $\ldots$, K_n have pairwise no common edges. Consequently:

> Theorem 3 A subgraph of an arbitrary graph G can be composed
> of cycles of G, if and only if it is an Euler graph.

From this theorem and Theorem II.1 we get the following theorem,
which contains Theorem I.9 as a special case:

> Theorem 4 If a composition of cycles is not the null graph,
> then it contains a cycle.

Now we prove

> Theorem 5 Every fundamental system of cycles of a graph is a
> composition basis for the totality of cycles of this graph.

Proof We must first show that from certain cycles of a fundamental
system F_s, which corresponds to an arbitrary frame S of G, a further
cycle of the same fundamental system cannot be composed. Since a
composition contains only those edges which occur in one of the
component parts this follows immediately from Theorem IV.29.

On the other hand we must show that every cycle K of the graph is
the composition of certain cycles of the fundamental system F_s. Let
k_1, k_2, $\ldots$, k_ν be those edges of K which do not belong to the frame S
and let K_i be (see Theorem IV.28) the only cycle which contains k_i, but
which otherwise has only edges of S (i = 1, 2, $\ldots$, ν). We form the
composition

$$N = K + K_1 + K_2 + \ldots + K_\nu.$$

N cannot contain any of the edges k_i ($i = 1, 2, \ldots, \nu$), since these edges occur in exactly two component parts (namely K and K_i). Nor are the remaining edges which do not belong to S contained in N, since none of the component parts contains these edges. Thus N can contain only edges of S, so that, by the definition of a frame, N contains no cycle. On the other hand, N, if it contained any edge at all, would, by Theorem 4, have to contain a cycle. So N is the null graph; by Theorem 1 the equation

$$K = K_1 + K_2 + \ldots + K_\nu$$

follows, and hence Theorem 5 is proved.///

As a consequence of Theorem 3 every composition basis B for the system of cycles, as well as (Theorem 5) every fundamental system of cycles yields at the same time a composition basis for the totality of Euler subgraphs. It also follows from the property of the basis, already mentioned, that every Euler subgraph E of G can be composed in only one way from elements of B. If, therefore, to every Euler subgraph E of G is assigned that finite subset of B whose elements in composition give E, then a one-to-one correspondence between the set of finite subsets of B and the set of Euler subgraphs of G is defined, and these two sets are equivalent.

We would like to draw some inferences from this equivalence.

First, let B be finite. (This is not only the case if G is finite but also if and only if G contains finitely many cycles.) If B contains ν elements, then B (the null set included) has 2^ν (finite) subsets.

So, as we saw above, the number of Euler subgraphs of G is 2^ν. If now G is a finite graph with connectivity μ and if a fundamental system of cycles is chosen for B, by Theorem 5, since this fundamental system contains μ cycles (Theorem IV.30), we get

Theorem 6   A finite graph with connectivity μ has (the null graph included) 2^μ Euler subgraphs.

Since every cycle is an Euler graph, this theorem contains the theorem of Ahrens [1, p. 317], that a finite graph with connectivity μ (the null graph included) contains[2] at most 2^μ cycles.

If now any composition basis for the system of cycles of a finite graph (with connectivity μ) contains ν elements, then the number of Euler subgraphs is also 2^ν; consequently $2^\nu = 2^\mu$ and $\nu = \mu$.   So we have Skolem's Theorem [1, p. 304][3]

Theorem 7   In the case of a finite graph with connectivity μ every composition basis for the system of cycles has μ elements.

Second, let the basis B be infinite and let b be its cardinality. We consider the set of finite subsets of B; its cardinality is likewise b, as can be seen by the Well-ordering Principle (from the relations $b^\nu = b$, $\aleph_o b = b$.)   As we have proved, the set of Euler subgraphs also has cardinality b.   If now m is the cardinality of the cycles, then $b \le m$, because B is a subset of the set of cycles, and $m \le b$, because every cycle is an Euler graph.

By the equivalence theorem we therefore have that b = m. Consequently
we have proved, with an eye to Theorem 7, the following theorem:

> Theorem 8   The composition bases for the system of cycles are
> equivalent to one another and have -- when there are infinitely
> many cycles -- the same cardinality as the set of cycles and
> the same cardinality as the set of Euler subgraphs.

To the connectivity of the finite graph (see Theorem 7)
corresponds, therefore, for graphs with infinitely many cycles the
cardinality m of the cycles (cf. Theorem IX.4).

§3. Composition of Stars

We now wish to study the stars of a graph with respect to
composition in a way similar to the way we did for cycles in §2. But
here -- in contrast to Chapter IX -- we shall consider the star as an
undirected graph. We do not restrict ourselves here to graphs of finite
degree, so that the stars, and therefore also the composition of stars
-- in contrast to the compositions of cycles -- may also be infinite
graphs. To the Euler subgraphs there correspond here those (finite or
infinite) subgraphs G' of the original graph G such that every[4] cycle of
G (and not only every cycle of G') contains an even number of edges of
G'. They will be designated as p-subgraphs of G. For an acyclic graph
G, therefore, every subgraph is a p-subgraph.

To Theorem 2 corresponds

> Theorem 9   The composition of p-subgraphs is itself a p-
> subgraph.

The theorem can be proved like Theorem 2 and, in fact, both proofs
given for that theorem can be carried over. It follows that the p-
subgraphs form a group with respect to composition. The analog of

Theorem 3 is:

> **Theorem 10**   If a graph G has finitely many vertices, then a
> subgraph G' of G can be composed from stars of G if and only if
> G' is a p-subgraph of G.

Proof   As for the necessity of this condition, it is clear that a star B
of G is a p-subgraph of G, since a cycle of G either contains two edges
of B (if namely the center of B belongs to the cycle) or no edge of B.
Then by Theorem 9 every composition of stars is also a p-subgraph.

In order to show that the condition is also sufficient we will
prove first that the graph is allowed to have infinitely many vertices.

> **Theorem 11**   If G' is a p-subgraph of G, then the vertices of G
> can be divided into two classes such that the endpoints of any
> edge of G belong to different classes if and only if this edge
> belongs to G'. This is possible only if G' is a p-subgraph of
> G.

Proof   First let G' be a p-subgraph of G. Let G_1 be any component of G
and let P_0 be any vertex of G_1. Now we place a vertex P of G_1 into the
first or second class (I and II), depending on whether a path which
joins P with P_0 contains an even or odd number of edges from G' (<u>not</u>
from G_1!). By Theorem I.8 (where the edges of G' should be regarded as
the distinguished edges) this determination is independent of the choice
of path. That this division (in two) satisfies our claims can be seen
in the following way. By Theorem I.6 there is a path W_1, which begins
at P_0 and ends with the edge AB of G_1. By removal of AB there results
from W_1 a second path W_2 (or the null graph). The one joins P_0 with A,
the other P_0 with B. If we consider these paths at the division of the
vertices A and B, we can see that A and B belong to different classes if
and only if AB is an edge of G'. (If, for example, G_1 is the

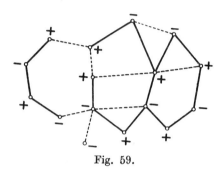

Fig. 59.

graph represented in Fig. 59, where the solid edges are the edges of the p-subgraph G', there results the division of the vertices into two classes indicated in the figure by + and - signs.)

If this division of the vertices into classes I and II is carried out for every component of G and all classes I and II are combined, we get the first part of the theorem for arbitrary graphs. (The choice of the vertices P_0 here, in case G has an infinite number of components, requires the Axiom of Choice. It can be immediately seen that the division of vertices of G_1 into two classes is independent of the choice of P_0; without the specification of P_0 it is indefinite which is Class I and which Class II, and this indefiniteness requires the application of the Axiom of Choice again.)

For the converse, let a division (I,II) of the kind mentioned be given for the vertices of G. In traveling around a cycle of G, the classes I and II to which the vertices belong must change an even number of times, since we return to the starting point, so that the cycle contains an even number of edges. In this way Theorem 11 is completely proved.///

With the help of Theorem 11 it can be easily proved that the condition expressed in Theorem 10 is also sufficient. Let G' be a p-subgraph of G with finitely many vertices. We consider a division for the vertices of G into two classes I and II corresponding to Theorem 11, and we prove that G' is identical to the composition K_I of those stars of G whose centers belong to I. Since every edge of G belongs to exactly two stars of G, then by the definition of composition an edge of G is contained in K_I if and only if it is contained in one and only one of those stars which arise as composition K_I and therefore if and only if exactly one endpoint of this edge belongs to I. These are the edges of G'. With this Theorem 10 is completely proved.///

This proof does not hold for a graph with an infinite number of vertices, since Class I can then contain an infinite number of vertices and the concept of composition was introduced only for a finite number of component parts. The condition in Theorem 10 is not sufficient for graphs with an infinite number of vertices, as can be shown in simple examples.

If every cycle of a graph G contains an even number of edges, then G is a p-subgraph of itself, and in this case G can only be a p-subgraph of itself. The following theorem is a special case of Theorem 11 which will be of use to us later (Chapter XI):

> Theorem 12 The vertices of a graph can be divided into two classes in such a way that the endpoints of every edge belong to different classes if and only if every cycle of the graph contains an even number of edges [5].

The concept of p-subgraphs is further illustrated by the following theorem, which we would now like to prove on the strength of Theorem 11.

> Theorem 13 A subgraph G' of G is a p-subgraph if all cycles of

a fundamental system for G contain an even number of edges of G'.

Proof Let S be any frame of G and F_s the corresponding fundamental system of cycles. We assume that every cycle of F_s contains an even number of edges of G'. The common edges of S and G' form a p-subgraph of S (since every subgraph of any acyclic graph is a p-subgraph). We can, by Theorem 11 (first part), divide the totality of vertices of S (and this agrees, by Theorem IV.21, with the totality of vertices of G) into two classes I and II so that the endpoints of an edge of S belong to different classes if and only if this edge belongs to G'. We show that this is the case for every edge AB of G, even if it does not belong to S. Let AB not belong to S.

By the definition of fundamental system there is in F_s a cycle K = A Q_1 Q_2 ... Q_r B A of G which contains only edges of S, with the exception of AB. By assumption K contains an even number of edges of G', and so the path A Q_1 Q_2 ... Q_r B, which belongs completely to S and which we get from K by removal of the single edge AB, contains an odd or even number of edges of G', depending on whether AB belongs to G' or not. In the first case A and B belong to different classes, since membership of the vertices A, Q_1, Q_2, ..., Q_r, B to the classes I or II changes an odd number of times. Likewise in the second case A and B belong to the same class. So the division (into two classes) of the vertices of G satisfies the condition mentioned for all edges of G'. By Theorem 11 (second part) G' is therefore a p-subgraph of G.///

Theorem 14 If a graph has a finite number α_o of vertices and ν components, then it has (the null graph included) $2^{\alpha_o - \nu}$ p-subgraphs.

Proof First, let the graph G be connected, $\nu = 1$. In this case a p-subgraph G' uniquely determines the division corresponding to it by

Theorem 11, into classes I and II (except for the order) of vertices of

G. If A and B are two arbitrary vertices of G and $A\ P_1\ P_2\ \ldots\ P_\nu\ B$ is a

path which joins them and contains r edges of G', then A and B belong to

the same or different classes I and II depending on whether r is even or

odd, since membership of the vertices to I or II in the sequence

$A,\ P_1,\ P_2,\ \ldots,\ P_\nu,\ B$ changes r times. So whether two vertices belong

to the same or different classes depends only on G'. If, on the other

hand, an arbitrary division into two classes of the set of vertices is

given, then the subgraph G' formed by those edges which join vertices of

different classes is the only one with the property that the endpoints

of an edge belong to different classes if and only if this edge belongs

to G'. By Theorem 11 (second part) G' is also a p-subgraph.

Consequently we have, for the p-subgraphs of G and for the divisions

into two classes of the vertices of G, a one-to-one correspondence (in

which two divisions are considered different if there are two vertices

which are in the same class in one division and in different classes in

the other division). Now the number of these divisions is

$$\frac{2^{\alpha_o}}{2} = 2^{\alpha_o - 1}$$

by which our theorem is proved for $\nu = 1$.

In the general case let $\alpha_o^{(i)}$ be the number of vertices of the

component G_i of $G = \sum G_i$. If now G_i' is a p-subgraph of G_i for every i,

then $\sum_i G_i'$ is a p-subgraph of G and every p-subgraph of G is such a

sum. The number of p-subgraphs, therefore, is the product

$$2^{\alpha_o^{(1)}-1} \cdot 2^{\alpha_o^{(2)}-1} \cdot \ldots \cdot 2^{\alpha_o^{(\nu)}-1} =$$

$$2^{\alpha_o^{(1)} + \alpha_o^{(2)} + \ldots + \alpha_o^{(\nu)} - \nu} = 2^{\alpha_o - \nu}.///$$

Since every edge belongs to exactly two stars, the system of all stars (in case this system is infinite) yields the null graph as a composition. If, however, a (finite) system of stars does not contain all stars of the connected graph G, then its composition K cannot be the null graph. Otherwise by Theorem I.25, G would have an edge P Q, such that the star with center P belonged to the system and the star with center Q did not; therefore P Q is contained in K. So we have the result:

Theorem 15 For a connected graph with a finite number of vertices the composition of certain stars yields the null graph if and only if all stars are in the composition. For a connected graph with an infinite number of vertices the stars are independent of one another with respect to composition.

This theorem shows how in the case of a connected graph a composition basis for the totality of stars (in fact, every such composition basis) can be obtained. The result can be immediately extended to disconnected graphs and yields

Theorem 16 If in every component with a finite number of vertices a vertex is chosen, then those stars whose centers are not chosen vertices form a composition basis for the totality of stars. Every composition basis can be obtained in this way.

§4. Linear forms modulo 2

The material with which we have concerned ourselves in this chapter shows a striking similarity with the material of Chapter IX. This similarity is explained if the linear forms $L = \sum c_\alpha x_\alpha$ considered in Chapter IX are reduced modulo 2. In doing so we have to restrict ourselves to whole number coefficients c_α. Reducing the linear form L

208

modulo 2 means that we replace all coefficients c_α by 0 or 1, depending on whether c_α is even or odd. Since $-1 \equiv 1$ (mod 2), the L (mod 2) is independent of the directions assigned to the edges of the graph and we can consider the original graph as well as its subgraphs as undirected graphs. The linear form (reduced mod 2) $\sum c_\alpha x_\alpha$, where $c_\alpha = 1$ or 0, corresponds to an undirected finite subgraph G' of G, depending on whether the edge k_α of G belongs to the subgraph G' or not. The linear forms $L' \equiv \sum c_\alpha x_\alpha$ and $L'' = \sum d_\alpha x_\alpha$ (mod 2) correspond in this way to the subgraphs G' and G", respectively, and so it is immediately seen that to the composition G' + G" there corresponds the linear form

$$L' + L'' \equiv \sum (c_\alpha + d_\alpha) x_\alpha \qquad \text{(mod 2)}$$

and similarly for arbitrarily many component parts[6]. The linear forms considered mod 2 can be made useful in this way for the theory of graph composition, and several theorems of this chapter follow from the corresponding theorems of Chapter IX -- or vice versa. We shall not go into this approach, which would make possible a shorter presentation of the results of both of these chapters.

Notes on Chapter X

[1] This composition is a concept of the general calculus of classes: in Boole - Schröder terminology of the algebra of logic $A\bar{B} + B\bar{A}$ is the composition of A and B. We restrict ourselves, however, to classes of edges, and of graphs, respectively. The composition of finite graphs was introduced, essentially in the way given here, by Skolem [1, pp. 303-305]. The basic idea of this concept formation stems, however, from Veblen, to which we shall return in §4.

[2] The proof which Ahrens gave for this theorem is not complete. Besides his Theorem VIII he uses also the theorem, which in his terminology (where "composition" does not mean the same as it does here) goes as follows: no two distinct cycles can be produced from the same cycles of a fundamental system by "composition" (where the two "compositions" would be distinguished only in the sequence of algebraic signs). This true, but not evident theorem is in Ahrens neither proved nor stated. -- Ahrens does not count the empty graph. For this reason the number $2^{\mu} - 1$ appears in his paper, and not 2^{μ} .

[3] The proof by Skolem is at a certain point not valid; ZZ_{σ} (line 9, p. 305) is not necessarily a cycle (Skolem writes Zyklus (directed cycle) for Kreis (cycle) and Fundamentalsystem (fundamental system) for Kompositionsbasis (composition basis)), since the cycles -- in contrast to Euler subgraphs -- do not form a group with respect to composition. Precisely for that reason it is advisable, in studies about composition with cycles to consider also the remaining Euler subgraphs, by which this gap in Skolem's proof can easily be filled. -- Since it is a matter of finite abelian groups, for which the order of each element is a prime number, namely 2, we get the result that for finite graphs two bases contain the same number of elements, even from general group theoretic considerations.

[4] Furthermore we show below that it is sufficient to postulate this only for the cycles of a fundamental system of cycles (Theorem 13).

[5] König [5, p. 454].

[6] The idea, to consider the coefficients of cycle forms mod 2, and, in particular, to set 0 for 1 + 1, stems from Veblen [1, §3]; it was then further developed by Veblen and Alexander [1] and forms an essential component of the theory, which Veblen [2], in connection with Poincaré, created for n-dimensional combinatorial topology. In another connection reduction mod 2 was introduced earlier by Tietze into the topology of n-dimensional manifolds (Über die topologischen Invarianten mehrdimensionaler Mannigfaltigkeiten, Monatshefte für Mathematik und Physik, 18, 1908; §9). Recently Alexander [1, §12] also introduced a reduction mod π into topology, where π stands for an arbitrary whole number. In connection with the Four Color Problem (see Chapter XII, §5) Heawood [1] made use of congruences mod 3 in graph theory.

Chapter XI

Factorization of regular finite graphs

§1. Factors of regular graphs

If every two of the finite or infinite graphs G_1, G_2, ... -- in
finite or infinite number -- have the same vertices but pairwise no
common edge, then the graph G which consists of all edges contained in
any one of the graphs G_1, G_2, ..., is called the <u>product</u> of the graphs
(<u>factors</u>) G_1, G_2, ... In this case we set

$$G = G_1 \ G_2 \ \dots \ \text{or} \ G = \Pi \ G_\alpha.$$

By specifying the graphs G_1, G_2, ... the graph G is split into
factors, and sometimes it should be understood that there are at least
two factors involved. We shall be interested in the factorization of a
graph in the case where the factors, and therefore also the factored
graph, are regular graphs. We designate a graph as regular if all its
vertices have the same degree. This constant number (cardinality) is
called the degree of the regular graph. If we add the degrees of the
factors, we get the degree of the product.

A regular graph of degree 1 is the sum of graphs, which consist of
a single edge. A regular graph of degree 2 is (Theorem I.28) the sum of
cycles and doubly infinite paths. Every component of a regular graph of
degree g is itself a regular graph of degree g.

The multiplication defined here is commutative and associative. If
a graph is the product of more than two regular graphs, then it is also
the product of two regular graphs.

In Fig. 60 the system of edges of the icosahedron is shown. The

edges marked with the same numbers each form a factor of degree 1. So
the figure[1] shows how this regular graph of degree 5 can be factored
into five factors of degree 1.

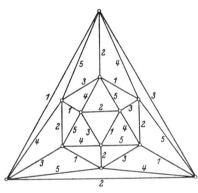

Fig. 60.

The concepts and nomenclature introduced here come (for the case of
finite graphs) from Petersen [1], who stated the problem of
factorization for regular finite graphs and partially solved it.

Petersen's starting point was the following problem in connection
with a problem of the theory of invariants[2] treated by Gordan and
Hilbert. Given a function of n variables x_1, x_2, ..., x_n of the form

$$F = (x_1 - x_2)^{\alpha_{12}} (x_1 - x_3)^{\alpha_{13}} \ldots (x_2 - x_3)^{\alpha_{23}} \ldots (x_{n-1} - x_n)^{\alpha_{n-1,n}}$$

where the α_{ij} stand for nonnegative whole numbers and the degree of F is
the same positive number k in every one of the n variables, we would
like to represent F as a product of functions of the same kind, with
smaller constant numbers k. If the x_i are vertices of a graph G, in

213

which x_i is joined with x_j by α_{ij} edges, then G is regular and of degree
k. Our problem is then equivalent to the problem of factoring graph G
into regular factors. This article by Petersen, in which Sylvester took
part[3], is certainly one of the most significant works of graph theory
but appears to have remained pretty much unnoticed for over 25 years.
Its most important results are proved without exception in this chapter
(§2 and §3) and in the next chapter. We would like to consider at the
same time non-regular graphs in order to extend certain results
concerning regular graphs to more general classes of graphs, and to be
able to apply the method of induction for finite graphs (with respect to
the number of edges). With restriction to regular graphs this method
does not lead to the goal, since, in removing an edge of a regular
graph, the graph loses its regularity in general (if there are more than
two vertices and the graph is not of degree 1).

   As a transition from regular graphs to arbitrary ones the following
theorem, which holds for both infinite graphs and infinite degrees, can
be used to advantage:

   Theorem 1 Let g be a positive integer. Let G be an arbitrary
   graph, in which every vertex P_i has degree g_i, which is less
   than or equal to g. Then by addition of certain vertices and
   edges G can be completed to a regular graph of degree g (and
   this graph is finite if G is finite)[4].

Proof Corresponding to each vertex P_i of G we introduce a new vertex
Q_i, and join Q_i with Q_j by as many (new) edges as join P_i with P_j in
G. Now we join all P_i with the corresponding Q_i by $g-g_i$ (new) edges.
(For infinite g, let $g-g_i$ denote any cardinality, for which the equation
$g_i + x = g$ holds.) It is clear that by this process we get a regular
graph of degree g, in which G is contained as a subgraph.///

   A regular graph cannot always be attained without introducing new vertices but merely adding new edges, if loops are not allowed. (see, for example, Fig. 61 with g = 3).

In this and the next chapter we would like to treat factorization of finite regular graphs. Complete graphs with an even number of vertices show a particularly simple behavior. (A graph with an odd number of vertices cannot, of course, have any factor of degree 1.) We now prove

Fig. 61.

<u>Theorem 2</u>   The complete finite graph G with 2n vertices splits into 2n-1 factors of degree 1.

<u>Proof</u>   G is, of course, a regular graph of degree 2n-1. We denote its vertices by P_0, P_1, P_2, ..., P_{2n-2}, and Q and split the set K of the edges of G into 2n-1 pairwise disjoint subsets K_0, K_1, K_2, ..., K_{2n-2} in the following way. Every edge is either of the form

$P_\mu P_\nu$ (μ, ν = 0, 1, 2, ..., 2n-2; $\mu \neq \nu$) or of the form

$P_\rho Q$ (ρ = 0, 1, 2, ..., 2n-2). The edge $P_\mu P_\nu$ ($\mu \neq \nu$) is assigned to K_r if

$$\mu + \nu \equiv r \ (\text{mod } 2n-1)$$

and $P_\rho Q$ is assigned to K_r if

$$2\rho \equiv r \ (\text{mod } 2n-1)$$

(r = 0, 1, 2, ..., 2n-2). Then every edge of G is assigned to one and only one of the sets K_r of edges. Now we show that the edges of K_r form a factor of degree 1 for every r; i.e. every vertex of G is the endpoint of one and only one edge of K_r. First, for the vertices P_ρ, either $P_\rho P_x$ or $P_\rho Q$ is the only edge of K_r going to P_ρ, depending on whether the only solution x ($0 \leq x \leq 2n-2$) of the congruence

$$\rho + x \equiv r \ (\text{mod } 2n-1)$$

satisfies $x \neq \rho$ or $x \equiv \rho$ (mod 2n-1). Second, we consider the vertex Q.

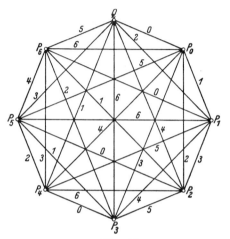

Fig. 62.

K_0	K_2	K_4	K_6	K_1	K_3	K_5
$P_0 Q$	$P_1 Q$	$P_2 Q$	$P_3 Q$	$P_4 Q$	$P_5 Q$	$P_6 Q$
$P_1 P_6$	$P_2 P_0$	$P_3 P_1$	$P_4 P_2$	$P_5 P_3$	$P_6 P_4$	$P_0 P_5$
$P_2 P_5$	$P_3 P_6$	$P_4 P_0$	$P_5 P_1$	$P_6 P_2$	$P_0 P_3$	$P_1 P_4$
$P_3 P_4$	$P_4 P_5$	$P_5 P_6$	$P_6 P_0$	$P_0 P_1$	$P_1 P_2$	$P_2 P_3$

If x (where $0 \leq x \leq 2n-2$) is the only solution of the congruence

$$2x \equiv r \pmod{2n-1}$$

216

then $Q\,P_x$ is the only edge of K_r going to Q. Consequently Theorem 2 is proved.///

For n = 4, using this method, we get the following factorization of the set K of edges: (see Fig. 62)

(For the order in which the K_r and the edges of K_r are enumerated compare the following footnote.)

The combinatorial fact[5] expressed in Theorem 2 is well known to the managers of chess or boxing tournaments. There it is a problem of bringing together 2n participants 2n-1 times into n pairs in such a way that everyone is paired with everyone else exactly once. The participants correspond to the vertices of G, and the problem is solved by decomposition of G into factors of degree 1.

§2. Graphs of even degree

We treat the general problem of factorization first for regular graphs of even degree. For degree 2 we have

> Theorem 3 A finite regular graph of degree 2, and consequently
> (see Theorem I.28) a finite sum of cycles is the product of two
> regular graphs of degree 1, if and only if every one of its
> cycles has an even number of edges[6].

Proof If all cycles K_i of $G = \sum K_i$ have an even number of edges, then (see Theorem I.12) every K_i is the product of two factors of degree 1: $K_i = K_i{}'\,K_i{}''$; likewise $G = (\sum K_i{}')\,(\sum K_i{}'')$. If, on the other hand, a cycle K_i contains an odd number of edges, then this cycle is not the product of two factors, and neither is G, since a factor of a regular graph determines a factor (of the same degree) for all its components.

For degree 4 we have

> Theorem 4 A finite regular graph G of degree 4 splits into two

217

factors of degree^7 2.

Proof It suffices to prove this for every component B_i of the graph $G = \sum B_i$. If every B_i is split into two factors of degree 2, $B_i = B_i' \, B_i''$, then the same holds for G; that is, $G = (\sum B_i')\,(\sum B_i'')$. and so G can be assumed to be connected in the proof of the theorem. By Theorem II.2, G is a closed trail and contains, since $4\alpha_o$ is double the number of edges, $2\alpha_o$ edges (α_o is the number of vertices of G). Since this number is even, every second edge can be chosen in the cyclic order determined by the trail. It is easily seen that the chosen edges and the remaining ones give a regular graph of degree 2. Their product is G.///

For connected graphs we get the following generalization: a finite regular graph of even degree g = 2k, with α_o vertices, is the product of two factors of degree k if $k\alpha_o$, the number of edges of the graph, is even. (For disconnected graphs the theorem does not hold: for example, the sum of two triangles, where k = 1, α_o = 6, cannot be factored.) The condition that $k\alpha_o$ is even is also necessary here, since, if the graph is split into two factors of degree k, then $k\alpha_o$ is double the number of edges of the factors.

Theorem 4 admits the following generalization for non-regular graphs.

Theorem 5 If each vertex of a finite graph G has degree at most 4, then all the edges of G can be divided into two classes in such a way that at most two edges of the same class go to any vertex.

Proof By Theorem 1 there is a finite regular graph H of degree 4, in which G is contained as a subgraph. By Theorem 4, H splits into two factors of degree 2. If an edge of G is put into one or the other of

the classes, depending on whether it belongs to one or the other of
these factors of H, then we have a partition of the required kind.///

There is also the quite general

> **Theorem 6** If at most 2g edges go to any vertex of a finite
> graph G, then all the edges of G can be divided into g classes
> in such a way that at most 2 edges of the same class go to any
> vertex.

Proof If the number of edges $\alpha_1 \leq 2g$ then the theorem is true: it
suffices to assign arbitrarily to each of the g classes at most 2 edges.
We can therefore use induction and assume the theorem proved for graphs
with $\alpha_1 - 1$ edges. Let PQ be any edge of G; let the rest of the edges
form the subgraph G' of G. Since G' contains only $\alpha_1 - 1$ edges, we can
by induction assumption divide the edges of G' into g classes K_1, K_2,
..., K_g in such a way that at most two edges of the same class go to the
same vertex. At most 2g-1 edges of G' go to P, and there is therefore a
class K_p which contains at most one edge of G' going to P. Likewise
there is a class K_q which contains at most one edge of G' going to Q.
If $K_p = K_q$ and the removed edge PQ is put into this class K_p ($=K_q$),
without changing the class membership of the rest of the edges, then the
required condition is fulfilled for all edges of G. We can therefore
assume that $K_p \neq K_q$. In the graph which is formed by the edges of K_p
and K_q at most three edges go to P, at most three edges go to Q, and at
most four edges go to the rest of the vertices. If the edge PQ is added
to the graph, there results a subgraph G_1 of G in which at most four
edges go to any vertex; so by Theorem 5 the edges of G_1 can be divided
into two classes $K_p{}'$ and $K_q{}'$ in such a way that at most two edges of the
same class go to the same vertex. So if in the class sequence
K_1, K_2, ..., K_g the class K_p is replaced by $K_p{}'$ and K_q by $K_q{}'$, then the
g classes of edges obtained in this way fulfill the desired condition

for G, and Theorem 6 is proved.///

We consider now the special case where G is a regular graph of degree 2g. Then every one of these classes must contain exactly two edges which go to P for each vertex P of G, since otherwise at least one of the remaining g-1 classes would have to contain three such edges.

We have thus arrived at the following theorem, which expresses[8] one of the chief results of the Petersen article:

> Theorem 7 Every finite regular graph of even degree (i.e. every regular Euler graph) splits into factors of degree 2.

In order to illustrate the logical connection of Theorems 6 and 7, let us make the following comment. If we had given for Theorem 7 a proof which does not depend on Theorem 6 -- as Petersen does -- we could, with the help of Theorem 1, deduce Theorem 6 from Theorem 7, as we did above for g = 2, when we deduced Theorem 5 from Theorem 4. In this way the finiteness of the graph, to which we shall appeal in Chapter XIII, does not need to be assumed. We give two examples for Theorem 7 (g = 2), i. e. for Theorem 4. In both Fig. 63 and 64 the edges labeled with the same number form a factor of degree 2. For the graph of Fig. 63 (this is the system of edges of the octahedron) these two factors split further into two factors of degree 1. For the graph of Fig. 64 this is, however, not the case, since both its factors are triangles of degree 2. By the way, this graph cannot have any factor of degree 1, since it contains an odd number of vertices.

Theorem 7 can be applied to a variety of combinatorial questions; as an example we use it to prove the following:

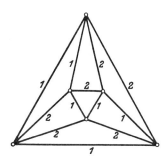   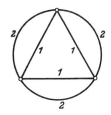

Fig. 63.   Fig. 64.

If in a group of people each person has exactly 2g

acquaintances, then the whole group can be arranged around a

certain number of round tables in such a way that everybody

knows both his neighbors. This can be done in g different ways

such that each person sits once and only once next to each of

his acquaintances. (The number of tables may be changed here

from case to case.)

Proof We let a vertex correspond to each person present and we join two

vertices by an edge if and only if the people corresponding to the

vertices are acquainted with each other, and so we get a regular graph

of degree 2g. By Theorem 7 it has a factor of degree 2; this is

(Theorem I.28) a sum of cycles and therefore furnishes a proof for our

first assertion. As for the second assertion we split G into g factors

of degree 2. These determine g arrangements of the required kind. It

is clear that they have the desired property.///

We shall give other applications of Theorem 7 in Chapter XIII,

since they are of particular interest for infinite graphs.

Some problems of combinatorics lead to the question: when can all factors of degree 2 into which a finite regular graph of even degree splits by Theorem 7, be chosen as connected subgraphs? That is, as Hamiltonian cycles of the graph? (In the interpretation just given this means that only a single table is permitted.) We do not wish to treat this difficult problem in general and shall content ourselves in citing some examples in which this is the case. (See also Figures 63 and 64):

1. The complete graph with 2n+1 vertices.

2. The graph which comes from the complete graph with 2n vertices by either removing or doubling the edges of a factor of degree 1.

3. The graph which consists of vertices P_1, P_2, ..., P_{2n}; Q_1, Q_2, ..., Q_{2n}; and edges $P_i Q_j$ (i, j = 1, 2, ..., 2n).

For the proof of the assertion that these four graphs have the desired property we refer to the treatment by Lucas.[9]

We would now like to carry over our results to graphs in the broader sense. Then there are two equally valuable interpretations to be strictly distinguished from each other. In counting the degree of a vertex, a loop can be counted as two edges (first way of counting) or as one edge (second way of counting). The degree of a vertex and hence also the regularity of a graph with loops naturally depends on the chosen way of counting. So, for example, the graph of Fig. 65 is regular only by the first way of counting, and the graph of Fig. 66 only by the second way.

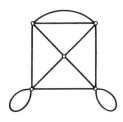

Fig. 65.   Fig. 66.

We can easily convince ourselves that Theorems 1 and 3-7 of this

chapter and Theorem II.2 (on which we have based the proof of Theorems 4

through 7) as well as the proofs given for these theorems remain

unaltered for graphs in the broader sense if we take as our basis the

first way of counting. For the second way of counting, however, this is

not the case. In the graph of Fig. 67 two edges go to each vertex, and

yet the graph obviously does not have an Euler line; it is also not the

sum of cycles. The second part of Theorem 3 is also not satisfied for

this graph: it is the product of two factors of degree 1 (1 and 2 in

Fig. 67), in spite of the fact that it contains cycles with an odd

number of edges (a loop is a cycle with a single edge). Nevertheless it

can be proved that, by the second way of counting, Theorems 4-7 remain

valid unaltered for graphs in the broader sense.[10] It suffices to prove

this for the most general of these theorems, Theorem 6. Instead of

carrying over the given proof, we prefer to reduce the general case to

Theorem 6.

Let G be a graph in the broader sense, in which at most 2g edges go

to each vertex, and in which every loop is counted as one edge. We

introduce a new vertex P'
corresponding to each vertex P of G
and join two new vertices (not
necessarily distinct) by as many new
edges as joined the two corresponding
G-vertices. We then remove all loops
(old and new), but introduce one new

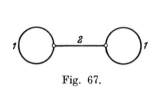

Fig. 67.

edge P P' each, corresponding to every loop going to P. In the
resulting graph H the same number of edges go to each P as in G, and
just as many edges go to P' in H as went to P in G (and go to P in H).
Consequently, at most 2g edges of H go to each vertex of H. On the
other hand, H has no loops. So Theorem 6 can be applied to H: its
edges split into g classes in such a way that at most two edges of the
same class go to the same vertex of H. Every edge of G which is not a
loop is also an edge of H, and we would like to assign it, as an edge of
G, to the same one of the g classes to which it belongs as an H-edge.
But every loop PP of G is put into the same class to which the new edge
PP' of H corresponding to this loop belongs. The partition of the edges
of G into g classes thus specified obviously satisfies the stated
condition.

Let it be noted that in this reduction of the general case to the
case of graphs in the stricter sense the finiteness of the graph did not
have to be assumed.

Summarizing, we can say:

Theorem 8 Theorems 6 and 7 also hold for graphs in the broader
sense and do not depend on whether the first or second way of
counting (loops) is used.

§3. Primitive graphs

We designate a regular graph as <u>primitive</u> if it cannot be split into regular factors. The material already presented (Theorems 3 and 7) gives the following result:

> <u>Theorem 9</u> A finite regular graph of even degree is primitive if and only if it is of degree 2 and contains an odd number of edges in at least one of its cycles.

It turns out to be much more difficult to find a simple characteristic for a regular graph of odd degree to be primitive, and this -- with the exception of degree 3, which will be treated in Chapter XII -- has not to date been accomplished.

Petersen [1, §13] proved the following theorem, which is quite general:

> <u>Theorem 10</u> For every odd number $2\alpha + 1$ there are primitive regular finite graphs of degree $2\alpha + 1$.

<u>Proof</u> Petersen gives the following example of a connected graph G_α: The following $3(2\alpha + 1) + 1 = 6\alpha + 4$ points are chosen as vertices of the graph:

$$P; \ A_1, \ B_1, \ C_1; \ A_2, \ B_2, \ C_2; \ \ldots; \ A_{2\alpha+1}, \ B_{2\alpha+1}, \ C_{2\alpha+1}.$$

For $i = 1, 2, \ldots, 2\alpha+1$ we join the point A_i with both B_i and C_i by α edges, B_i with C_i by $\alpha+1$ edges, and finally A_i with P by one edge. Of these $(2\alpha+1)(3\alpha+2)$ edges, there are $2\alpha+1$ edges going to each vertex of the graph G_α so defined.

It can be seen immediately that none of the edges PA_i ($i = 1, 2, \ldots, 2\alpha+1$) is contained in a cycle of G_α; but then P also cannot belong to any cycle of G_α. From this it follows that G_α cannot have any factor of degree 2, since every factor must contain every vertex. If G_α were not primitive and therefore the product of two factors, one of these factors would have to be of even degree; this would either have to be of degree 2 itself or would by Theorem 7 have a factor of degree 2, so that in any case G_α would contain a factor of degree 2. But this has just been proved to be impossible.///

(This proof shows that a finite regular graph of odd degree is always primitive, if it contains a vertex P, which is not contained in any cycle.)

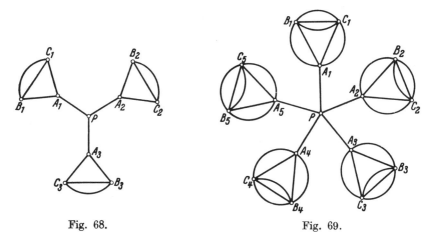

Fig. 68.   Fig. 69.

In Fig. 68 the graph G_α just defined is shown for $\alpha = 1$. It was designated by Petersen [1, Fig. 11] as the Sylvester graph. Fig. 69 shows the graph G_α for $\alpha = 2$. Fig. 70 gives a second example of Petersen's [1, Fig. 14] of a primitive regular graph of degree 3; here every vertex is contained in a cycle. (We leave it to the reader to

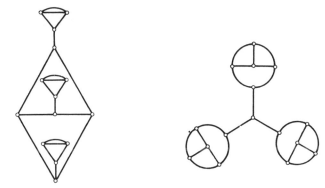

Fig. 70.   Fig. 71.

convince himself that this graph is primitive.) The example of the graph of Fig. 71 shows that a primitive graph of degree 3 does not necessarily have to have 2-cycles.

The factorization of regular graphs into primitive factors shows an analogy with the factorization of the natural numbers into prime factors, through the fact that every non-primitive graph of finite degree can be composed of primitive factors, which can be done directly stepwise. There are also essential differences with elementary number theory. The factorization of regular graphs into primitive factors is not unique. If the graph G is the sum of the quadrilaterals 12341 and 1'2'3'4'1', then it is the product of the primitive factors (12, 34, 1'2', 3'4') and (14, 23, 1'4', 2'3') as well as the product of the primitive factors (12, 34, 1'4', 2'3') and (14, 23, 1'2', 3'4'). (As we shall soon see, a connected graph can also be given as an example of

this non-uniqueness.) In fact, the degrees of primitive factors into
which a graph splits are never uniquely determined by the graph. In
order to see this, we consider any primitive regular graph of odd
degree, such as the Sylvester graph S of Fig. 68. If, corresponding to
each of its edges, a second edge is introduced, which joins the same two
vertices, then the resulting graph S' of degree 6 is certainly the
product of two primitive factors of degree 3. On the other hand, by
Theorem 7, S' is also a product of three graphs of degree 2. If these
three factors, or some of them, are not primitive, then they split
further into two factors of degree 1. So S' is the product of primitive
factors of degree 3 as well as the product of primitive factors of
degrees 1 and 2. In this sense, Petersen's assertion [1, p. 194] that
in the case of even degree, "all primitive factors are of degree 1 or 2"
needs to be emended. The question of factorizations of a regular finite
graph of even degree into primitive factors is in no way taken care of
by Theorem 7 if factors of odd degree are also allowed. Problems arise
which cannot now be dealt with. The simplest are: When does a regular
graph of degree 4 have factors of degree 1? When is a regular graph of
degree 6 the product of two (primitive) factors of degree 3? etc.

     For primitive graphs we now prove

     Theorem 11 If isomorphic graphs are regarded as not distinct,
     then there are for every natural number ν only finitely many
     primitive finite regular graphs with ν vertices.[11]

We base our proof[12] on the following number theoretical
consideration. An ordered system $(a_1, a_2, \ldots, a_n)$ will be designated
as an n-termed complex if all a_i are nonnegative whole numbers and at
least one a_i is > 0. For two such complexes we set

$$(a_1, a_2, \ldots, a_n) < (b_1, b_2, \ldots, b_n)$$

if, for every i, $a_i \leq b_i$ and the "equals" sign does not hold for every

i. If neither of two distinct complexes is greater than the other, we shall call them incomparable. We now prove the lemma

Lemma 1. If M is a set of n-termed complexes and any two elements of M are incomparable then M is finite.

Proof The lemma is trivial for n = 1, since then there are no incomparable complexes; we can then, using induction, assume that n > 1 and assume the theorem as proved when the number of terms is n-1.

Let $a = (a_1, a_2, \ldots, a_n)$ be an element of M. If $x = (x_1, x_2, \ldots, x_n)$ is any other arbitrary element of M, then there must be an i for which the inequality $x_i < a_i$ holds, since otherwise x > a and x would be comparable with a. If M were infinite then infinitely many (namely all distinct from a) elements of M would satisfy at least one of the finitely many conditions $x_i < a_i$ (i=1, 2, ..., n). But then there would have to be an i such that the inequality $x_i < a_i$ would hold for infinitely many elements of M. Without loss of generality we may assume that i = 1. Since only finitely many x_1 are smaller than a_1, there would have to be a number b_1 ($<a_1$) with the property that infinitely many elements of M have the form $(b_1, x_2, x_3, \ldots, x_n)$. To each of these elements corresponds an n-1 - termed complex $(x_2, x_3, \ldots, x_n)$; these are, of course, likewise pairwise incomparable, and therefore, since we assume the theorem for n-1, their number is finite. This contradiction establishes the correctness of Lemma 1.///[13]

We now return to Theorem 11. Since the graphs to be considered have the same number ν of vertices, we may assume that they all have the same vertices $P_1, P_2, \ldots, P_\nu$. A graph is completely determined by its structure, if for every pair $(P_i, P_j) = P_{ij} = P_\rho$ (where i, j = 1, 2, ..., ν; i ≠ j and $\rho = 1, 2, \ldots, \binom{\nu}{2}$) the number of

edges a_ρ joining P_i with P_j is given. The structure of G is
consequently determined by a $\binom{\nu}{2} = \mu$ - termed complex
$(a_1, a_2, \ldots, a_\mu)$. If two primitive graphs G_a and G_b are determined by
complexes $a = (a_1, a_2, \ldots, a_\mu)$ and $b = (b_1, b_2, \ldots, b_\mu)$ then these
complexes are incomparable; for if $a > b$, then G_a would be the product
of two regular graphs, whose structures are determined by the complexes
$(b_1, b_2, \ldots, b_\mu)$ and $(a_1 - b_1, a_2 - b_2, \ldots, a_\mu - b_\mu)$, and G_a would not
be primitive. By Lemma 1 there are then only finitely many primitive
finite graphs of distinct structure with ν vertices.///

We should mention here that if primitivity is not required, infinitely
many finite regular graphs of distinct structure exist, with the same
vertices $P_1, P_2, \ldots, P_\nu$. If each of the ν pairs of vertices (P_1, P_2),
$(P_2, P_3), \ldots, (P_{\nu-1}, P_\nu), (P_\nu, P_1)$ are joined by k edges, then we get a
regular graph of degree 2k for arbitrary k. But if the degrees remain
under a bound, then there are only finitely many graphs with ν vertices
of distinct structure. Theorem 11, thus proved, is completely
equivalent to the following theorem:

> Theorem 11* The degrees of finite primitive regular graphs
> with ν vertices have an upper bound dependent on ν.

If only even degrees are allowed, then by Theorem 9 the number 2 is
always an upper bound independent of ν. If we wished to prove the
existence of such a bound also for the case of nothing but odd degrees
we would get a second proof for Theorems 11* and 11. This method can be
pursued: Petersen [1, §13] proved not only the existence of such a
bound, but also determined this bound as a function of ν; he showed that
a finite regular primitive graph of odd degree with ν vertices can have
at most degree $\dfrac{\nu}{3} + 1$. We shall not go into the proof of this theorem.

§4. Bipartite graphs

With respect to factorization one class of graphs, which also plays [14] an important role also in other parts of graph theory and its applications, shows a particularly simple behavior. This is the class of bipartite graphs. A finite or infinite graph is called <u>bipartite</u> if each of its cycles contains [15] an even number of edges, as, for example, the system of edges of the cube or the universal infinite plane square lattice. Every subgraph of a bipartite graph is a bipartite graph. The simplest bipartite graphs are acyclic graphs. A bipartite graph can, of course, have no loops, and so it is a graph in the narrower sense.

A characteristic property of bipartite graphs is expressed in the following theorem, which has already been proved as Theorem X.12.

> <u>Theorem 12</u> A graph is a bipartite graph if and only if its vertices can be divided into two classes in such a way that only vertices of different classes are joined by an edge.

For connected bipartite graphs this partition of the vertices into two classes is uniquely determined (apart from the order of the classes). If the graph is finite and regular then these two classes always contain the same number of vertices, for the following reason. If one contains m vertices and the other contains n vertices, and if g is the degree of the graph, then the number of edges is gm and gn. From gm = gn it follows that m = n.

We now prove the following general theorems [16]:

> <u>Theorem 13</u> Every finite bipartite regular graph has a factor of degree 1.

> <u>Theorem 14</u> Every finite bipartite regular graph of degree g splits into g factors of degree 1.

231

<u>Theorem 15</u> If the degree of any vertex of a finite bipartite graph G is at most g, then all the edges of the graph can be divided into g classes in such a way that two edges which meet at a vertex always belong to different classes.

We would like to give three different types of proofs for these theorems. The first[17] concerns the proof of Theorem 15; this proof will be analogous to the proof which we gave for Theorem 6. The assertion of Theorem 15 can be formulated so that it is possible to assign one of g indices to the edges of the graph G[18] in such a way that two edges which meet at a vertex always receive different indices.

If the number of edges $\alpha_1 \leq g$, then Theorem 15 is, of course, true. We use induction and assume that the theorem is proved when the number of edges is $\alpha_1 - 1$. If an arbitrary edge AB is removed from G, we get a graph G' (still bipartite), whose edges can be assigned indices from the g numbers in a way corresponding to the theorem. If an index occurs neither at A (i.e. in an edge going to A) nor at B, then this index can be assigned to edge AB, and our goal is reached. So we may restrict ourselves to the case where every index occurs at either A or B. Since the degree of A (and also of B) in G' is at most g-1, then at each of these vertices there is at least one index missing. Let "1" denote an index missing at B (and therefore surely occurring at A), and "2" an index missing at A (and therefore surely occurring at B). Now let AA_1 be the edge with index "1". Possibly there is an edge A_1A_2 with index "2", and then perhaps an edge A_2A_3 with index "1", an edge A_3A_4 with index "2", etc. We form this "alternating" sequence of edges $A\ A_1\ A_2\ A_3\ \ldots$ as far as possible. It is -- according to the choice of indices "1" and "2" -- uniquely determined. In the sequence of vertices $A, A_1, A_2, \ldots$ no vertex can occur twice; for if A_i were the first vertex, which had already occurred once, then there would have to occur

at A_i two edges with index "1" or two edges with index "2". This is
impossible. We also cannot return to A again, since "2" does not occur
at A and "1" was already used (with edge A A_1). Finally, we cannot get
to B on this path, for this could happen only with an edge of index "2"
("1" does not occur at B), and then the path from A to B would contain
an even number of edges; combined with the omitted edge AB, this path --
in contradiction to our assumptions -- would give a cycle of G with an
odd number of edges.

So the path A A_1 A_2 ... A_r is therefore a path of G' which does not
contain the vertex B and whose edges have the indices "1" and "2"
alternately. We switch now the indices "1" and "2" on the edges of this
path,[19] without changing the indices of the other edges. The
distribution of the indices also satisfies the requirements after this
switching. This is seen not only for the first vertex A, from which no
edge with index "2" went out originally, but for the interior vertices
A_i as well, and for the endpoint A_r, from which no second edge with
index "1" (or "2", respectively) can go out, since otherwise our path
could not end in A_r. But we have now arranged it so that the index "1"
no longer occurs at A, and it can be assigned to the omitted edge AB.

Theorem 15 is thus proved.///

We consider now the special case where G is a regular graph of
degree g. Then each of the g classes of Theorem 15 must for every
vertex P of G contain exactly one edge going to P, since otherwise one
of the remaining classes would contain at least two such edges. So each
of these classes forms a factor of degree 1 of G. With this Theorem 14
and Theorem 13 are proved as a consequence of Theorem 15.///

The second type of proof[20] which we would like to give for Theorems
13, 14, and 15 begins with the proof of Theorem 13 and is based on
Petersen's[21] Theorem (Theorem 7). Let G be an arbitrary regular

233

bipartite graph of degree g. Corresponding to every edge k of G we introduce a new edge k', which joins the same two vertices to each other as k does. So out of G we get the regular graph G_{2g} of degree 2g. G_{2g} is also a bipartite graph; if K is a cycle of G_{2g}, then either K is a 2-cycle, or by replacing every new edge k' by the corresponding edge k of G, we get from K a cycle of G with the same number of edges. By Theorem 7, G_{2g} has a factor of degree 2. As a subgraph of a bipartite graph this factor is also a bipartite graph and by Theorem 3 splits into two factors of degree 1. If all the new edges k' in one of these factors of degree 1 are replaced by the corresponding edges of G, we obviously get a factor of degree 1 of G, and Theorem 13 is proved.///

If G_1 is a factor of degree 1 of G, then $G = G_1 G_{k-1}$, where G_{k-1} is a regular bipartite graph of degree k - 1, and so by Theorem 13 also has a factor G_1': $G_{k-1} = G_1' G_{k-2}$; likewise $G_{k-2} = G_1'' G_{k-3}$, etc., where G_1', G_1'', ... are also factors of degree 1 of G. Finally, therefore, $G = G_1 G_1' G_1'' \ldots G_1^{(k-1)}$ splits into factors of degree 1. With this Theorem 14 is proved as a consequence of Theorem 13.///

Finally, Theorem 15 comes from Theorem 14 in the following way. Let G be a finite bipartite graph, in which every vertex has degree $\leq g$. In the proof of Theorem 1 we constructed a regular finite graph H of degree g, in which G is contained as a subgraph. We show that if G is a bipartite graph, then H must also be bipartite. By Theorem 12 the vertices of G can be divided into two classes I and II in such a way that every edge of G joins vertices of different classes. If the new vertices (not belonging to G) of H are put into Class I or II, depending on whether the corresponding vertex of G belongs to II or I, then all edges of H join vertices of different classes. So by Theorem 12, H is a bipartite graph, and H by Theorem 14 can be split into g factors of degree 1. If the edges of G are now divided into g classes, depending

234

on which of these g factors of H the edge belongs to, this partition satisfies our requirements, and Theorem 15 is proved.///

The third proof we shall give for Theorem 13 comes -- in combinatorial formulation -- from E. Sperner.[22]

Let P_1, P_2, ..., P_m, Q_1, Q_2, ..., Q_m be the vertices of the regular finite graph $G^{(m)}$ of degree g whose edges (their number is gm) join a vertex P with a vertex Q. The theorem is true for m = 1, when the number of vertices is α_0 = 2m = 2. By using induction we may assume the theorem proved for $\alpha_0 \leq 2m$ (and arbitrary g) and prove it only for α_0 = 2 (m+1) > 2. For the same m (and arbitrary g) the following also holds:

> α) If the degree of every vertex of a bipartite graph G, whose edges join one of the vertices P_1, P_2, ..., P_m with one of the vertices Q_1, Q_2, ..., Q_m, is $\leq$ g, and if the number of edges of the graph is mg - k, where $0 \leq k \leq g - 1$, then G has a factor of degree 1.[23]

It is clear that G can be supplemented by the introduction of k new edges of the form $P_\alpha Q_\beta$ to a regular bipartite graph, $G^{(m)}$, of degree g, with the same 2m vertices. By induction assumption $G^{(m)}$ has a factor G_1 of degree 1: $G^{(m)} = G_1 G_1'$; for the same reason G_1' also has a factor G_2 of degree 1, etc. So we get g factors of degree 1 of $G^{(m)}$ which have pairwise no common edges. Since k < g, there is certainly one factor among these g factors -- say, G_1 -- which contains none of the k new edges, and so consists of edges of G. Then G_1 is a factor of degree 1 of G.

Now the existence of a factor of degree 1 for the regular graph $G^{(m+1)}$ of degree g with 2(m+1) vertices can be proved as follows. If the edges of $G^{(m+1)}$ join each vertex P_i with a vertex Q_j (i, j = 1, 2, ..., m+1), then we can choose the designation of the vertices so that an edge $P_{m+1} Q_{m+1}$ is contained in $G^{(m+1)}$. Now we remove from $G^{(m+1)}$ the vertices P_{m+1} and Q_{m+1} and all edges going to P_{m+1} or Q_{m+1}. For the number ℓ of removed edges we have the following inequality:

$$g \leq \ell \leq 2g - 1, \text{ i.e. } 0 \leq \ell - g \leq g - 1,$$

and there remain $(m + 1) g - \ell = mg - (\ell - g)$ edges which form a bipartite graph $G^{(m)}$ with $2m$ vertices. So with $k = \ell - g$ the assumptions in our assertion (α) hold for $G^{(m)}$, and so $G^{(m)}$ has by (α) a factor of degree 1; if the edge $P_{m+1} Q_{m+1}$ is added to this factor, we get a factor of degree 1 of $G^{(m+1)}$. With this we have completed the third proof[24] of Theorem 13.///

On the one hand, we showed above that Theorem 13 follows from Theorem 14, and Theorem 14 follows from Theorem 15; on the other hand, we have also proved that Theorem 14 can be deduced from Theorem 13, and Theorem 15 from Theorem 14. For later reference it is important to note that in this proof of the equivalence of Theorems 13, 14 and 15 the finiteness of the graphs under discussion did not have to be assumed anywhere. If we also wish to prove these three theorems for infinite graphs (where the degree g remains finite), it is sufficient to

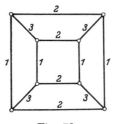

Fig. 72.   Fig. 73.

generalize any one of these theorems for the case of infinite graphs. We shall come back to this in Chapter XIII.

As an example for Theorem 14, Fig. 72 shows how the system of edges of the cube splits into three factors of degree 1. The edges labeled with the same number form each such factor.

By Theorem 3 the assumption that G be a bipartite graph is also necessary for the validity of Theorems 13 and 14 for degree g = 2. This is, however, not the case for g > 2, as Fig. 73 shows (system of edges of the tetrahedron) (see also Fig. 60).

§5. Applications

Theorem 13 can be applied to different problems of combinatorics. Instead of speaking of general combinations, binary relations, etc. it will be clearest if we show this in three concrete examples.

I. Application

In a mathematical journal there are n problems stated. Solutions arrive from n contributors, who each send in k solutions, (among which there may be several solutions to the same problem), in such a way that altogether exactly k solutions arrive for each of the n problems. The editorial staff is always able to publish, for each of the n problems, exactly one of the solutions sent in, so that exactly one solution by each of the n contributors is published.

II. Application

At a dance there are n ladies and n gentlemen who do not enjoy making new acquaintances. It turns out that every lady knows k gentlemen and every gentleman k ladies. It is possible to arrange everyone present into pairs -- each lady with a gentleman -- in such a way that only acquaintances are joined in each pair.

III. Application

At an officers' party there are n regiments and n ranks represented, and exactly k members of the party belong to each regiment and to each rank. A delegation can always be chosen from the party in such a way that every regiment and every rank are represented by exactly one soldier.

From these three assertions we get the following by Theorem 13. We consider the problems in I (the gentlemen in II and the regiments in III, respectively) as vertices A_1, A_2, ..., A_n and the contributors in I (the ladies in II and the ranks in III, respectively) as vertices B_1, B_2, ..., B_n. The vertices A_i and B_j are joined by r_{ij} edges (i, j = 1, 2, ..., n). For I, r_{ij} denotes the number of solutions which are sent in for the problem A_i by the contributor B_j; for II, r_{ij} = 1 or 0, depending on whether or not A_i and B_j know each other, and for III, r_{ij} is the number of officers who belong to the regiment A_i and at the same time to the rank B_j. The requirements imply in all three cases that the corresponding graph is regular and of degree k. It is true also that G is a bipartite graph, since every edge joins an A-vertex with a B-vertex (see Theorem 12). In all three cases the existence of a factor of degree 1 by Theorem 13 furnishes the desired result.

It is easily seen that the statements I, II, III are equivalent to Theorem 13 with the single exception that formulation II expresses Theorem 13 only for the case where the same two vertices in the graph are joined[25] with each other by at most one edge.

If not only Theorem 13 but also Theorem 14 were used in the three applications, further results could be formulated in all three cases. We shall not go into this here; cf. the second assertion in the application which we gave above for Theorem 7.

It is clear that assertion III can be expressed, without being put

238

in concrete form, by the following theorem.

> Let two partitions of a finite set be given. Let one of
> them split the set into n mutually disjoint classes
> A_1, A_2, ..., A_n with k elements each. Let the other also split
> the set into n mutually disjoint classes B_1, B_2, ..., B_n with k
> elements each. Then there is a system of n elements
> x_1, x_2, .., x_n such that every A-class, as well as every
> B-class, is represented among the x_i's by an element.

This formulation of Theorem 13 comes from van der Waerden [1]. His
motivation was to show that a theorem of G. A. Miller's[26] about abstract
finite groups "is basically not a group theoretical theorem but a set
theoretical one"; we would say that it is a graph theoretical theorem.

As a further application we shall deduce a corollary of Theorem 13
(14, resp.), which gives information concerning the structure of the
symmetric group.

Let a_{ij} (i = 1, 2, ..., m; j = 1, 2, ..., n) be m n distinct
elements, which we collect in the form of the matrix $A = ((a_{ij}))$ (with m
rows and n columns). Let $\{b_{ij}\}$ be an arbitrary permutation of the
sequence of elements $\{a_{ij}\} = (a_{11}, a_{12}, ..., a_{1n}, a_{21}, ..., a_{mn})$. This
determines a matrix $B = ((b_{ij}))$ which also has m rows and n columns. A
graph can be assigned to this permutation in the following way. We let
the vertices P_1, P_2, ..., P_m correspond to the rows of A and the
vertices Q_1, Q_2, ..., Q_m correspond to the rows of B and we introduce an
edge $P_i Q_\beta$ corresponding to each element a_{ij}, if a_{ij} is contained in the
β-th row of B. The graph G thus obtained is, as can be immediately
seen, a regular bipartite graph of degree n with 2m vertices. So by
Theorem 14 it splits into n factors of degree 1:
$G = G_1 G_2 ... G_n$. To the m edges of any of these factors correspond m
elements a_{ij}, which belong to the m different rows in A as well as in B.

A permutation inside the rows in A can be arranged so that in the resulting matrix A' the first column contains the elements corresponding to the edges of G_1, the second column contains the elements corresponding to the edges of G_2, etc. Likewise, a permutation inside the rows of B can be arranged so that the resulting matrix B' has the same property. So A' and B' contain the same elements in every two corresponding columns. Consequently we can get B' from A' by permuting only inside the columns. Since the inverse of a permutation which permutes only inside the rows (columns) is also such a permutation, we have proved the following theorem of R. Rado.[27]

Every permutation of the m n elements a_{ij} (i = 1, 2, ..., m; j = 1, 2, ..., n) can be obtained by permuting inside

Step 1. the rows of the matrix $((a_{ij}))$,

Step 2. the columns of the resulting matrix, and finally

Step 3. the rows of the last matrix.

As an added remark to this theorem we mention that in general none of the three steps can be dispensed with, since two distinct elements of the same row cannot, in general, be brought to the same given column by two of these steps.

An application of Theorem 13 to the theory of determinants will be treated in Chapter XIV, §3. Other applications and formulations of this theorem will be discussed in Chapter XIII, since these are of particular importance for infinite graphs (but not exclusively).

Notes on Chapter XI

[1] Ahrens [2, Vol. II, p. 222]. A method is found in G. Kowalewski [2, p. 92] to get to a five class partition of the set of edges of this special graph.

[2] Already before Petersen Sylvester [3], Clifford, Spottiswoode, Buchheim, and Kempe made use of the concept of graphs in the theory of invariants (partly in connection with chemical atomistics); see the information on the literature in W. Fr. Meyer, Invariantentheorie, Encyklopädie der math. Wissenschaften, Vol. I, Footnotes 239-242.

[3] Concerning this Peterson writes [1, p. 194] as follows: "Mr. Sylvester at the same time took up with me the question of basic factors, and I communicated with him frequently concerning them. Although we sought the answer in quite different ways, I owe to his communications an excitement without which I would have been worn down by the great difficulties which arose at every step.

[4] König [5, p. 464].

[5] This was first proved by M. Reiß : Über eine Steinersche kombinatorische Aufgabe ..., Journal für reine u. ang. Mathematik, 56, 1859, p. 326; §2. An elegant geometrical solution, which stems from Walecki, is given by Lucas [1, Vol. II, p. 176] (Les promenades du pensionnat). It can be shown that this solution of Walecki's agrees with the solution given here. It is easily seen, for example, that out of the first column (K_0) in the table given above for n = 4 the other columns can be obtained by leaving Q alone and permuting the indices 0, 1, 2, 3, 4, 5, 6 of P six times cyclically in this order. -- Let it be further mentioned that our combinatorial problem represents one of the simplest cases ($\mu = \nu = 2$) of the general and partly unsolved problem, which asks for the producing of complete (μ ; ν) - systems of n things,

where by a "complete (μ ; ν) - system of n things" a system is
understood, whose elements are themselves systems, which each contain
ν of the n things, while arbitrary μ of the n things are contained in
one and only one of these systems of n elements; cf. Skolem, Note 16 in
Netto: Lehrbuch der Combinatorik, 2nd ed., Leipzig and Berlin 1927, p.
321.

[6] Petersen [1, §3].

[7] Petersen [1, §6].

[8] Petersen [1, §§3-9]. The proof given here, which uses the
basic idea of Petersen's, is simpler than that of Petersen and also has
the advantage of leading directly to the generalization, which was
expressed in Theorem 6. This generalization will make possible for us
the transition to infinite graphs (in Chapter XIII). -- Reidemeister
also gives a proof of Theorem 7 [1, pp. 116-119].

[9] The decomposition of this graph into the product of
Hamiltonian cycles is equivalent to the problems, which are solved with
the help of the above mentioned geometrical method of Walecki in Lucas
[1, Vol. II, pp. 162-170] under the titles: 1. "Les rondes
enfantines", 2. "Les rondes paires", 3. "Les rondes alternées". --
Also the problems "Les rondes à centre" and "Les files indiennes" dealt
with there can be interpreted as questions of graph theory.

[10] Theorem 1 remains true together with its proof given above.

[11] In the introduction to his article Petersen [1] stated a
theorem equivalent to this theorem and noted that it follows from a
theorem of Gordan. This theorem of Gordan's (Vorlesungen über
Invariantentheorie, Leipzig, 1885-1887, Vol. I, pp. 196-201) which forms
the foundation of its proof (ibid. Vol. II, pp. 231-236) relating to the
finiteness of the system of invariants for fundamental binary forms,
goes as follows: A system of arbitrarily many linear and homogeneous

Diophantine equations has a finite number of nonnegative solutions, from which every nonnegative solution can be obtained by addition. This theorem can be derived immediately from Lemma α , which follows here, see König [10]. -- Contrary to a remark in the Encyklopädie der mathematischen Wissenschaften (Vol. I_1, p. 365, cf. also p. 342) Petersen's work contains no proof of Gordan's theorem.

[12] König [10].

[13] My original proof was not so simple; the simplification I owe to a conversation with Mr. E. Egerváry. -- Mr. L. Fejér communicated to me another proof of Lemma α , which shows that the lemma remains true if transfinite numbers are allowed as elements of the complexes.

[14] The graphs, for example, which were treated in Chapter VI, §3, and in Chapter VIII, §4, belong to this class of graphs.

[15] This concept (graphe à circuits pairs) was introduced by König [3 and 5]. It already, however, plays a certain role in Kempe [1, p. 200]. Sainte-Laguë [1, 2, and 4] uses for "bipartite" the designation bipartie or de rang 2, where he understands by the "rang" of a graph the minimal number of classes into which its set of vertices can be split so that the endpoints of any edge always belong to different classes (cf. Theorem 12 following here).

[16] The author stated Theorems 13 and 14 for the first time in 1914 at the Congress for Mathematical Philosophy in Paris [3]. He then proved all three theorems in his work [5]. Later these theorems were treated by Frobenius, Sainte-Laguë [1 and 2], van der Waerden [1], Sperner, Skolem [1], and Egerváry. We shall come back to the treatment of van der Waerden and Sperner shortly. Frobenius and Egerváry treated the determinant theoretical formulation of Theorem 13; this is discussed in Chapter XIV, §3. In Sainte-Laguë's terminology Theorem 14 goes as

follows: "The class of a regular bipartite graph is equal to its degree," where by class he means the minimal number of classes, into which its edges can be split so that two edges which meet at a vertex always belong to different classes.

[17] König [5, §1].

[18] In similar circumstances often "colors" are spoken of instead of indices.

[19] Here an idea is used which was first used by Kempe [1, p. 194]. In Kempe it is a matter, not of edges, but of "districts."

[20] This was indicated in König's work [5, p. 455, Note **].

[21] As we shall see in Chapter XIII, Theorem 7 can be derived conversely from Theorem 14.

[22] Sperner, Note zu der Arbeit von Herrn B. L. van der Waerden: "Ein Satz über Klasseneinteilungen von endlichen Mengen," Abhandlungen aus dem mathematischen Seminar, Hamburg, 5, 1927, p. 232.

[23] It should be remembered here that we have introduced the concept of factor of degree 1 also for nonregular graphs already (in Chapter I, §2).

[24] In Chapter XIV, §2, we shall give still a fourth proof.

[25] This exception could be remedied by assigning to each "acquaintance" (say, according to the degree of intimacy of the acquaintance) a natural number as multiplicity, and, instead of requiring that every person present have k acquaintances, postulating that for each of the people present the sum of the multiplicities of his acquaintances be k.

[26] On a method due to Galois, Quarterly Journal of Mathematics, 41, 1910, pp. 382-384. The theorem under consideration is: If Q is a subgroup of index μ of a finite group, then there is a common system of representatives x_1, x_2, ..., x_μ for righthand and

lefthand subgroups of Q. -- Also a generalization of this theorem

stemming from G. Scorza (A proposito di un teorema del Chapman,

Bolletino della Unione Mathematica Italiana, 6, 1927, pp. 1-6) is an

immediate consequence of Theorem 13. This theorem of Scorza's goes as

follows: If Q_1 and Q_2 are two subgroups of the same index μ of a finite

group, then there is a common system of representatives

x_1, x_2, ..., x_μ for the righthand subgroups of Q_1 and the lefthand

subgroups of Q_2.

27 Bemerkungen zur Kombinatorik im Anschluß an Untersuchungen

von Herrn D. König. (Remarks on Combinatorics in connection with

studies of Mr. D. König), Sitzungsberichte der Berliner Mathematischen

Gesellschaft, 32, 1933, pp. 60-75. This work also contains other

combinatoric and group theoretical applications of Theorem 13.

Chapter XII

Factorization of regular finite graphs of degree 3

§1.   Bridges and Leaves

As preparation for the main subject matter of this chapter we must first deal with material which relates not only to regular graphs of degree 3 but also to completely arbitrary graphs.

An edge b of the graph G which is not an end edge and does not belong to any cycle of G is called a bridge of G (see b in Fig. 74).[1]   It is clear that if G' is any subgraph of G which contains the bridge b of G, b is either an end edge or a bridge of G'. Since an end edge of a

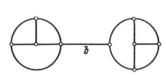

Fig. 74.

component of G is at the same time an end edge of G, it follows that a bridge of G is at the same time a bridge of that component B of G to which it belongs.

Theorem 1   If a bridge b = PQ of a connected graph G is removed, the resulting graph G' consists of exactly two components.

Proof   If the graph G' were connected, then it would contain a path from P to Q, and we would get from this, together with PQ, a cycle of G. If, on the other hand, A is any vertex of G, then there is a path

$$A\ B_1\ B_2\ \ldots\ B_n\ P$$

of G which joins A with P. If $B_n\ P \neq QP$ then this is a path in G', but if $B_n\ P = QP$, then $B_n = Q$, and

$$A \ B_1 \ B_2 \ \ldots \ B_{n-1} \ Q$$

is a path in G'. For every vertex A of G' there is therefore in G' either a path from A to P or from A to Q. So those two components of G' which contain P and Q, respectively, contain all vertices and therefore also all edges of G'.///

These two components U_1 and U_2 of G' will be designated as the two shores of the bridge PQ (with respect to G); one "originates from the endpoint P of the bridge", the other from the other endpoint Q. We shall also use these expressions if G is not connected. If the bridge PQ is contained in the component B of G, both shores of PQ with respect to B will also be called shores of PQ with respect to G. A shore will be called a leaf of the graph if it does not contain[2] any bridge of G.

From the definitions given above, the following theorems are immediately obtained:

Theorem 2   If the bridge PQ is not contained in the path P P_1 P_2 ... (i.e., $P_1 \neq Q$), then the vertices and edges of this path belong to the shore of PQ originating from P.

Theorem 3   If G is connected, then every connected subgraph of G which does not contain the bridge b of G is a subgraph of one of the two shores of b.

Everything said here about bridges and shores holds for both finite and infinite graphs. The following theorem, however, holds only for finite graphs:

Theorem 4   Each of the two shores of every bridge of the finite graph G contains a leaf of G[3] as subgraph.

Proof   In proving this we may assume that G is connected. If U is one shore of the bridge b of G, then U is either itself a leaf of G or it contains a bridge b_1 of G. Let U_1 be that shore of b_1 which does not contain b. Since U_1 is connected, a path goes out of every one of its vertices to one endpoint of b_1. If b_1 is added to this path, we get a path W. By Theorem 3, W belongs completely to one shore of b; but since W contains the edge b_1, this shore can only be U. But since -- also by Theorem 3 -- U_1 is a subgraph of a shore of b and U_1 has a vertex in common with W, this shore is also U. Consequently U_1 is a subgraph of U. Now either U_1 is bridgeless, in which case our goal is reached, or U_1 contains a bridge b_2 of G. Let U_2 be that shore of b_2 which does not contain b_1. By repetition of the above reasoning U_2 is a subgraph of U_1, etc. So we get the sequence U, U_1, U_2, ..., where every element is a subgraph of the preceding element and (since the edge b_i of U_{i-1} is not contained in U_i) contains at least one less edge than the preceding element. Since the number of edges is finite this sequence must end, and we thus reach a shore U_r contained in U as subgraph which is a leaf of G.///

So if a finite graph has a bridge, then it has at least two leaves.

In order to show that the theorem just proved does not hold for infinite graphs, let G be a doubly infinite path; then every edge is a bridge. All shores are singly infinite paths, and therefore have bridges, so that G contains no leaf at all. We now prove

Theorem 5 If a finite graph H contains exactly two leaves, B_1 and B_2, then every path W of G which joins a vertex of B_1 with a vertex of B_2 contains all the bridges of G.[4]

Proof First, let G be connected. We suppose that there is a bridge b of G not contained in W. It is clear that the subgraph G' of G, which consists of the edges of B_1, B_2, and W, and therefore does not contain b, is connected, so that G' -- by Theorem 3 -- belongs completely to a shore of b. But the other shore of b contains -- by Theorem 4 -- a leaf. And this would be a third leaf.

If G is not connected, then both leaves must belong to the same component G_1 of G, since -- by Theorem 4 -- a finite graph can never have a single leaf. So the remaining components of G have no leaves and therefore (Theorem 4) also have no bridges. So the proof has been reduced to the case of a connected graph.///

It is a good idea now to introduce another definition. Let R_1 S_1 and R_2 S_2 be two edges of G. We remove these two edges and introduce two new vertices, P and Q, and five new edges R_1 P, P S_1, R_2 Q, Q S_2, and P Q. We shall say that the graph H thus obtained from G comes from "joining the edges R_1 S_1 and R_2 S_2" (Fig. 75).

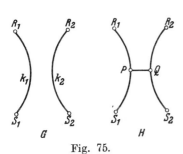

Fig. 75.

Theorem 6   If a finite graph G has exactly two leaves, B_1 and B_2, and an edge k_1 of B_1 is joined with an edge k_2 of B_2, then the resulting graph H is bridgeless.[5]

The second part of the proof of Theorem 5 shows that we may restrict ourselves to connected graphs here. Then there is a path W of G which joins an endpoint of k_1 = R_1 S_1 with an endpoint of k_2 = R_2 S_2, and which therefore -- by Theorem 5 -- contains all bridges of G. We may assume that W contains neither k_1 nor k_2, since this can be achieved by the possible removal of one or both endpoints of W. Then W, which joins, for instance, R_1 with R_2, is also a path of H. If the path R_1 P Q R_2 is added to it, then we get a cycle of H which contains all bridges of G and also PQ. So neither P Q nor the bridges of G are bridges of H. But neither can an edge k of G which is not a bridge of G be a bridge of H. The edge k is contained in a cycle of G, and this is either a cycle also of H, or becomes a cycle of H if R_1 S_1 is replaced

249

by R_1 P S_1 and R_2 S_2 is replaced by R_2 Q S_2. Finally, R_1 P, P S_1, R_2 Q,

and Q S_2 are not bridges of H. This can be seen, for example, for R_1 P

as follows. As an edge of a leaf, R_1 S_1 is not a bridge of G and so is

contained in a cycle of G; if in this cycle the substitution just

mentioned is carried out, then we get a cycle which contains R_1 P. So

Theorem 6 is proved.///

Theorems 1 - 6 remain valid also for graphs in the broader sense.

§2. Splitting of an edge
Frink's Theorem

This section is a preparation for the main contents (§3) of this

chapter.

Let x = M N be an edge of the arbitrary graph G, which 1) joins two

vertices of degree 3, and 2) is not contained in a 2-cycle (MN is the

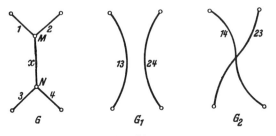

Fig. 76.

only edge of G which joins M with N). The edges 1 and 2 in G go to M

and edges 3 and 4 go to N. We remove from G the vertices M and N and

the five edges 1, 2, 3, 4, and x, and introduce in their place two new

edges, which we designate 13 and 24; the first joins the endpoints of

the path 1 x 3, and the second the endpoints of the path 2 x 4. Let G_1

be the resulting graph. If, on the other hand, M and N and 1, 2, 3, 4, and x are removed from G, and 14 and 23 are introduced as new edges, which join the endpoints of the paths 1 x 4 and 2 x 3, respectively, then we call the resulting graph G_2. Both G_1 and G_2 have two fewer vertices than G. We shall say that G_1 and G_2 come from G by <u>splitting</u> <u>the edge</u> x.[6] In the case where 1 and 3 have a common endpoint

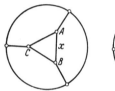

Fig. 77.   Fig. 78.

C, it is to be understood that the new edge 13 is a loop of G_1 at C. We get the graph of Fig. 78 with loop at C by splitting the edge x in the graph of Fig. 77. The result of the splitting of an edge x is possibly a graph only in the broader sense (if, namely, x is the edge of a triangle.) So in splitting edges, graphs in the broader sense must be considered.[7] By splitting an edge we get from a regular graph of degree 3 another such graph. If a loop results this loop ought always to be counted double. Let it be once again emphasized that splitting is done only of edges which do not belong to a 2-cycle.

Now we have Frink's[8] Theorem:

> <u>Theorem 7</u> Let G be a connected bridgeless regular graph of degree 3 and let x be any edge of G not contained in any 2-cycle of G. Let G_1 and G_2 be the regular graphs (in the broader sense) resulting from the two ways of splitting x. Then either G_1 or G_2 is at the same time connected, bridgeless, and a graph in the narrower sense.

251

<u>Proof</u> We divide the proof into five steps and use the notation (see
Fig. 76) introduced above. Let P, Q, S, and R be the endpoints,
distinct from M and N, of 1, 2, 3, and 4, respectively (Fig. 79).

First we prove:

a) If G_1 is not connected,
then it has exactly two
components; one containing
13, and the other 24.

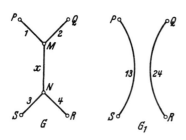

Fig. 79.

We assume a component B of G_1
contains neither 13 nor 24. Then B
is a subgraph of G. Let A ($\neq$ M) be
any one of its vertices. A path
joins A with M in G; this must

contain P, Q, S, or R. If P is the first of these four vertices on this
path going out from A, then a part of this path is a path which joins A
with P and does not contain the edges 1, 2, 3, 4, or x, and so is a path
of G_1. But then P (and therefore P S = 13) belongs to B. If it is not
P, but Q, S or R that is the first vertex with the property mentioned,
then either 13 or 24 belongs to B. This contradicts our assumption.
But now 13 and 24 can be contained in only separate components, so that
(a) is proved.

b) Neither 13 nor 24 is a bridge of G_1.

We assume that 13 is a bridge of G_1. Then there is no path from P
to S in G_1 which does not contain 13 (this path with 13 would give a
cycle containing 13). Since 1 is not a bridge of G, there is a path in
G not containing 1 and going from P to M; this must contain Q, R, or S.
Going out on this path from P, either Q, R, or S can be the first of
these vertices, which we meet. If S, a part of this path would also be
a path of G_1 which joins P with S and does not contain 13, and 13 would

not be a bridge of G_1. If Q or R, a subpath contained in G_1 and not containing 13 and 24 goes from P to Q or to R. In like manner we get a path from S to Q or to R. Certain edges of these two paths generate (Theorem I.5), a path of G_1, possibly by adding QR = 24, which does not contain 13 and which goes from P to S, and again 13 would not be a bridge of G_1. In the same way the assumption that 24 is a bridge leads to a contradiction.

      c)   If $b = A_1 A_2$ is a bridge of G_1, then one shore of b

            contains the edge 13, the other shore the edge 24.

By (b) the edge b belongs to G and is therefore contained in a cycle K of G, since end edges are excluded. The cycle K must contain 1, 2, 3, or 4 and therefore P, Q, S, or R, since otherwise it would not also contain x and would be a cycle of G_1 which contains b. Proceeding from A_1 in the direction on K which does not lead directly to A_2, let P be the first of the vertices P, Q, R, and S reached. The path $A_1 \ldots$ P so described is also a path of G_1 and does not contain the bridge b. By Theorem 2, P (and therefore also 13) belongs to the shore of b originating from A_1. Proceeding from A_2 in the other direction on K, the vertex Q or R is reached first (otherwise, if P or S were reached sooner, this part of K together with b and the path $A_1 \ldots$ P -- possibly with the addition of the edge P S = 13 -- would give a cycle of G_1 containing b). But then 24 also belongs to the other shore of b.

By (b) there is a cycle K_1 of G_1 which contains 13, and a cycle K_2 of G_1 which contains 24. We prove:

      d)   If G_1 is not both connected and bridgeless, then the cycles

            K_1 and K_2 are always mutually disjoint.

If G_1 is not connected, then (Theorem I.26) K_1 and K_2 each belong completely to a component of G_1. By (a) these components for K_1 and for K_2 are distinct from each other and therefore disjoint.

If G is connected and has the bridge b, then it follows from
Theorem 3, since b is not contained in any cycle, that K_1 and K_2 each
belong completely to a shore of b. By (c) these shores for K_1 and for
K_2 are distinct and therefore disjoint.

Now we come to the fifth step of our proof. We assume that G_1 is
not both connected and bridgeless. If the edges 13 and 24 are removed
from K_1 and K_2, respectively, then we get the paths W_1 = P ... S and
W_2 = Q ... R of G_1 which by (d) are disjoint from each other. These are
also paths of G and also of G_2. The paths W_1 and W_2 give, with the

edges 14 and 23 of G_2, a cycle of G_2,
which contains 14 and 23 (Fig. 80).
If (d) is applied to G_2 instead of to
G_1 (where 14 and 23 replace 13 and
24), then G_2 is both connected and
bridgeless. So we have proved that
if G_1 is not both connected and
bridgeless, then G_2 must have both
these properties.

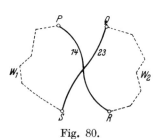

Fig. 80.

It now remains to be shown only that if G_1 is connected and
bridgeless, then G_1 is a graph in the narrower sense. If there were a
loop at a vertex A of G_1, then -- using the first way of counting loops
-- there would have to be exactly one edge of G_1 (not a loop) going to
A, and this edge would have to be a bridge of G_1, since it cannot be
contained in any cycle of G_1.

Frink's Theorem (Theorem 7) is thus proved.///

§3. Petersen's Theorem

In the case of finite regular graphs of degree 3 Petersen succeeded in finding a simple necessary condition for primitivity. This most complicated result of Petersen's paper goes as follows:

> Theorem 8 If a finite regular graph of degree 3 has at most two leaves, then it splits into two factors.[9]

Frink's Theorem (Theorem 7) forms an essential basis for the proof. If a regular graph of degree 3 splits into two factors, then one is of degree 1 and the other of degree 2. If all components split into factors, this also holds for the original graph. On the other hand every component of a graph which has at most two leaves has likewise at most two leaves. So we may assume that the graph is connected. If such a factorization is given, then the edges of the factor of degree 1 are designated as red (r), and the edges of the factor of degree 2 as blue (b). (Since -- by Theorem I.28 -- the blue edges form cycles, a bridge must always be red.) If the edges in a cycle of a factored graph are alternately red and blue, then we shall call the cycle an alternating cycle.[10] If the colors of all edges of an alternating cycle are switched without altering the color of the other edges, then the new coloring determines a factorization, since one red and two blue edges also go to each vertex according to this new coloring. This observation of Petersen's plays an important role now.

In order to get to Petersen's Theorem we must first prove:

> Theorem 9 Let G be a finite, connected, bridgeless regular graph of degree 3, which is split into two factors, b and r. Then every edge x of G is contained[11] in an alternating cycle of G.

Proof We suppose that the theorem is not generally true, and we let G be a graph with a minimum number of vertices for which the theorem does not hold.

We consider first the case where G contains a 2-cycle A B A. If the third edges AC and BD, which go to A and B, respectively, are identical, then the entire graph -- since it is connected -- consists of these three edges, and all three edges are contained in an alternating cycle ABA (Fig. 81). So we may assume that AC and BD are distinct. Then C ≠ D, since otherwise the third edge going to C = D would be a bridge. We remove

Fig. 81.

from G both edges AB and the edges AC and BD (as well as the vertices A and B), and introduce a new edge CD. The resulting graph G'[12] is, of course, regular of degree 3. It is also connected, for if P and Q are any two vertices of G', then there is in G a path W which joins them; now either W is also a path in G' or it contains a subpath CABD; if this is replaced by the new edge CD, then we get a path in G' which joins P with Q. Finally, G' is also bridgeless; it can be easily seen that CD is not a bridge of G'; but if PQ is an arbitrary edge of G', which is distinct from CD, then it is contained in a cycle K of G; now either K is also a cycle of G' or it contains a path CABD; if this is replaced by CD, then we get a cycle of G'. It follows from "bridgelessness" for regular graphs G' of degree 3 -- as we saw above -- that G' is a graph in the narrower sense. Now there are two possibilities, depending on whether both edges AB have different colors or the same color b:

α) AC and BD are both b and

β) AC and BD are both r (Fig. 82).

In case (α) we give the new edge CD the color b, in case (β) the color r, while we give the remaining edges of G' the same color which they had as edges of G. Then G' is clearly split into two factors. So

256

summarizing, we can say that G' is also a finite, connected, bridgeless, regular graph of degree 3 which was split into two factors b and r. Since G' has two fewer vertices than G, every edge of G' is -- by assumption -- contained in an alternating cycle K of G'. If CD is not contained in K, then K is also an alternating cycle of G; but if CD is contained in K, then we replace CD in K by CA, $(AB)_1$, and BD [where we understand by $(AB)_1$ an arbitrary one

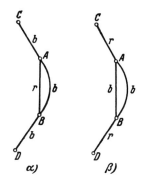

Fig. 82.

of the edges AB in case (β) and that one of the edges AB in case (α) whose color is different from the color of AC and BD, see Fig. 82], by which we get an alternating cycle of G containing CABD.

Those edges of G which at the same time belong to G' are therefore contained in an alternating cycle of G. But this holds also for the edges AC and BD; we get such a cycle if in the alternating cycle of G' which contains the edge CD this edge is replaced by CA, $(AB)_1$, and BD. Here an arbitrary edge of the two edges AB can be chosen for $(AB)_1$, in case these (case (β)) have the same color (b), so that in this case both edges AB are contained in an alternating cycle of G. But this is also the case if they have different colors (case (α)), since then ABA is such a cycle. So Theorem 9 would hold for G in contradiction to our assumption.

We may therefore assume that G does not contain a 2-cycle. Let x = ST be an arbitrary b-edge of G, which we wish to prove is contained in an alternating cycle of G. Let y = SA be the other b-edge which goes

to S and z = AB be the r- edge which
goes to A (Fig. 83). The third edge
going to A we denote by PA, the
second and third edges going to B by
QB and RB, respectively. Since z is
not contained in any 2-cycle, it can
be split. Let G_1 (which contains the
edges PQ and SR and not PR and SQ) be
that one of the two graphs which we
get from this splitting which --
corresponding to Frink's Theorem

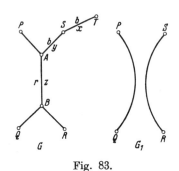

Fig. 83.

(Theorem 7) -- is connected and bridgeless (and does not contain a
loop). We now give each edge of G_1 the same color (b or r,
respectively), which it has as an edge of G, while the new edges PQ and
SR are colored blue. Since all adjacent edges of z in G are blue, it
can be immediately seen that G_1 is split into two factors by this
coloring.

Since the vertices P, Q, R, S and T are not necessarily completely
distinct from one another, it is possible -- this case ought to be
considered first -- that x is identical to one of the five edges AB, PA,
SA, QB, and RB, which are not contained in G_1. The edge x cannot be
identical to AB, since it is red; by definition of y, x = SA is also
impossible. If it were true that x = AP (and therefore T = A), then x
and y would form a 2-cycle. If it were true that x = BR, then we would
have that S = R (since, in case S = B, the edges y and z would form a 2-
cycle), and SR would be a loop of G_1. So there remains only the fifth
possibility x = BQ. Since S ≠ B, in this case T = B (Fig. 84). Since
G_1 has two fewer vertices than G, there is in G_1 an alternating cycle K
which contains the b-edge PQ. Since two b-edges of K cannot be

258

adjacent, SR (≠PQ) is not contained in K. So if the b-edge PQ in K is
replaced by PA, AB, and BQ, which are in order b, r, b, then we get an
alternating cycle of G containing the edge x = BQ. (This idea holds
also in the case where P = R along with Q = S; Fig. 85 shows that this
case can actually happen.)

   We consider now the second case, in which the b-edge x of G is an
edge of G_1. By our assumption about G there is an alternating cycle K
of G_1 which contains x. This cannot contain the blue edge SR adjacent
to x (see Fig. 83). So either K is also an alternating cycle of G or,

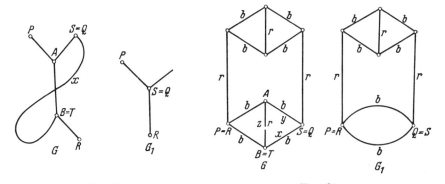

Fig. 84.   Fig. 85·

if it contains PQ, we get such a cycle from it if PQ is replaced by PA,
AB, and BQ. So we have proved that every b-edge x of G is contained in
an alternating cycle of G.

   There would then have to be an r-edge x = AB in G not contained in
any alternating cycle, but this also leads to a contradiction. If
y = BC is a blue edge adjacent to x, then, as we saw above, y is

contained in an alternating cycle. But this must also contain x, since
x is the only r-edge which ends in the endpoint B of y.

So Theorem 9 is proved.///

Following Frink's order of proof, we now prove the following
theorem, which expresses the most essential particular case of
Petersen's Theorem.

> Theorem 10   Every finite bridgeless regular graph G of degree 3
> splits into two factors.

Proof Let G be the graph with the minimum number of vertices for which
the theorem does not hold. This graph is, of course, connected, so that
there cannot be three edges in G which join the same pair of vertices,
since otherwise G would consist only of these three edges (see Fig. 81
above) and would split into two factors. But then there is in G one
edge which is not contained in any 2-cycle. If this is not the case for
x = AB (see Fig. 86), then there is a second edge y which joins A with
B, and so let z be the third edge which goes to A. The other endpoint
of z is then distinct from B. Since there is no fourth edge that goes
to A, z is an edge not contained in any 2-cycle.

Let x = MN be an edge of G not contained in any 2-cycle. Let the
other edges going to M and N be 1 = PM, 2 = QM, 3 = SN, and 4 = RN. We
split the edge x; of the two graphs which we can get out of G, let G_1
(which contains the edges 13 = PS and 24 = QR) -- corresponding to
Theorem 7 -- be connected, bridgeless, and a graph in the narrower
sense. Since G_1 is also regular of degree 3 and has two vertices fewer
than G, it splits, by assumption, into two factors, b and r. We show
now that this, in contradiction to our assumption, is then also the case
for G. We temporarily give those edges which also belong to G_1 the same
color b or r in G, which they have as edges of G_1. As for the colors of
1, 2, 3, 4, and x, three cases are to be distinguished:

260

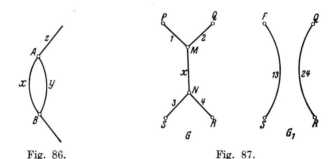

Fig. 86.   Fig. 87.

α) 13 and 24 are both blue; then let 1, 2, 3, 4, and x be, in order, b, b, b, b, and r.

β) one of the edges 13 and 24 is blue, the other red; if, say, 13 is blue and 24 red, let 1, 2, 3, 4, and x be, in order, b, r, b, r, and b.

It is immediately seen that in both cases the given coloring -- in contradiction to our assumption -- defines a factorization of G into two factors.

γ) 13 and 24 are both red.

By Theorem 9, the edge 13 is contained in an alternating cycle of G_1. If the colors of the edges of this cycle are switched without altering the colors of the remaining edges, then the new coloring also defines a factorization of G_1 where 13 is blue. Depending on whether or not 24 changes its color in this color switch, we have reduced the situation to case (α) or (β), already taken care of above.

So Theorem 10 is proved.///

Now we prove Petersen's Theorem (Theorem 8) in the following way. If

261

the graph G is bridgeless, then Theorem 10 implies its correctness. A
graph cannot have a single leaf (Theorem 4). There remains only the
case where G has exactly two leaves. Let AB be an edge of one leaf and
let CD be an edge of the other. We
"join" these two edges with an edge
PQ (Fig. 88). We get a finite
connected regular graph G' of degree
3. By Theorem 6 it is also
bridgeless. Theorem 10 then implies
that it splits into two factors, b
and r. For G, then, a factorization
can be given as follows. We
temporarily give those edges of G,

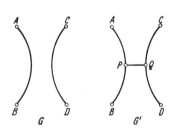

Fig. 88.

which also belong to G', the same color, b or r, which they have as
edges of G'. As for the colors of AB and CD, we distinguish two cases.
α) PQ is red (and so AP, PB, CQ, and QD are blue); then let both AB and
CD in G be blue. It is immediately seen that a factorization for G is
defined by this. β) PQ is blue; by Theorem 9 PQ is contained in an
alternating cycle of G'; we switch the colors of the edges of this
cycle, and so PQ becomes red, and we are reduced to case (α).

Peterson's Theorem is proved.///

Recently a new proof of Petersen's Theorem by Schönberger [1]
appeared; using Frink's Theorem he proved first, in a way similar to the
way Theorem 9 was proved by Frink (and also here), the following
theorem:

Every finite connected bridgeless regular graph of degree 3 can
be split into two factors in such a way that two arbitrarily
chosen edges are blue.

Petersen's Theorem is then deduced from this theorem in a way similar to the way it was deduced from Theorem 9. The proof by Schönberger has the advantage of not having to work with alternating cycles; through this it is shown that the necessary color changes in the reduction occur in a (previously determined) restricted part of the graph.

§4. Supplements to Petersen's Theorem

Petersen's Theorem states a condition by which a graph of degree 3 splits into two factors. If this is the case, and all cycles of the factor of degree 2 contain an even number of edges, then it also splits into three factors. The latter happens if the factor of degree two is one cycle, and if the graph therefore has [13] a Hamiltonian cycle. If the graph has α_0 vertices, then $3\alpha_0$ is double the number of edges, so that α_0 -- the number of edges of the Hamiltonian cycle -- is even (cf. Theorem II.3).[14] But not every graph of degree 3 splits into factors of degree 1, even if it has at most two leaves. Fig. 89 shows the simplest example.

No finite[15] regular graph which has a bridge can split into factors of degree 1. If in a factorization into factors of degree 1, G_1 is that factor which contains a bridge g, and G_2 is any second factor of this factorization, then the product $G_1 G_2$ is a factor of degree 2 which contains g; so g would have to be (Theorem I.28) contained in a cycle. On the other hand we have seen (Theorem XI.14) that a finite regular bipartite graph splits into factors

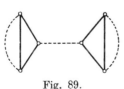

Fig. 89.

of degree 1, and so a finite regular bipartite graph G cannot have[16] a bridge. This theorem can also be proved directly in the following way.

<u>Proof</u> Let g = PQ be a bridge of G and let U be the shore of g

originating from the vertex P, so that Q does not belong to the shore

U. As a subgraph of G, U is also a bipartite graph, and so the vertices

of U (Theorem XI.12) split into two classes I and II in such a way that

the two endpoints of every edge of U belong to different classes.

Let μ and ν be the number of vertices of I and II, respectively, and let

P belong to I. If now d (> 1) is the degree of G, then d μ - 1 and d ν

must be the number of edges of U. But an equation d μ - 1 = d ν is

impossible.///

That a finite regular graph of even degree cannot have a bridge can

be proved not only using Theorem XI.7, but also directly -- with the

help of Theorem II.3. We have the more general theorem that an Euler

graph can have no bridges. (Theorem II.1 can be sharpened in the

following way: every edge of an Euler graph is contained in a cycle of

the graph.)

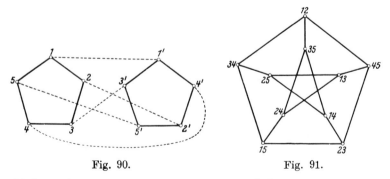

Fig. 90.   Fig. 91.

It is an important discovery of Petersen [2] that bridgeless

264

regular graphs of degree 3 also exist which do not split into factors of degree 1. His example is shown[17] in Fig. 90. It is easy to convince oneself of the impossibility of such a factorization. This graph splits -- according to Petersen's Theorem -- into two factors: in Fig. 90 the red edges are drawn dotted.

The Petersen graph (just like the system of edges of the regular solids) shows noteworthy and varied properties of symmetry, so that it would be profitable to study it more closely, particularly its group -- as this concept was defined in Chapter I (at the end of §1). In this study we shall use -- referring to the color order theory given by A. Kowalewski [2] -- the following idea: we let each vertex correspond to one of the ten combinations 12, 13, ..., 45 which can be formed from the five elements 1, 2, 3, 4, and 5 and join two vertices by an edge if and only if the two corresponding combinations have no element in common (see Fig. 91, where a somewhat modified form was chosen for the graph). That the resulting graph -- as is easily seen -- has the same structure as the Petersen graph was recognized by G. Kowalewski [2, p. 88]. We shall return to other representations of the Petersen graph in the next section (see Figures 96-100). Nothing is known about how the Petersen graph can be extended to regular graphs of degree 5, 7, 9, ... Petersen [1, §14] expressed the conjecture that these graphs can be primitive if and only if they contain bridges, but he "found difficulties that were too great and restricted his study to graphs of degree 3." At the end of his paper Petersen wrote: "It appears that the way followed here can lead to the goal" (i.e. for odd degrees > 3). In spite of the more than 40 years that have elapsed since then and in spite of simplifications (in particular, the ones found by Frink), which have certainly made the Petersen way more practicable, this goal has not yet been reached.

Certainly certain results concerning the factorization or
primitivity of regular finite graphs of arbitrary degree can easily be
proved[18] in connection with the bridges of the graph.

We single out especially the following theorem:

Every factor of odd degree of a regular finite graph G
contains every bridge of G.

Proof We may assume in the proof that the degree of G is odd, since an
Euler graph, as we have seen, has no bridges. Let F be an arbitrary
factor of odd degree of G; so G = FF', where F' is of even degree.
Since F' cannot have a bridge (i.e. every edge of F' is contained in a
cycle of F' and therefore of G) all bridges of G are contained in F.///

The theorem just proved admits the following conclusions. If a
regular finite graph with at least one bridge can be factored, then at
most one of these factors can be of odd degree. Otherwise two factors
of odd degree would have to contain the bridge. If ν bridges go to the
same vertex of a regular finite graph, then the factor of odd degree has
degree at least ν. The ν bridges must be contained in this factor. In
particular, we get from this the following result proved in Chapter XI,
§3:

If nothing but bridges goes to the same vertex of a
regular finite graph of odd degree, then the graph is primitive
(cf. Fig. 68 and 69).

§5. Connection with relative graph theory
---- The Four Color Theorem.

The factorization of regular finite graphs of degree 3 is very
closely connected with important studies of relative graph theory.
These lie outside the scope of this book, so that we shall indicate this
connection only briefly and without proofs. If we speak of graphs on

surfaces, it is to be understood that we are interpreting the vertices
as geometrical points and the edges as sets of points and, in fact, as
topological images of intervals (of the line). We restrict ourselves
here to finite graphs and to surfaces of finite cardinal number (of
finite genus), and we assume the basic concepts of the topology of
surfaces are known.

Of fundamental significance here is the generalized Euler Theorem
on polyhedra[19] :

> If the surface F with genus g is divided into α_2
> elementary surfaces by a graph G, which has α_0 vertices
> and α_1 edges, then the following equation holds:
>
> $$\alpha_0 - \alpha_1 + \alpha_2 = 2 - g.$$

(If F is a closed orientable surface of genus p, then it is well
known that g = 2p.)

An important problem of graph theory on surfaces is:

If a graph G and a surface F are given, how can it be decided
whether or not[20] G can be drawn on F, i.e. whether or not a subset of F
is homeomorphic to G?[21] This question is of particular significance if
F stands for the (closed) orientable surface of genus 0, the (function
theoretical or Euclidean) plane (the surface of a sphere), which is to
be distinguished topologically from the projective plane.

In this case we can show that every tree can be drawn on the
surface of a sphere and, in fact, on any graph which does not have a
"θ-shaped" subgraph; a graph is called θ-shaped if it is homeomorphic to
the graph which has two vertices with three edges joining them (Fig.
81).[22]

We give now the two simplest examples of graphs which cannot be
drawn on the sphere. The first (M) is obtained by joining in the system
of edges A_1 A_2 B_1 B_2 of the tetrahedron two opposite edges A_1 A_2 and

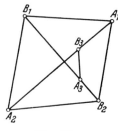

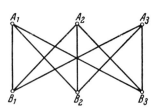

Fig. 92a.　　　　　　　　Fig. 92b.

B_1 B_2 by a new edge A_3 B_3 (Fig. 92a). This graph can also be obtained
as follows: two point triples A_1, A_2, A_3 and B_1, B_2, B_3 are chosen as
vertices, and each point of the first triple is joined to each point of
the second triple by an edge (Fig. 92b). The same graph can also be
defined as a hexagon A_1 B_2 A_3 B_1 A_2 B_3 A_1 with its three diagonals
A_1 B_1, A_2 B_2, and A_3 B_3.[23] The second example (K) is obtained by
joining every pair of five vertices by an edge (Fig. 93).[24]

1) The following two theorems[25]
can easily be proved:

If p is chosen large enough an
arbitrary graph can be drawn[26] on
a (closed) orientable surface of
genus p.

2) If F is a closed orientable
surface of smallest possible
genus on which a given connected

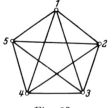

Fig. 93.　　　　graph can be drawn, then it

splits this surface into elementary surfaces.

If p is the smallest number with the property that the graph G can
be drawn on an orientable surface of genus p, then p can be called the

genus of G. In particular, a graph is of genus zero if it can[27] be
drawn in the plane (on a sphere).

With the help of the generalized Euler theorem on polyhedra, a
graph which can be drawn on an orientable surface of genus p, whose
genus is therefore $\leq$ p, can be easily characterized[28] by internal
(absolute) properties. This condition is, even for p = 0, too
complicated to be used to advantage. But for the special case, where
p = 0, Kuratowski [1] succeeded in solving the problem in a very elegant
way. Kuratowski's Theorem goes as follows:

> A graph is of genus zero if and only if it does not have
> a subgraph homeomorphic to either the graph M of Fig. 92 or the
> graph K of Fig. 93.

If the degrees of all its vertices are at most 3, so that the graph
is regular and of degree 3, then it can never contain a subgraph which
is homeomorphic to K. Such a graph is, by Kuratowski's Theorem, of
genus zero if and only if it does not contain a subgraph homeomorphic to
M. This partial result of Kuratowski's discovery was also found at the
same time by Menger [3]. Whitney [3, Theorem 29] gave a different kind
of necessary and sufficient condition for a graph to be of genus zero.
He defines the "duality" of two graphs by interior (combinatorial)
properties and shows that a graph is of genus zero if and only if it has
a dual graph. With the help of this result Whitney [4] gave a new proof
for Kuratowski's Theorem.

It would be interesting to study whether a theorem analogous to
Kuratowski's holds for arbitrary genus numbers p.

We content ourselves here with illustrating Kuratowski's Theorem by
two examples. The first example is obtained by putting four cubes
together. This graph, which represents a part of the three-dimensional

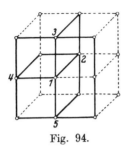

Fig. 94.

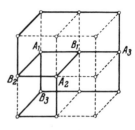

Fig. 95.

cubic lattice, contains a subgraph homeomorphic to K. This is represented by the heavy lines of Fig. 94. The same graph also has a subgraph homeomorphic to M; see the heavy lines of Fig. 95. For the second example we consider the Petersen graph P (Fig. 90, 91). It can easily be seen that it cannot be drawn on the sphere, and so it must have a subgraph homeomorphic to M. This can be verified in the following way. We divide the three edges $A_1 B_1$, $A_2 B_2$, and $A_3 B_3$ of M each into two edges by C_1, C_2, and C_3, respectively, and join these three vertices with a tenth vertex D by one edge each. The resulting graph (Fig. 96) has the same structure as P. This is seen immediately if we let the vertices

$$A_1, A_2, A_3, B_1, B_2, B_3, C_1, C_2, C_3, D$$

correspond, in order, to the vertices

1, 3', 2', 5', 2, 1', 5, 3, 4', 4

of P in Fig. 90.

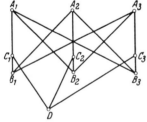

Fig. 96.

We would also like to give simple examples of surfaces on which the Petersen graph can be drawn. Such a surface is the projective plane, shown in Fig. 97, where all five sides of the pentagon 1' 3' 5' 2' 4' 1'

go to infinity. This figure is nothing more than the projection of the
(regular) dodecahedron from its center to the plane (considered as
projective) of one of its surfaces. This connection between the
Petersen graph P and the system of edges of the dodecahedron[29] is
obtained from the following property of the latter, found by A.
Kowalewski [1]: each of the ten pairs 12, 13, ..., 45 can be assigned
to the 20 vertices of the dodecahedron in such a way that

1. the same pair is assigned to two opposite vertices and only to
such vertices, and

2. two vertices are joined by an edge if and only if the two
corresponding pairs have no common element.

It is easily seen that the graph (which has α_o = 10 vertices and
α_1 = 15 edges) divides the projective plane into α_2 = 6 pentagons (I,
II, III, IV, V, VI); this corresponds to the Euler Theorem on polyhedra,
which says that for the projective plane (which has genus 1)

$$\alpha_o - \alpha_1 + \alpha_2 = 1.$$

If an elementary surface is cut out of the projective plane, a
well-known surface is obtained homeomorphic to the Möbius strip; so the
graph P must also be able to be drawn on the Möbius strip. Fig. 98
shows this, where MN is to be identified with M' N', M coinciding with
M', A with A', etc. Here the rectangle M N M' N' does not belong to the
graph. This representation can be accomplished in such a way that the
edge of the strip completely belongs to the graph; see Fig. 99, where
the pentagon 1 2 3 4 5 1 is the center line and the other pentagon

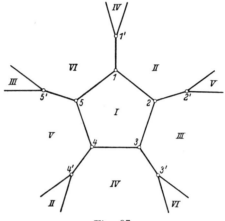

Fig. 97.

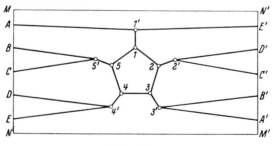

Fig. 98.

1' 3' 5' 2' 4' 1' is the edge of the Möbius strip. In perspective

representation -- for which I thank Mr. E. Egerváry -- this is shown in

Fig. 100. Here the Möbius strip is divided into five pentagons (I-V).

272

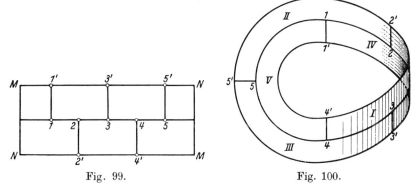

Fig. 99.   Fig. 100.

Our graph P (like the graphs of Fig. 92,93, and 94) can also be drawn on the surface of a ring; see Errera [1, Fig. 29].

An important and much treated problem of graph theory on surfaces is the problem of map coloring. This is the problem of relative graph theory most closely connected with the factorization of regular finite graphs of degree 3, and particularly with Petersen's Theorem. The Four Color Theorem, found empirically and still unproved,[30] is:

> If the surface of a sphere is divided into countries by an arbitrary (finite[31]) graph, then one of four colors can be assigned to each country in such a way that any two countries, which have a common border edge, have different colors assigned to them.

The difficulty of the problem was first recognized[32] in 1878 by Cayley. Cayley also showed that a proof may[33] be restricted to regular

graphs of degree 3. It may also be assumed that the graph does not have
a bridge; if a graph is drawn on a surface, the same country lies on
both sides of a bridge, so that the system of borders of a real map
cannot have a bridge. Now Tait has proved [1, §10 and 2, §15] the
theorem:

> If the surface of a sphere is divided into countries by a
> bridgeless regular graph G of degree 3, then these countries
> can be colored with four colors if and only if G splits into
> three factors.

With this Tait proved that the Four Color Theorem is equivalent to
the following theorem (called Tait's Theorem):

> Every regular bridgeless graph of degree 3 and of genus 0
> splits into three factors of degree 1.

This theorem was proved neither by Tait nor anyone else. If the
condition that the graph be of genus 0 is not included -- Tait
originally expressed it in this generality [1] and believed he had
proved it -- then the theorem becomes false. The example of the
Petersen graph P [Fig. 90] shows this. The Petersen graph, however,
does not contradict Tait's Theorem formulated here, and therefore does
not contradict the Four Color Theorem, since the Petersen graph, as was
already mentioned, cannot be drawn on the surface of a sphere.

These remarks let the close connection between the Four Color
Problem and Petersen's Theorem be recognized; even today this theorem
can be considered[34] the fundamental step in the direction of the
solution of the map coloring problem. Certainly the above stated
theorem of Kuratowski awakens new hopes, since it makes possible the
Four Color Theorem without the concept of surfaces, as a theorem of
absolute graph theory.

These remarks let the close connection between the Four Color
Problem and Petersen's Theorem be recognized; even today this theorem

This formulation is:

274

If a bridgeless regular (finite) graph of degree 3 does not have a subgraph homeomorphic to the graph M (Fig. 92), then it splits into three factors of degree 1.

Notes on Chapter XII

[1] Tait appears to have been the first to consider such edges, without giving them any particular designation [1, Fig. 2]. A proper designation (isthme und chemin isolant) was introduced by Lucas [1, Vol. IV, p. 235]. In English the word isthmus is used.

[2] This definition stems from Brahana [1] (leaf). In Petersen [1], who introduced the designation "Blatt," every shore (Ufer) is designated as a leaf (Blatt). The French designation is feuille.

[3] Errera. [1, p. 14].

[4] Errera [1, p. 15]. There also is found the following converse: If a path contains all bridges, then the graph has at most two leaves.

[5] Errera [1, p. 14].

[6] This concept, without any designation, stems from Frink [1].

[7] In every reduction process which may be chosen for the material of this chapter it turns out that the graphs cannot be restricted to graphs in the narrow sense: "on crée parfois des boucles, qu'on le veuille ou non" -- as was stressed by Errera [1, p. 13]. The consideration of graphs in the broader sense will not, however, affect the validity of the following proofs, since these proofs depend on the theorems of the first chapter and on the preceding theorems of this chapter, and these theorems -- as was already mentioned -- retain their validity for graphs with loops. Let it be also mentioned that the splitting of an edge will be used only for graphs in the narrower sense.

[8] Frink [1, Theorem I]. Frink's condition, that the graph have more than two vertices, is superfluous, if -- as is necessary -- the splitting of edges is restricted to such edges, which are not contained in a two-cycle. Frink's proof seems to us to be much too short to be convincing. The proof following here I owe to the

assistance of Mr. Tibor Schönberger, cf. his work which has meanwhile appeared:

Schönberger [1]. -- Frink says of his proof: "it is unfortunate that such a trivial and obvious (?) theorem should require such a long proof." Of course, this holds to an even greater degree for the proof following here. Frink's Theorem in itself would probably not deserve the trouble which is necessary for its proof; but we shall see that it forms the foundation for the simplest (Frink's) proof of an important theorem (see further under Petersen's Theorem).

[9] Petersen's proof [1, §§14-20] is long and extraordinarily complicated. It was simplified first by Brahana [1] and then by Errera [1, pp. 15-20 and 2]. Cf. also Sainte-Laguë [4, pp. 27-29 and 3]. Finally Frink [1] succeeded in reducing the theorem in an elegant and comparatively easy way to Theorem 7, which stems from him, in which -- in contrast to all his predecessors -- he was able to avoid any counting process. The proof following here essentially agrees with Frink's proof. That it is longer and more complicated is partly explained by the fact that Frink did not take into account sufficiently the circumstance that the "splitting of an edge" actually loses its meaning when the edge is contained in a 2-cycle (cf., nevertheless, the footnote following below on p. 188). Also it seems to us indispensable to treat separately certain possibilities, which involve the edge x of Theorem 9. -- We shall return to the most recent proof of Petersen's Theorem 8, which was given by Schönberger [1].

[10] In Petersen it is called a Wechselpolygon.

[11] Frink [1, Theorem II].

[12] We could say: "G' results from G by splitting one of the edges AB." If the concept of splitting would be extended in this way also to 2-cycle edges, then the proof following here would show that

Frink's Theorem is also true in this setting. It is to be understood
that in the splitting of a 2-cycle edge both kinds of splitting produce
the <u>same</u> graph $G_1 = G_2 = G'$. Even if this generalization corresponds to
a certain degree with the concrete interpretation of "splitting," it
nevertheless seemed clearer to us to avoid it. Our proofs could be made
more uniform by this generalized concept of splitting, but not simpler.

[13] de Polignac [2].

[14] As for the converse, however, not every factorization into
three factors gives a Hamiltonian cycle; for every two of these three
factors disconnected products can occur, as Fig. 72 shows; there are, in
fact, graphs which split into three factors of degree 1 and have no
Hamiltonian cycle at all; see Figures 10 and 14, as well as Fig. 107 in
Chapter XIV.

[15] The theorem does not hold for infinite graphs.

[16] Sainte-Laguë [2, p. 42] -- This theorem also does not hold
for infinite graphs.

[17] A second similar example stemming from F. Fitting is given
in Ahrens [2, Vol. 2, pp. 362-363].

[18] I owe the results of this section, for the case where the
degree of the graph is odd, to Mr. T. Grünwald.

[19] Concerning its history see, for example, Dehn and Heegaard
[1, Footnote 99, p. 199].

[20] It is always to be understood that no new cut points arise.
[Translator's note: The edges must not cross one another.]

[21] For the case where G is the complete graph with n vertices
and F stands for an orientable surface of given genus p, Heffter [1]
dealt with this problem -- in connection with the studies of Heawood [1]
and with the so-called Steiner problem of triple systems. For certain
Cayley color groups Maschke [1], Hilton [1, p. 90, Ex. 16] and Baker [1]

stated and dealt with the problem of drawing it on the sphere, on the projective plane, and on the surface of a ring (p = 1).

[22] For the history of these theorems (Mazurkiewicz, Ważewski , Menger, Ayres) see Kuratowski [1, p. 271].

[23] That this graph M cannot be drawn in the plane implies the unsolvability of the following long-known problem: "In the plane are three houses and three wells: the problem is to join every house with every well by nine paths, in all, so that no two of these paths cross each other." --

Errera [3] proved the following generalization: "If the points A_1, A_2, ..., A_u; B_1, B_2, ..., B_v lie in a plane, then exactly 2u + 2v - 4 of the uv edges $A_i B_j$ -- and not more -- can be drawn without any two of the edges crossing each other in the plane (u > 1, v > 1).

[24] The assertion that this graph K cannot be drawn in the plane is the dual counterpart of the so-called Weiske Theorem, which states that it is impossible to give five regions in the plane such that any two of these regions touch each other along a line; see Baltzer [1]. (In the projective plane there are even six such regions; for more on this see Fig. 97 below.)

[25] See König [1].

[26] This theorem was already stated and applied in the (false, of course) work of Petersen [3, p. 420].

[27] Sainte-Laguë [4, p. 6] designates such graphs as _sphérique_, Menger [3] calls them _plättbar_, and Whitney [3] calls them _planar_; Kuratowski [1] refers to all other graphs as _gauche_.

[28] See König [2].

[29] In analogy to the designation "double surface" (Doppelfläche) (a nonorientable surface) this connection could be expressed in the following way: the system of edges of the dodecahedron

is the "double graph" (Doppelgraph) of the Petersen graph.

[30] (Translator's note) The Four Color Theorem was proved by Kenneth Appel and Wofgang Haken in 1976. See Appel and Haken, Every planar map is four colorable, Bulletin American Math. Soc. 82 (1976), 711-712 and also Illinois J. of Math. 21 (1977), 429-567.

[31] With the help of the Infinity Lemma (Theorem VI.6) it can be proved that the minimum number of necessary colors is not greater if infinite graphs are also allowed; see König [8, §3].

[32] Since then a large literature of the subject matter has come about. A bibliography of the Four Color Theorem is to be found in Errera [1]. It was added to, for the years 1921 through 1926 by Reynolds [1, p. 481]; cf. also Errera [4] and Sainte-Laguë [5]; the greatest part of this last named work is a report on the map coloring problem. -- That five colors are always enough for the surface of a sphere, could be proved (but is of little interest for graph theory); the proof of this assertion, stemming from Kempe [1] and Heawood [1] is also to be found in Errera [1, p. 38] and in Rademacher and Toeplitz [1, §12a].

[33] The assertion of Ahrens [2, Vol. 2, p. 215 and p. 221] that three colors are always enough in this case, is, of course, wrong.

[34] As a more recent statement on the solution of the problem we mention a work of Birkhoff's [1]. There the following theorem is treated: If $P_n(\lambda)$ is the number of ways in which a plane map of n (≥ 3) countries can be colored with λ colors, then

$$P_n(\lambda) \geq \lambda(\lambda-1)(\lambda-2)(\lambda-3)^{n-3}.$$

For $\lambda=4$ we get that $P_n(4) \geq 24$, and so, since $P_n(4) > 0$ we have the assertion of the Four Color Theorem. Birkhoff proved the stated theorem precisely with the exception of the critical case $\lambda = 4$.

Chapter XIII

Factorization of regular infinite graphs

§1. Graphs of finite even degree

In this chapter we take up the question of how the results of
Chapter XI can be extended to infinite graphs.

We would like to show first that Theorems XI.6 and XI.7 remain
valid also for infinite graphs of finite degree. In our proofs our
Infinity Lemma (Theorem VI.6) will form the basis.[1]

Let G be an infinite graph in which at most 2g edges go to each
vertex. We shall assume that g is finite. We may assume that G is
connected, because if a partition into classes of the kind required in
Theorem XI.6 exists for each component of G, then by summing we also get
such a partition into classes for G. (In general, the Axiom of Choice
is needed for this conclusion.) By Theorem VI.1 the edges of G can be
counted in a countable sequence k_1, k_2, ..., k_ν, ... Let G_ν be the
finite graph formed by the edges k_1, k_2, ..., k_ν (ν = 1, 2, 3, ... ad
inf.). As in G, at most 2g edges go in each G_ν to each of its
vertices. By Theorem XI.6 the edges of G_ν can be divided into g classes
such that at most two edges of the same class go to each vertex of G_ν.
Now it will be more convenient to formulate this property (similarly to
the way it was done in the first proof of Theorem XI.15) as follows:
one of the "colors" (indices) (1), (2), ..., (g) can be assigned to each
of the edges of G_ν in such a way that at most two edges of the same
color meet at the same vertex. Every such assignment will be called a
"coloring" of G_ν. Since the set of edges as well as the set of colors

281

is finite, the different colorings of G_ν form a <u>finite</u> set π_ν, and none
of these sets π_ν ($\nu = 1, 2, \ldots$ ad inf.) is empty. Since G_ν is a
subgraph of $G_{\nu+1}$ every coloring $F_{\nu+1}$ of $G_{\nu+1}$ uniquely determines a
coloring F_ν of G_ν "coming from $F_{\nu+1}$", in which every edge of G_ν is
colored the same color as it was colored as an edge of $G_{\nu+1}$ by the
coloring $F_{\nu+1}$.

We would like to consider the elements of the finite sets
π_ν ($\nu = 1, 2, \ldots$ ad inf.) as vertices of a graph Γ in which a point
F_ν of π_ν is joined with a point $F_{\nu+1}$ of $\pi_{\nu+1}$ by an edge, if the coloring
F_ν "comes out of" the coloring $F_{\nu+1}$. Γ has no other edges. The
conditions of the Infinity Lemma are fulfilled by the given definition
of π_1, π_2, $\ldots$, Γ. If the lemma is applied, we get a countably infinite
sequence

$$F_1^0, F_2^0, F_3^0, \ldots, F_\nu^0, \ldots \qquad (F)$$

with the property:

F_ν^0 is a coloring of G_ν (element of π_ν), and the same color is
assigned to an arbitrary edge k of G_ν by F_ν^0 and $F_{\nu+1}^0$. It follows that
the same color is also assigned to the edge k by $F_{\nu+2}^0$, $F_{\nu+3}^0$, etc., and
that therefore k is colored the same color by each of the colorings
F_1, F_2, F_3, $\ldots$, which assigns a color to this edge. Since every edge
of G is contained in a G_ν (then also in $G_{\nu+1}$, $G_{\nu+2}$, $\ldots$), a definite
color of the colors (1), (2), $\ldots$, (g) is therefore assigned by the
sequence (F) to each edge of G.

It remains to be shown only that three edges which meet at the same
vertex are never assigned the same color by (F). Clearly, if there were
three such edges k_p, k_q, and k_r, then they would be contained in G_ν for
large enough ν and would be colored the identical colors by F_ν^0, which
contradicts the definition of "coloring".///

So Theorem XI.6 holds for infinite graphs of finite degree; but it

follows from this -- exactly in the way we saw in the finite case --
that in Theorem XI.7 the finiteness of the graph does not need to be
assumed. From these results we would particularly like to stress the
following theorems. (The carrying over of Theorem XI.6 -- as we have
seen -- follows from these for infinite graphs with the help of Theorem
XI.1).

> Theorem 1 Every regular graph of even (finite) degree has a
> factor of degree 2.
>
> Theorem 2 Every regular graph of even (finite) degree splits
> into factors of degree 2.

We shall give another proof for this theorem.

For the graph of degree 4, represented by the plane square lattice,
the horizontal edges form in themselves a factor of degree 2, as do the
vertical edges. The same holds for the graph of degree 6 which is
represented by the corresponding spatial construct. For the infinite
graph of degree 6 of Fig. 101 (which should be thought of as continuing
to infinity in every direction) the edges of the same direction also
form a factor of degree 2. In these three examples the factors of
degree 2 split further into two factors of degree 1. Now we give an
example of a connected infinite graph of degree 4 where this is not the
case. In Fig. 102 (which should be thought of as continuing to
infinity) the similarly marked edges each form a factor of degree 2.
One of these factors contains a triangle, the other a heptagon, so that
they cannot be factored further. (This graph also splits into a factor
of degree 1 and one of degree 3; the latter is shown in Fig. 103.)

We have seen (Theorem XI.8) that Theorems 1 and 2 for finite graphs
remain unchanged even if graphs in the broader sense are allowed and, in
fact, for both ways of counting loops. The reduction given above of
Theorems 1 and 2 to the case of finite graphs shows -- without any

change -- that this is also the case for infinite graphs:

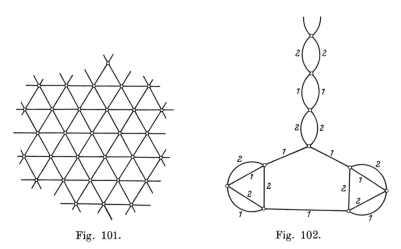

Fig. 101.   Fig. 102.

<u>Theorem 3</u> Theorems 1 and 2 also hold for graphs in the broader
sense and, in fact, in both cases of whether loops are counted
the first or second way.

§2. Applications

In order to give an application of Theorem 2 to set theory, we
prove the following theorem:

Let E be a point set of the plane symmetric with respect
to the line y = x and with the property that every straight

line parallel to one of the coördinate axes and containing a
point of E contains exactly 2g points of E (g finite). Then E
has a subset E_1, of which exactly two points lie on each of
these straight lines. There are, in fact, g subsets E_1, E_2,
..., E_g with this property which are pairwise disjoint and
together generate the set E.

The set M of the abscissas of the points of E is identical to the
set of ordinates of these points. We define a graph G, whose vertices
are the elements of M, in which two vertices x and y are joined by an
edge if and only if the point (x,y) belongs to E (and therefore the
point (y, x) also). We first assume that E does not contain a point of
the line x = y, and the graph so defined is a graph in the narrow sense;
each of its vertices has degree 2g. By Theorem 1, G has a factor G_1 of
degree 2. Let E_1 be that subset of E which consists of the points
(x, y) for which x and y are joined by an edge of G_1. Obviously E_1 has
the desired property. If G, by Theorem 2, now splits into factors of
degree 2: $G = G_1 G_2 \ldots G_g$, and we let a subset E_i of E correspond to
each G_i in the manner stated, then it is clear that the sets E_1, E_2,
..., E_g fulfill the requirements stated in the second part of the
theorem. By Theorem 3 the restriction that no point of E should lie on
the line y = x can be removed and, in fact, in both cases, no matter
whether the points of y = x are always counted as one or two points.///

A more general formulation can be given to the theorem proved,
corresponding to the situation where two vertices can be joined by more
than one edge, by assigning a multiplicity (positive integer) to the
points of E and, instead of requiring that 2g points of E should lie on
each of the lines involved, by requiring that the sum of the
multiplicities of the points of E, which lie on any one of these lines,
be 2g. Then the existence of a factorization $E = E_1 + E_2 + \ldots + E_g$ can

be shown which has the property that for every E_i the sum of the multiplicities (1 and 2) of the points of E_i lying on one of these lines is always 2. The factorization here is such that the E_i's are not necessarily pairwise disjoint, but that the sum of the multiplicities with which a point in E_1, E_2, ..., E_g is counted is equal to the multiplicity with which this point is counted in E.

Theorem 3 is particularly important for the second way of counting. This part of the theorem makes it possible to interpret Theorem 1, extended (in the sense of the second way of counting) to graphs in the broader sense, as a very general theorem of abstract set theory. In order to be able to express this briefly, a designation should first be given concerning certain multi-valued mappings.

A "reciprocal $(1, \nu)$ - mapping into itself" is given for a set M, in case ν elements of M are assigned to every element of M. But these ν elements need not be distinct. We let correspond to every assignment of an element a of M to an element b of M a multiplicity, and if we say that ν elements correspond to the element a, this shall mean: If b_1, b_2, ..., b_μ are the elements assigned to a and these assignments have, in order, the multiplicities s_1, s_2, ..., s_μ, then for each element a of M we have: $s_1 + s_2 + ... + s_\mu = \nu$. (It is often convenient to express the fact that the element b does not correspond to the element a by saying that the assignment of a to b has multiplicity zero; then we have the following equation: $\sum s_\alpha = \nu$, where the summation is to range over all elements of M. We require secondly that, if the assignment of a to b has multiplicity s, the multiplicity of the assignment from b to a likewise be s. So we have that if b is one of the elements corresponding to a, then a must be one of the elements corresponding to b. From now on, ν and the multiplicities will signify finite numbers.

A reciprocal $(1, \nu)$ - mapping into itself for a set M can be represented by a regular graph of degree ν, in which the vertices are the elements of M and a pair (a, b) of vertices is joined by as many edges (possible zero) as the multiplicity of the reciprocal assignment of a and b specifies. If in the $(1, \nu)$ - mapping an element a is assigned to itself, then to this assignment corresponds a loop of the graph; if the assignment of a to a has multiplicity s, then the graph has s loops at the vertex a. The graph is, in general, regular if and only if each of these loops is counted as one edge; for this reason we must use here the second way of counting.

Also, conversely, a reciprocal $(1, \nu)$ - mapping into itself is specified for the set of its vertices by every regular graph in the broader sense (with the second way of counting loops), in case it is of degree ν: to an element a are assigned those elements b which are joined with a by an edge, and the multiplicity of the reciprocal assignment of a to b is the number of edges which join a with b.

In set theoretic formulation Theorem 1, extended in the sense of Theorem 3, is:

> If there is for a set M a reciprocal $(1, 2g)$ - mapping Z_{2g} into itself (g finite), then there is also such a reciprocal $(1,2)$ - mapping Z_2 into itself, which assigns to one another only those elements which are also assigned to one another by Z_{2g}; the assignment of the two elements by Z_2 has at most the multiplicity which it has by Z_{2g}.

With the help of the concept of a function another form can be given to this theorem and to Theorem 2 (extended to graphs in the broader sense), which can be connected with certain questions of reducibility in algebra. Here, however, it is a matter of the most general (so-called logical) concept of a function. Let Φ be a single-

287

valued symmetric function of the variables x and y, which vary in a
domain B. A function Ψ of one variable is defined in this domain by the
equation Φ (x, y) = 0, for which the statement y = Ψ(x) (and likewise
the statement x = Ψ(y)) is equivalent to the statement Φ (x, y) = 0.
We assume that Ψ is a ν-valued function in B (this means that Ψ is for B
a reciprocal (1, ν) - mapping into itself), which we also express by
saying that Φ has the value ν. We assign to the vanishing of
Φ(x, y) the multiplicity with which x and y are assigned to each other
by the (1, ν) - mapping Ψ. With the notation introduced Theorem 2 can
be formulated:

If Φ (x, y) is a single-valued symmetric function of value 2g, then
there are single-valued symmetric functions $\Phi_1(x, y)$, $\Phi_2(x, y)$, ...,
$\Phi_g(x, y)$ of value 2 for which the statements $\Phi(x, y) = 0$ and
$\Phi_1(x, y)$ $\Phi_2(x, y)$ ... $\Phi_g(x, y) = 0$ are equivalent. (These two equations
with the same multiplicity hold, if, to the vanishing of a product,
there is assigned as multiplicity the sum of those multiplicities with
which the individual factors vanish.) The theorem can be applied to the
case where B signifies the complex number plane (with ∞ included)
and Φ is an arbitrary symmetric polynomial of degree 2g in x (and
therefore also in y). The $\Phi_i(x, y)$ can be chosen here as polynomials
only in special cases; in general they will be completely discontinuous
functions.

§3. Primitivity of graphs of finite degree

The question of the primitivity of infinite regular graphs of even
(finite) degree is essentially taken care of by the proof of Theorem
2. This theorem states that such a graph can be primitive if and only
if it is of degree 2. We know (Theorem I.28) that each of its

components is either a cycle or a doubly infinite path. But a graph of degree 2 can be primitive if and only if one of its components is primitive (Axiom of Choice). This primitive component can only be a cycle and must have an odd number of edges (Theorem I.12). Since a graph (of degree 2) is primitive if it has a primitive component, it is therefore proved that Theorem XI.9 also remains true for infinite graphs, that is, the following is true:

Theorem 4   A regular graph of even (finite) degree is primitive if and only if it is of degree 2 and one of its components is a cycle consisting of an odd number of edges.

Also Theorem XI.10 can also be carried over to infinite graphs. This is clear if the graph is not required to be connected; it is sufficient to assume that one of its components is a finite primitive graph. But a restriction may be made to connected graphs. We now prove

Theorem 5   For every odd number $2\alpha + 1$ there exist connected primitive infinite regular graphs of degree $2\alpha + 1$.

This can be shown by the following example. We choose as vertices of the graph the points

$$P;\ A_1,\ B_1,\ C_1;\ A_2,\ B_2,\ C_2;\ \ldots;$$
$$A_{2\alpha},\ B_{2\alpha},\ C_{2\alpha};\ Q_1,\ Q_2,\ Q_3,\ \ldots \text{ ad inf.}$$

We join P with each of the vertices A_i ($i = 1, 2, \ldots, 2\alpha$) and with Q_1 by an edge; the vertices A_i are also joined with B_i and with C_i by α edges each, while B_i is joined with C_i by $\alpha + 1$ edges each ($i = 1, 2, \ldots, 2\alpha$); finally each Q_{2i} is joined with Q_{2i+1} (for $i = 1, 2, \ldots$, ad inf.) by an edge, and Q_{2i-1} with Q_{2i} (for $i = 1, 2, \ldots$, ad inf.) by 2α edges each. It is easily seen that a connected infinite regular graph G of degree $2\alpha+1$ is thus defined, where each edge going to P is a bridge. So P cannot belong to any cycle of G, but since no singly infinite path exists in G beginning with

PA_i (i = 1, 2, ..., 2α), P also cannot belong to any doubly infinite

path of G. (Such a path certainly does not exist in G.) But then G

cannot have a factor of degree 2, since such a factor (Theorem I.28) has

only cycles and doubly infinite paths as components. Now if G were not

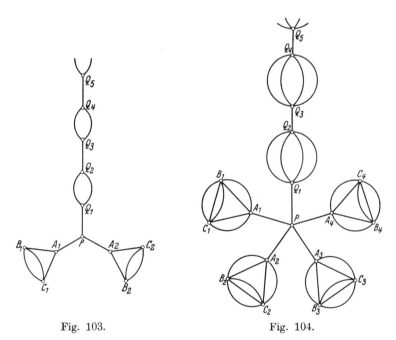

Fig. 103.   Fig. 104.

primitive, and therefore the product of two factors, then one of these

factors would have to be of even degree; this would either itself be of

degree 2 or -- according to Theorem 1 -- have a factor of degree 2, so

that in any case G would have a factor of degree 2, but this has been proved to be impossible.///

In Figures 103 and 104 this graph G is represented for $\alpha = 1$ and $\alpha = 2$ respectively. Both figures are to be thought of as continuing indefinitely. The graph of degree 3 of Fig. 103 has, as is easily seen, exactly two leaves; so Petersen's Theorem XII.8 loses its validity for infinite graphs. It would be interesting to be able to decide whether an <u>infinite</u> regular graph of degree 3 which has no bridges or leaves at all can be primitive. If this is the case then there must be a connected graph of this type.

§4. Bipartite graphs of finite degree

We shall now consider bipartite graphs. In fact, Theorem XI.15 will be extended to infinite graphs of finite degree analogously to the way in which we extended Theorem XI.6 to infinite graphs in §1. Again the Infinity Lemma forms the basis of the proof. We may assume, as we did there, that the infinite bipartite graph G, in which at most g edges go to the same vertex, is connected (Axiom of Choice), and therefore (Theorem VI.1) has countably many edges k_1, k_2, k_3, ... By Theorem XI.15 the edges of the finite graph G_ν ($\nu = 1, 2, 3, ..., $ ad inf.) consisting of the edges k_1, k_2, ..., k_ν can be colored with the g colors (1), (2), ..., (g) in such a way that two edges of the same color never meet at a vertex; these "colorings" of G form a finite nonempty set π_ν. We consider the elements of the set $\sum\limits_{\nu=1}^{\infty} \pi_\nu$ to be the vertices of a graph Γ, in which a point F_ν of π_ν is joined with a point $F_{\nu+1}$ of $\pi_{\nu+1}$ by an edge if the coloring F_ν assigns to every edge of G_ν the same color as is assigned to it by $F_{\nu+1}$. Γ has no other edges. So the conditions of the Infinity Lemma (Theorem VI.6) are fulfilled for π_1, π_2, ..., Γ; its application gives an infinite sequence

291

$$F_1^0, F_2^0, F_3^0, \ldots, F_\nu^0, \ldots, \tag{F}$$

in which F_ν^0 is a coloring of G_ν, and where two elements (colorings) which each assign a color to an edge k of G always assign the same color to it. By (F) every edge of G is thus assigned a definite color of the colors (1), (2), ..., (g), in such a way that our requirement which specifies that two edges with a common endpoint should always be assigned different colors is fulfilled; for a violation of this requirement would appear in one of the finite graphs G_ν. By this it is proved that Theorem XI.15 is also true for infinite graphs (g finite).///

That the same must then hold for Theorems XI.13 and XI.14 was already mentioned in Chapter XI. So we content ourselves here with formulating[2] the carrying over of Theorems XI.13 and XI.14:

> Theorem 6   Every regular bipartite graph of finite degree has a factor of degree 1.
>
> Theorem 7   Every regular bipartite graph of finite degree factors into factors of degree 1.

Theorem 6 can be deduced in an analogous way to the second method of proof we gave for Theorems XI.13, XI.14, and XI.15, and Theorem 7 can also be deduced in this way from Theorem 1. We shall not go into the details of this second method of proof here.

We would like, however, to show how conversely Theorem 2 can be deduced easily (both for finite and infinite graphs) from Theorem 7 with the help of Theorem II.12. Let G be a regular graph of degree 2g. Analogously to Theorem II.12 we transform it into a directed graph in such a way that each of its vertices is an initial vertex as well as terminal vertex of exactly g edges. We now define a graph H in the following way. To every vertex P_α of G we let two vertices P'_α and

P''_α correspond and to each (directed) edge $\overrightarrow{P_\alpha P_\beta}$ of G an (undirected) edge $P'_\alpha P''_\beta$ (so $P''_\alpha \ P'_\beta$ corresponds to an edge $\overleftarrow{P_\alpha P_\beta}$). The graph H defined by the vertices and edges introduced is a bipartite graph, since neither two P'-vertices nor two P"-vertices are joined by an edge (see Theorem XI.12). Since P_α is the initial vertex (terminal vertex) of g edges of G, g edges in H go to P'_α (to P''_α). So by Theorem 7, H splits into g factors of degree 1: $H = H_1 H_2 \ldots H_g$. Each of these factors contains <u>one</u> edge going to P'_α and one going to P''_α, and so there are among those edges of G which correspond to the edges of H_ν ($\nu = 1, 2, \ldots$ or g) exactly two edges going to P_α. So factors of degree 2 correspond in G to the factors H_ν. Since the correspondence between the edges of G and of H is one-to-one, these factors of degree 2 have pairwise no common edge, and their product yields the graph G. So Theorem 2 is proved.///

This proof shows how directed graphs can be applied in the theory of undirected graphs.

In order to give a geometrical application of Theorems 6 and 7, we prove the following theorem[3]:

> Let E be a set of points in the plane such that every line parallel to one of the two coördinate axes and containing a point of E contains exactly g points of E (g finite). Then E has a subset E_1, such that exactly one point of E_1 lies on each of these lines. There are g subsets of this kind, $E_1, E_2, \ldots,$ E_g, which are pairwise disjoint and together make up the set E.

<u>Proof</u> Let X and Y be the projections of E onto the coördinate axes. We consider X + Y as the set of vertices of a graph G, in which a point x of X is joined with a point y of Y by an edge if and only if (x, y) belongs to E. No other edges are introduced. G is a regular graph of

degree g and is also a bipartite graph, since all its edges join an X-point with a Y-point. By Theorem 7, G splits into factors of degree 1: $G = G_1 G_2 \ldots G_g$. Those points (x, y) of E for which the x and y are joined by an edge of G_1 then form a subset E_1 of E with the desired property. The sets E_1, E_2, ..., E_g, which correspond in the same way to the factors G_1, G_2, ..., G_g have the property required in the second part of the theorem.///

If Theorem XI.15 is applied instead of Theorem 7, so that the graph does not need to be assumed to be finite, we get more generally:

> If E is a set of points of the plane such that every line
> parallel to one of the coördinate axes contains at most g
> points of E (g finite) then E can be partitioned into g
> disjoint parts in such a way that each of these lines contains
> at most one point of each of these sets.

Just like the point set theorem, which we proved in §2 as an application of Theorem 2, these theorems can also be generalized by assigning multiplicities to the points. Another way of generalization is possible if more general concepts are used in place of plane and line. But it can be shown that these theorems lose[4] their validity for three dimensional space -- and also for finite sets E.

The application of Theorem 7 given here is connected very closely with those applications of determinants and matrices, which we shall deal with, for finite graphs, in Chapter XIV, §3 (see, in particular, Theorems A and B there).

§5. Set theoretical formulations

Theorem 6 expresses a very general theorem of abstract set theory. In order to show this, we would first like to give an exact

definition of the concept of a reversible $(1, \nu)$ - relation
(correspondence, mapping) (where ν can also stand for an infinite
cardinality). We say that the set M stands in a $(1, \nu)$ - relation to
the set N if to every element of M there correspond ν elements of N;
these, however, do not need to be ν distinct elements of N, since we let
a multiplicity correspond to the assignment of an M-element to an N-
element. In saying that there correspond ν elements of N to the element
c of M it ought to be understood that if the elements b_1, b_2, b_3, ... of
N correspond to the M-element c and these assignments have, in order,
the multiplicities s_1, s_2, s_3, ... then, for every element c of M, we
have that $s_1 + s_2 + s_3 + ... = \nu$. Such a relation from M to N is called
reversible if the converse of this relation, which assigns the elements
of M to the elements of N, is likewise a $(1, \nu)$ - relation. If this
converse has the property that if one of the N elements which correspond
to the element c of M is b, then c is one of the M-elements which
correspond to the element b of N by the converse relation, and these two
correspondences have the same multiplicity.

As a consequence the reversible $(1, \nu)$ - relation of two (element
disjoint) sets M and N can be represented by a regular bipartite graph
G: let the elements of M and N themselves be the vertices of the graph
G; an M-point is joined to an N-point by s edges, where s stands for the
multiplicity of the (reciprocal) assignment of these two elements. Two
M-points, as well as two N-points, are never joined with each other. So
the resulting graph G is, in fact, a bipartite graph, and it is also
regular, since exactly ν edges go to each of its vertices.

Now let ν be finite. By Theorem 6 this graph G has a factor of
degree 1; this represents a reversible single-valued relation between M
and N. So we have proved

Theorem 6* If there exists for two sets [5] M and N a reversible

$(1, \nu)$ - relation R_ν (ν finite), then

    α) there is a reversible, single-valued relation R for M
and N;

    β) two elements are assigned to each other by R only if
they also correspond to each other by R_ν.

Since (see Theorem XI.12) each regular bipartite graph of degree ν
can be thought of as a graph which represents the reversible $(1,\nu)$ -
relation between two sets this theorem contains the full content of
Theorem 6. The assertion that two sets M and N of cardinality m and n,
respectively, stand in a reversible $(1, \nu)$ - relation to each other is
obviously synonymous with the assertion that the two sets which result
from M and N if all elements are replaced by ν distinct elements are
equivalent, and that therefore $\nu\, m = \nu\, n$.

So the first part α of Theorem 6* expresses the Bernstein Theorem[6],
that m = n always follows from the equation $\nu\, m = \nu\, n$ (ν finite).

The additional statement β gives an essential sharpening of the
Bernstein Theorem, just as the Equivalence Theorem was sharpened in
Chapter VI, §3.

The greater part of the preceding reflections is superfluous for
the proof of the Bernstein Theorem. This proof comes about quite
simply, graph theoretically. We represent the equation $\nu\, m = \nu\, n$, that
is, the existence of a reversible $(1, \nu)$ - relation between M and N, by
the graph G introduced above, whose vertices are the elements of
M + N. Let G_1 be an arbitrary component of G; its vertices should form
the subsets M_1 and N_1 of M and of N, respectively, and let the
cardinality of M_1 and N_1 be m_1 and n_1, respectively.
Then $\nu\, m_1 = \nu\, n_1$, since $\nu\, m_1$, as well as $\nu\, n_1$, stands for the
cardinality of the set of edges of G_1. But here (by Theorem VI.1) m_1
and n_1 are finite numbers or $\aleph_o$, and since for this case the

Bernstein Theorem, of course, holds, it follows from this that $m_1 = n_1$. If this is done for each component of G, and these are added, the equation $m = n$ is the result.

In order to be able to give still another set theoretical formulation for Theorem 6, we wish to introduce here a commonly used notation. If ϕ is a reversible single-valued relation between the sets M and N, by which the element n of N corresponds to the element m of M, then we set $m = \phi(n)$ and $n = \phi(m)$. If the elements of the subset N' of N correspond by ϕ to the elements of the subset M' of M, then we also set $M' = \phi(N')$ and $N' = \phi(M')$. For $\phi(\psi(a))$ we set $\phi \psi(a)$ for any ψ. So we always have $\phi\phi(a) = a$.

Now we can express the formulation mentioned, which will turn out to be fully equivalent to Theorem 6, as follows[7]:

Theorem 6** If a set E is split in two different ways into ν pairwise disjoint parts (ν finite):

$$E = M_1 + M_2 + \ldots + M_\nu = N_1 + N_2 + \ldots + N_\nu,$$

and there are reversible single-valued relations ϕ_i between M_1 and M_i and reversible single-valued relations ψ_i between N_1 and N_i for $i = 1, 2, \ldots, \nu$, and therefore

$$M_i = \phi_i(M_1),$$

$$N_i = \psi_i(N_1),$$

then M_1 as well as N_1 can be split into ν^2 pairwise disjoint subsets

$$M_1 = \sum_{i=1}^{\nu} \sum_{j=1}^{\nu} M_{ij},$$

$$N_1 = \sum_{i=1}^{\nu} \sum_{j=1}^{\nu} N_{ij}$$

in such a way that the relation

$$\phi_i \ (M_{ij}) = \psi_j \ (N_{ij})$$
$$\text{i.e. } M_{ij} = \phi_i \ \psi_j \ (N_{ij})$$

holds for i,j = 1, 2, ..., ν.

Let E' be a set equivalent to E and disjoint from E; furthermore, let a one-to-one correspondence between E and E' be determined. The element in this correspondence corresponding to the element x of E and the subset of E' corresponding to the subset X of E are designated by x' and X', respectively. So, for example, $E' = \sum_{i=1}^{\nu} N_i'$, and from x' = y' it follows that x = y. We complete the definition of the functions ψ_i for arguments from E', by setting y' = ψ_i(x') in case y = ψ_i(x); so we have that ψ_i(x') = (ψ_i(x))'.

We form the graph H, which contains as vertices the elements of E + E', and the edges of which join element P of E with the corresponding element P' of E' and only with this element. H is a regular graph of degree 1. If in H the ν points ϕ_i(P) of E (i = 1, 2, ..., ν) which belong to a point P of M_1 are[8] identified as one point and if likewise the ν points ψ_i(Q') of E' (i = 1, 2, ..., ν) where Q is any point of N_1 are identified as one point, then from H we get the regular graph G of degree ν. G is a bipartite graph, since each of its edges joins an E-point with an E'-point (Theorem XI.12). So by Theorem 6, G has a factor of degree 1: G_1.

Now we can define the sets M_{ij} and N_{ij} under consideration. A point P of M_1 belongs to M_{ij} if and only if a point Q of N_1 exists such that the points ϕ_i(P) and ψ_j(Q') are joined by an edge of G_1. Since only one edge of G_1 goes to each vertex of G, one and only one of the ν points ϕ_i(P) (i = 1, 2, ..., ν) is the endpoint of an edge of G_1. The other endpoint of this edge belongs to a certain one of the sets N_j' ($j = 1, 2, \ldots, \nu$) and can therefore be unambiguously designated by $\phi_j(Q')$, where Q' belongs to N_1. So here i and j (and also Q', and so also Q) are determined by P, and indeed uniquely determined, so that each of the points P of M_1 belongs to one and only one of the ν^2 sets M_{ij} (i, j = 1, 2, \ldots, \nu); so $M_1 = \sum_{i=1}^{\nu} \sum_{j=1}^{\nu} M_{ij}$ is a decomposition into pairwise disjoint parts.

In an analogous way we now define the set N_{ij}' so that it contains those points Q' of N_1' for which a point P of M_1 exists, such that the points $\phi_i(P)$ and $\psi_j(Q')$ are joined by an edge of G_1. It follows as before that $N_1' = \sum_{i=1}^{\nu} \sum_{j=1}^{\nu} N_{ij}'$, and so $N_1 = \sum_{i=1}^{\nu} \sum_{j=1}^{\nu} N_{ij}$ are decompositions into pairwise disjoint parts. We have seen that to each point P of M_{ij} one and only one point Q of N_{ij} (and likewise to each point Q of N_{ij} one and only one point P of M_{ij}) belongs, with the property that $\phi_i(P)$ and $\psi_j(Q')$ are joined by an edge of G_1. Since all edges of H, and so also of G_1, have the form A A', we have in this case $(\phi_i(P))' = \psi_j(Q')$ and consequently also $= (\psi_j(Q))'$. Out of this it follows that $\phi_i(P) = \psi_j(Q)$. If all elements of M_{ij} go through P, and therefore those of N_{ij} through Q, then $\phi_i(M_{ij}) = \psi_j(N_{ij})$ and Theorem 6** is proved.///

Now we show conversely that Theorem 6 can be deduced from Theorem 6**. Let G be an arbitrary regular bipartite graph of degree ν, all edges of which join a point of Π_1 with a point of Π_2. Each of the ν indices (1), (2), \ldots, (ν) can be assigned to the ν edges of G, which meet at the same point of Π_1. This is carried out for all points of Π_1 (Axiom of Choice). If M_i is used to designate the set of those edges of G which have received the index (i) (i = 1, 2, \ldots, \nu), then we have the decomposition $K = M_1 + M_2 + \ldots + M_\nu$ into pairwise disjoint subsets for the set K of edges of G. If to an edge of M_i is assigned that edge of M_1 which has a common endpoint with it from π_1, then we have a one-to-one correspondence ϕ_i between M_1 and M_i: $M_i = \phi_i(M_1)$ for $i = 1, 2, \ldots, \nu$. If we now consider π_2 instead of π_1 we get a second decomposition $K = N_1 + N_2 + \ldots + N_\nu$ into disjoint parts and a one-to-one correspondence ψ_i for which the equation $N_i = \psi_i(N_1)$ holds. By Theorem 6** there are then decompositions into ν^2 pairwise disjoint subsets:

$$M_1 = \sum_{i=1}^{\nu} \sum_{j=1}^{\nu} M_{ij}, \qquad\qquad N_1 = \sum_{i=1}^{\nu} \sum_{j=1}^{\nu} N_{ij},$$

so that the equations

$$\phi_i(M_{ij}) = \psi_j(N_{ij})$$

hold for $i, j = 1, 2, \ldots, \nu$.

We now define a subset F of the set of edges of G in the following way. An edge k* of G shall belong to F if and only if for two numbers i and j from the number sequence $1, 2, \ldots, \nu$ an edge k in M_{ij} and an edge ℓ in N_{ij} exist with the property that $k* = \phi_i(k) = \Psi_j(\ell)$.

We show first that an edge of F goes to each vertex P of π_1. Let k be an edge of M_1 going to P; let it belong to M_{ij}, which is one of the sets $M_{11}, M_{12}, \ldots, M_{\nu\nu}$. Since it is true that $\phi_i(M_{ij}) = \psi_j(N_{ij})$, there is an edge ℓ of N_{ij} such that $\phi_i(k) = \psi_j(\ell)$. The edge $k* = \phi_i(k) = \psi_j(\ell)$ also going to P has the desired property. We show secondly that not more than one edge of F goes to any point P of π_1. A single edge of M_1 ends in P. So the edge k is uniquely determined by the condition that the edge $k* = \phi_i(k) = \psi_j(\ell)$ of F (where k belongs to M_{ij} and ℓ belongs to N_{ij}) should end in P; furthermore the edge k uniquely determines the number i (and the number j) for which k is an element of M_{ij}. So by stating P the edge of F ending in P is uniquely determined.

Since the set F of edges was defined symmetrically with respect to π_1 and π_2, a point of π_2 also belongs to one and only one edge of F as endpoint. Thus the edges of F form a factor of G of degree 1, and so Theorem 6** as well as Theorem 6* is proved equivalent to Theorem 6.///

Theorem 6** gives rise to a question, which could be formulated only with difficulty in connection with Theorem 6* or 6. Now only the following component statement from the content of Theorem 6** will be considered.

If

$$E = \sum_{i=1}^{\nu} M_i = \sum_{i=1}^{\nu} N_i,$$

$$M_i = \phi_i(M_1), \ N_i = \psi_i(N_1) \ (i = 1, 2, \ldots, \nu);$$

then ν^2 elements $x_1, x_2, \ldots, x_{\nu^2}$ can be chosen from the group, which is generated[9] by ϕ_i and ψ_i in such a way that

$$M_1 = \sum_{i=1}^{\nu^2} A_i \ , \quad N_1 = \sum_{i=1}^{\nu^2} B_i,$$

$$A_i = x_i(B_i) \qquad (i = 1, 2, \ldots, \nu^2).$$

The question is whether the number ν^2 here can be replaced by a smaller number. This question was posed and answered for $\nu = 2$ by Ulam [1]. He proved by a simple finite example, which he represented by a graph, that the number $\nu^2 = 4$ cannot be replaced here by 3. It would probably be worth the trouble to look into this question also for $\nu > 2$.

§6. Graphs of infinite degree

For the sake of completeness we should now investigate how the factorization of regular graphs of infinite degree into regular factors can be done. Neither Theorems 1 and 2 nor Theorems 6 and 7 can be carried over without any restriction in this case. This is shown by the example of the graph which has three vertices A, B, C and in which A is joined both with B and with C by $\aleph_o$ edges (there is no edge BC). This is a bipartite regular graph of degree $\aleph_o$. It has, as can easily be seen, neither a factor of degree 1 nor a factor of degree 2. A second essential difference we would like to point out here is that in the case of a finite degree number as well as in the consideration of graphs of even degree, as with bipartite graphs, we can infer from the existence of a factor of degree 2 (1) the splitting up of the graph into factors of degree 2 (1). This successive separating of the factors (compare the second proof of Theorem XI.14 in Chapter XI, §4) does not lead to any result in the case of graphs of infinite degree.

By application of Zermelo's Well Ordering Theorem (or by the assumption that all cardinalities being considered are alephs) we can with the help of a restrictive condition carry over our factorization results -- and indeed in a still more general form -- to graphs of infinite degree if an upper bound is postulated for the cardinality of the edges which join the same two vertices. It would suffice to assume that two vertices are joined always only by one edge or by a bounded finite number of edges, but we wish quite generally and without having to distinguish between bipartite and non-bipartite graphs to prove the following theorem, where it will be a matter of transfinite calculation, while special graph theoretical considerations, like those we used in the case of graphs of finite degree, are quite superfluous here.

<u>Theorem 8</u> Let G be a regular graph of infinite degree g; if
there is a cardinality g' < g with the property that any two
vertices in G are joined[10] by fewer than g' edges, then G
splits into factors of degree 1.

<u>Proof</u> If this is proved for every component G_α of $G = \sum\limits_{(\alpha)} G_\alpha$ (which are
likewise regular graphs of degree g), and if therefore for every
α: $G_\alpha = \prod\limits_{(\beta)} G_\alpha^{(\beta)}$, where the $G_\alpha^{(\beta)}$ are regular graphs of degree 1 and
these products consist of g factors for every α, then $G = \prod\limits_{(\beta)} \sum\limits_{(\alpha)} G_\alpha^{(\beta)}$
is also split (Axiom of Choice) into factors $G^{(\beta)} = \sum\limits_{(\alpha)} G_\alpha^{(\beta)}$. So we
may assume that G is connected. Since the cardinality of the vertices
which have distance ν from a fixed vertex (compare the proof of Theorem
VI.1) is at most $g^\nu(\nu = 1, 2, \ldots$ ad inf.), the cardinality m of all the
vertices of G is at most equal to

$$\sum_{\nu=0}^{\infty} g^\nu = \sum_{\nu=0}^{\infty} g = \aleph_0\, g = g.$$

So $m \leq g$.

Now let I be any set of cardinality g, the elements of which we
shall designate as indices. Our task can be formulated thus: we need
to assign to each edge of G an index from I in such a way that for each
vertex P of G every index is assigned to one and only one edge going to
P. We order the set of vertices of G as a well-ordered sequence:

$$P_1, P_2, \ldots, P_\omega, P_{\omega+1}, \ldots P_\alpha, \ldots, \tag{π}$$

in such a way that the ordinal number of this sequence is an initial
number, so that every segment of π has a cardinality less than m. The
assignment of indices to the edges which go to P_1, then, if necessary,
to the edges which go to P_2, etc., can be carried out in the desired
manner. We prove with the help of transfinite induction that this can
be continued transfinitely arbitrarily far. Let P_α be an arbitrary but

fixed vertex of π. We assume that we have already assigned each index
to all edges which go to the points P_β, where $\beta < \alpha$, each index
according to our condition, and in such a way that two edges meeting at
the same vertex P_γ received different indices. We prove that this can
also be carried through for the edges which go to P_α.

If m_α $(< m)$ is the cardinality of the vertices P_β, for
which $\beta < \alpha$, and if P_α is joined with P_β $(\beta < \alpha)$ by $n_{\alpha\beta}$ $(< g')$ edges,
then an index was assigned to the $\sum\limits_{\beta<\alpha} n_{\alpha\beta}$ edges which go to P_α before
we reached the vertex P_α. But it now follows from $n_{\alpha\beta} < g'$ for this
m_α- termed sum:

$$\sum_{\beta<\alpha} n_{\alpha\beta} \leq m_\alpha \, g'.$$

Here (according to whether $m_\alpha \geq g'$ or $m_\alpha \leq g'$) $m_\alpha \, g'$ is either
$= m_\alpha < m \leq g$ or $= g' < g$. So in any case $\sum\limits_{\beta<\alpha} n_{\alpha\beta} < g$. But the
equation $\sum\limits_{\beta<\alpha} n_{\alpha\beta} + x = g$ has the single solution $x = g$. It follows from
this that on the one hand the set of those edges going to P_α which have
not yet received an index (i.e.: do not go to P_β, $\beta < \alpha$), and on the
other hand the set of those indices, which were not yet used for edges
going to P_α have cardinality g.

These two equivalent sets may be ordered as well ordered sequences:

$$k_1, \, k_2, \, \ldots, \, k_\omega, \, k_{\omega+1}, \, \ldots, \, k_\mu, \, \ldots \tag{1}$$

and

$$i_1, \, i_2, \, \ldots, \, i_\omega, \, i_{\omega+1}, \, \ldots, \, i_\mu, \, \ldots \tag{2}$$

respectively.

The ordinal number of both sequences is the same initial number;
the segments are therefore again of smaller cardinality than the whole
sequence. We assign the index i_1 to the first such edge from (1) which
does not have a common endpoint with any edge which has already received

the index (1). We assign the index i_2 to the first such edge of (1) which so far has received no index and which does not have a common endpoint with any edge which has already received the index i_2, etc. We show that this assignment process can be carried through for all elements from (2) transfinitely.

We assume that we have assigned in our process all indices up to i_μ. The set of those elements from (1) which so far have already received indices is equivalent to a segment of (2) and so has a cardinality less than g. On the other hand, the set of those elements of (1) which have a common endpoint with one such edge, to which was already earlier assigned the index i_μ, is of cardinality less than g. (Each of these edges goes to a vertex of the segment of the sequence (π) determined by P_α; but this segment has a cardinality $< m \leq g$; the set of edges under consideration therefore has a cardinality $< g \, g' = g$.) So there are edges in (1) which do not belong to either of these sets. The index i_μ is assigned by our process to the first of these edges from (1).

In order to show that our induction assumption is also fulfilled for P_α, we need only still prove that every edge from (1) receives an index by our assignment process. We assume that an edge k_μ does not receive an index. Since the cardinality of the vertices $P_\beta (\beta < \alpha)$ is smaller than $m \leq g$, the cardinality of the edges which have a common endpoint with k_μ and have already received an index is smaller than $g' \, g = g$. The indices "denied" for k_μ are therefore also of smaller cardinality than g. The indices not denied therefore have the cardinality g. That none of these indices was assigned to the edge k_μ implies that an edge from (1) which precedes the edge k_μ in (1) was assigned to each of these indices. But this is impossible, since the

305

segment of (1) determined by k_μ has smaller cardinality than g. Theorem 8 is thus completely proved.[11] ///

This Theorem 8 also implies, for example, that every infinite complete graph splits into factors of degree 1. Consequently we have carried Theorem XI.2 over to the infinite.

In the same way that we were able to give Theorem 6 the formulation of Theorem 6*, there follows from Theorem 8, if we consider only the statement concerning the existence of one factor of degree 1, the following result, where we are dealing with a bipartite graph.

If there is a reversible (1, g) - relation R_g for two sets M and N, such that the multiplicity of the reciprocal correspondence of two elements is always less than a cardinality g' < g (if all these multiplicities are equal to 1), then there is a one-to-one correspondence between M and N which assigns only such elements to each other which also correspond to each other in R_g.

In order to give a special application for g = c (the cardinality of the continuum) we mention that from this it follows that:

There is a one-to-one correspondence between the points of a plane[12] and the lines of the same plane which has the property that every point lies on the line corresponding to it.

The bipartite graph G of degree c corresponding to this correspondence arises thus: we let a vertex of G correspond to each point and to each line and join two vertices by one edge if and only if one vertex corresponds to a point and the other to a line, and the point lies on the line. Every factor of degree 1 of G (Theorem 8) gives rise to a correspondence of the desired kind.

Notes on Chapter XIII

[1] The proof given for the finite case cannot be carried over, since the proof, which we gave for Theorem XI.6, by using induction, must assume that the number of edges is finite.

[2] Theorem 6 and Theorem 6* following below, which express the set theoretical equivalent of Theorem 6, were first stated by the author in 1914 in his address [3] mentioned above. In his paper [5] he proved these theorems for the case where the degree (g) is a power of 2 and reduced the general case to the case where the set of vertices is countable and g is a prime number. The general case was proved, however, ten years later with the assistance of St. Valkó; see König and Valkó [1]. This theorem could be simplified on the one hand by generalizing to non-regular graphs and on the other hand by dissection and isolation of the Infinity Lemma; see König [7]. The proof given above agrees essentially with the proof in this paper; a difference consists in the fact that in the paper in Fundamenta Mathematicae -- to be appropriate to the orientation of this journal -- the terminology of graph theory was avoided.

[3] König [7, §6].

[4] See König [6, §1 and 7, §6]. In the first mentioned reference this spatial problem is connected, on the one hand, with cubic matrices, and on the other hand with surface complexes -- this is the two dimensional analog to the (one dimensional) graph.

[5] It is clear that M and N do not have to be assumed to be disjoint here.

[6] F. Bernstein, Untersuchungen aus der Mengenlehre, §2; Mathematische Annalen, 61, 1905, pp. 117-155 (originally appeared as a Goettingen dissertation in 1901). The Bernstein proof is extraordinarily complicated even for the case where $\nu = 2$ (actually the

proof was completely done by Bernstein only for this case). The very
simple graph theoretical proof given here is given by König [3 and 5].
Sierpiński gave a proof for the case where $\nu = 2$ without using the Axiom
of Choice: Sur l'égalité 2m = 2n pour les nombres cardinaux, Fundamenta
Mathematicae, 3, 1922, pp. 1-6. Such a proof was assumed to be
impossible by König [3]. The Bernstein Theorem follows naturally
directly from Zermelo's Well Ordering Principle (since the equation
$\nu m = m$ has this as a consequence); for the additional statement β this
is not at all the case. The attempt to apply the methods of proof,
which J. König gave for the equivalence theorem (see Chapter VI, §3) to
the proof of the Bernstein Theorem, was the starting point in 1914 of my
studies in infinite graphs.

[7] König [7, §5]. For the case where $\nu = 2$ this formulation
stems from Kuratowski, Une propriété des correspondances biunivoques,
Fundamenta Mathematicae, 6, 1924, p. 240. Kuratowski's proof, apart
from the terminology, agrees essentially with the proof which was given
earlier for Theorem 6 in the case where $\nu = 2$ by König [3 and 5]. With
respect to the clarity of the argument the superiority of the graph
theoretical terminology over that of group theory cannot be denied.
Kuratowski gave a generalization for the case where $\nu = 2$ in the
reference cited, where more general binary relations figure instead of
set theoretical equivalences. This generalization can, of course, be
extended to arbitrary ν, which can be useful for certain applications --
as were given, for example, in a paper by Banach and Tarski (p. 254)
mentioned in Chapter VI, §3. Certainly it is possible to manage in the
case where $\nu = 2$ or $\nu = 2^n$ without having to use Theorems 6, 6*, or 6**
for arbitrary ν. (cf. J. von Neumann, Zur allgemeinen Theorie des
Maßes, Fundamenta Mathematicae, 13, 1929, particularly pp. 82 and 103).
-- Concerning further generalization possibilities and consequences, cf.

also Tarski, Über Äquivalenz der Mengen in bezug auf eine beliebige

Klasse von Abbildungen, Atti del Congresso Internationale dei

Matematici, Bologna 1928, Vol. 2 (Bologna, 1930), pp. 243-252.

[8] It is clear how this intuitive expression is to be

understood and how it could be rigorously defined (cf. the proof of

Theorem II.11). Also we may consider the edges of G as identical to the

edges of H, without having to say "the edge of H corresponding to the

edge k of G."

[9] It is to be understood that $\phi_i(a)$ = a is put for an element

a of E, if a does not belong to either M_1 or M_i, and likewise for ψ_i .

[10] It is questionable whether the theorem still holds if this

condition is replaced by the following weaker condition: "every two

vertices are joined by fewer than g edges."

[11] I owe this proof to the kind assistance of Messrs. G. Hajós

and L.Kalmár.

[12] It does not matter here whether the Euclidean, function

theoretical, or the projective plane is meant.

Chapter XIV

Separating Vertices and Sets of Vertices

§1. Cut points and Blocks

We would like now to consider a type of vertex which is characterized by a property analogous to the one that characterizes bridges among edges. A vertex A of a graph G is called a <u>cut</u> <u>point</u> of G if G has two edges, AP and AQ, going to A with the property that every path of G which joins P with Q also contains A. (Fig. 105). In other words, there is no cycle of G containing all three vertices A,

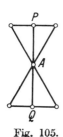

Fig. 105.

P, and Q at the same time. From this definition it follows immediately that: an endpoint of the graph is never a cut point; the endpoint of an end edge which is not an endpoint of the graph is a cut point; a vertex is a cut point if and only if it is a cut point of the component to which it belongs; every vertex of an acyclic graph is either an endpoint or a cut point of the graph; the endpoints of a bridge are cut points.

It is a good idea to introduce the following notation here. If one endpoint of a path or an edge belongs to the set π_1 and the other endpoint to the set π_2, it is designated as a <u>$\pi_1\pi_2$-path</u> or a <u>$\pi_1\pi_2$-edge</u>, respectively.

Theorem 1 A vertex A of a connected graph G is a cut point of G if and only if the remaining vertices of G can be divided into two nonempty sets π_1 and π_2, such that each $\pi_1\pi_2$-path of G contains the vertex A.

310

Proof Let A be a cut point; AP and AQ are two edges such that every

path from P to Q contains A. We consider all paths W_1 = A P ... X,

which begin with the edge AP. The second endpoints X of all these paths

form the set π_1. Then P belongs to π_1, and Q belongs to the

complementary set π_2. If there were a path W going from a point X

of π_1 to a point Y of π_2 which did not contain A, then by Theorem I.5 a

path going from A to Y could be composed out of the edges of paths W_1

and W. This would begin with the edge AP (since no other edge of W or

of W_1 ends in A), and Y would not belong to π_2 but to π_1.

   Now we assume that the sets π_1 and π_2 have the above-named

property. Let X ... R_2 R_1 A S_1 S_2 ... Y be a $\pi_1\pi_2$-path of G. A is a

cut point, since every path W which goes from R_1 to S_1 ($\neq R_1$) contains

A; in the opposite case the three paths X ... R_2 R_1, W, and S_1 S_2 ... Y

(none of which contains A) would give rise to a $\pi_1\pi_2$-path from X to Y

which does not contain A, by two applications of Theorem I.5.///

   The concept of cut point (as "articulation") was introduced[1] by

Sainte-Laguë [2, p. 3] by the property expressed in Theorem 1. By its

cut points the graph is "split into its blocks." How this expression is

to be understood is described in the following paragraph.

   If a cycle of the graph G contains the edges k_1 and k_2, this is

temporarily expressed as "k_1 o k_2". Moreover, for every edge k of G we

write k o k, even if it is not contained in any cycle of G. The binary

relation o, so defined for the set K of edges of G, has the following

properties:

   α) k o k is always true (Reflexivity).

   β) k_1 o k_2 implies k_2 o k_1 (Symmetry).

   γ) k_1 o k_2 and k_2 o k_3 imply k_1 o k_3 (Transitivity).

The validity of this last property is implied by Theorem I.11. It

follows from these three properties that the set K of edges of G can be

divided into pairwise disjoint sets K_α (in only one way) such that two
edges k_1 and k_2 belong to the same set if and only if $k_1 \circ k_2$ is true
(compare Chap. I, §4). The edges which belong to the same set of these
sets form a <u>block</u> of G.[2]

Every block is a connected subgraph of G. The end edges and
bridges each form -- since they are not contained in any cycle -- a
block in themselves. All other blocks contain at least two edges.
Every edge belongs to one and only one block. Two edges which join the
same pair of vertices belong[3] to the same block. The concepts of block
and cut point are connected by:

> <u>Theorem 2</u> A vertex belongs to more than one block or to one
> block, depending on whether it is a cut point or not,
> respectively.

<u>Proof</u> If A is a cut point, then there are two edges AP and AQ which do
not simultaneously belong to a cycle, and therefore belong to different
blocks; both blocks contain A. If, on the other hand, A belongs to two
blocks, one block must contain AP, and the other block must contain AQ,
whereby no cycle contains AP and AQ simultaneously; but then A is a cut
point.///

Now we prove four theorems about graphs without cut points[4].

> <u>Theorem 3</u> If a graph does not have a cut point, then its
> blocks are identical to its components.

<u>Proof</u> If the edges k_1 and k_2 of G belong to the same block of G, then
they also belong to the same component of G, since a block is always
connected, by Theorem I.26. If k_1 and k_2 belong to the same component,
there is, by Theorem I.19, a path $P_1 P_2 P_3 \ldots P_n$, where $P_1 P_2 = k_1$ and
$P_{n-1} P_n = k_2$. Since P_2 is not a cut point, $P_2 P_3$ belongs by Theorem 2
to the same block as $P_1 P_2 = k_1$. Since $P_3, P_4, \ldots$ are also not cut
points, $P_3 P_4, P_4 P_5, \ldots$ and finally $P_{n-1} P_n = k_2$ belong, in turn, to

the same block.///

The theorem thus proved gives rise to the following as a special
case (compare Theorem I.20):

> Theorem 4 If a connected graph has no cut point, then it is
> itself its only block.

Now we prove

> Theorem 5 If a connected graph does not consist of a single
> edge and it does not have a cut point, then for every two
> vertices P, Q of the graph there are two paths which join P
> with Q and have[5] no common vertex other than P and Q.

Proof Let k_1 be an edge which ends in P and k_2 an edge which ends in Q.
Since the graph does not consist of a single edge PQ, k_1 and k_2 can be
chosen to be distinct. By Theorem 4 there is a cycle which contains k_1
and k_2; this is split by P and Q into two paths of the desired kind.///

It can easily be seen that the converses of Theorems 3-5 are
valid. We give now a fourth characterization of graphs without cut
points:

> Theorem 5[a] A connected graph contains no cut point if and only
> if for every three vertices A, B, C (in arbitrarily determined
> order) there always exists a path which goes from A to C and
> also contains[6] B.

Proof If A is a cut point, there are two edges AB and AC with the
property that every path going from B to C also contains A. But if
there were a path from A to C containing B, then a subpath of this path
would go from B to C and would not contain A.

In order to show that the condition is also necessary, we assume
that the graph does not have a cut point. We may assume that the graph
does not consist of a single edge, since otherwise there would be only
two vertices and there would be nothing to prove. Let A, B, and C be

three arbitrary vertices. By Theorem 5 there is a cycle K_1 which
contains A and B and a cycle K_2 which contains B and C. If A also lies
on K_2, then one of the two paths into which K_2 is split by A and C is a
path of the desired kind. So let A not be contained in K_2, and likewise
let C not be contained in K_1. Let D be the first vertex of K_1 reached
(it is at the farthest D = B) going from C in one of the two directions
on K_2. We have described a path W = C ... D, which has the single
vertex D in common with K_1 (compare Theorem I.16). One of the two paths
into which K_1 is split by D and A contains B. This path
W' = D ... B ... A also has the single vertex D in common with W. So
the union of W and W' gives rise to a path which goes from A to C and
also contains B.///

Concerning this proof we also note that in a graph without cut
points every three vertices are not necessarily contained in a cycle.
The simplest example is furnished by the graph which consists of the
edges AP, BP, CP, AQ, BQ, and CQ. No cycle of this graph contains at
the same time A, B, and C.

Now we prove

Theorem 6   No vertex of a block G_α is a cut point of G_α.
Proof   If this were not true, there would exist in G_α two edges, AP and
AQ, which were not contained in the same cycle of G_α; but then AP and AQ
would also not be contained in any cycle of G, since all edges of a
cycle of G belong to the same block G_α. This contradicts the definition
of a block for G_α. ///

Theorem 7   Two blocks of a graph have at most one common
vertex.

Proof   We assume that P and Q are common vertices of the blocks
G_α and G_β of G. Then, since a block is always connected, there is a path
P R_1 R_2 ... Q of G_α as well as a path P S_1 S_2 ... Q of G_β from P to Q.

Let R_k be the first vertex in the sequence R_1 R_2, ..., Q, which is also contained in the second path, $R_k = S_\ell$ (possibly Q). Then the two paths P R_1 R_2 ... R_k and P S_1 S_2 ... S_ℓ have no common interior vertex and so form a cycle; this contradicts the fact that PS_1 and PR_1 belong to different blocks.///

If the edges P_{k-1} P_k and P_k P_{k+1} of the path $W = P_1$ P_2 ... P_k ... P_n ($1 < k < n$) belong to different blocks, then by Theorem 2, P_k is a cut point. It shall be designated as a transit cutpoint of W.

Theorem 8   If the endpoints of a path $W = P_1$ P_2 ... P_n do not belong to the same block, then at least one interior vertex of W is a transit cut point of W.

Proof   Since P_1 P_2 and P_{n-1} P_n belong to different blocks, there must be two adjacent edges P_{k-1} P_k and P_k P_{k+1} in the path W which likewise belong to different blocks; then P_k is a transit cut point of W.///

Theorem 9   If the path P_1 P_2 ... P_k ... P_n contains the transit cutpoint P_k, then P_k is contained in every path W which joins P_1 with P_n.

Proof   We suppose this is not the case. By two applications of Theorem I.5 the three paths P_{k-1} P_{k-2} ... P_2 P_1, $W = P_1$ ... P_n, and P_n P_{n-1} ... P_{k+2} P_{k+1} give rise to a path $\bar{W}$ going from P_{k-1} to P_{k+1} which, since none of these three paths contains P_k, also does not contain the vertex P_k. The paths $\bar{W}$ and P_{k-1} P_k P_{k+1} have therefore only their endpoints in common. Their union gives rise to a cycle which contains P_{k-1} P_k and P_k P_{k+1}. But this is impossible, since these two edges belong to different blocks, and so the theorem is proved.///

Thus W has the form P_1 Q_1 Q_2 ... Q_ν P_k R_1 R_2 ... R_μ P_n. We show that Q_ν P_k belongs to the same block as P_{k-1} P_k. This is clear if $Q_\nu = P_{k-1}$, but if $Q_\nu \neq P_{k-1}$, then the paths P_{k-1} P_{k-2} ... P_2 P_1 and

$P_1 Q_1 Q_2 \ldots Q_{\nu-1} Q_\nu$, neither of which contains P_k, give rise to a path W_0 going from P_{k-1} to Q_ν which likewise does not contain P_k and so has only the endpoints in common with the path $P_{k-1} P_k Q_\nu$. The union of the paths W_0 and $P_{k-1} P_k Q_\nu$ therefore forms a cycle which contains $P_{k-1} P_k$ and $Q_\nu P_k$, so that these two edges belong to the same block. In like manner $P_k P_{k+1}$ and $P_k R_1$ belong to one and the same block. But since $P_{k-1} P_k$ and $P_k P_{k+1}$ belong to different blocks, $Q_\nu P_k$ and $P_k R_1$ also belong to different blocks. So we have shown that P_k is also a transit cut point for W.///

We have thus obtained the following sharpening of Theorem 9:

Theorem 10   Two paths with the same endpoints have the same
            transit cut points.

The chain of reasoning just carried out shows that these transit cut points follow one another in the same sequential order on both paths.

Let P and Q be two arbitrary vertices of the graph G. If P and Q lie in the same block G_α of G, then two cases are possible: either G_α consists of a single edge PQ, so that PQ is not contained in any cycle (is an end edge or a bridge), and this edge PQ is the only path from P to Q, or since G_α is connected and (Theorem 6) has no cut point itself, by Theorem 5 there are two paths from P to Q which have no common vertex except for P and Q. If P and Q do not belong to the same block, then all paths which join P with Q have a common interior vertex, since (Theorem 8) every such path has a transit cut point and this cut point (Theorem 9) is also contained in every other path which goes from P to Q. Our result can be formulated without the concepts of cut point and block as follows:

Theorem 11   For every two vertices P and Q of a graph there are
            -- as long as an edge PQ is not the only path[7] going from P to

316

Q -- only the following two possibilities:

  1) There are two paths from P to Q which have no common vertex other than P and Q.

  2) All paths from P to Q contain a common interior vertex.

From this we get

Theorem 12 Let π_1 and π_2 be two disjoint subsets of the set of vertices of a graph G. Either there are two disjoint $\pi_1\pi_2$-paths of G or the graph G has a vertex which is contained in every $\pi_1\pi_2$-path of G.

Proof Let P_α be the elements of π_1 and Q_α the elements of π_2 (π_1 and π_2 are also allowed to be infinite sets and, in fact, of arbitrarily large cardinality). We augment G to make it the graph G* by adding to G two new vertices, P and Q, and the new edges $P\,P_\alpha$ and $Q\,Q_\alpha$, where P_α ranges over all elements of π_1 and Q_α ranges over all elements of π_2. Since G* has no edge PQ, Theorem 11 can be applied to G* and to the pair of vertices, P and Q. If (case 1) G* contains two paths from P to Q which have no common vertex except for P and Q, then, since P is joined by an edge only with vertices of π_1, and Q only with vertices of π_2, these paths have the form $P\,P_\alpha\,\ldots\,Q_\beta\,Q$ and $P\,P_{\alpha'}\,\ldots\,Q_{\beta'}\,Q$, and the subpaths $P_\alpha\,\ldots\,Q_\beta$ and $P_{\alpha'}\,\ldots\,Q_{\beta'}$ of these paths are two disjoint $\pi_1\pi_2$-paths of G. On the other hand (case 2) if all paths P ... Q of G* contain a common vertex which is distinct from P and Q and therefore belongs to G, then this vertex is contained in every $\pi_1\pi_2$-path $P_\alpha\,\ldots\,Q_\beta$ of G, since this path can be completed to form a path $P\,P_\alpha\,\ldots\,Q_\beta\,Q$ of G*.///

The properties of cut points and blocks treated here can be related to the theory of acyclic graphs and trees. For every graph G a graph H can be defined as follows. We let correspond a vertex to every block of

G and join two of these vertices by one edge if and only if the two
corresponding blocks of G have a common vertex. It can be shown that
the resulting graph H is acyclic; it is a tree if G is finite and
connected. We shall not go into this in detail here.

We would like to mention that the concepts and theorems treated in
this section, particularly Theorem 11, have their analogs in the
continuous concept of graphs (in the new theory of curves), which has
been widely investigated in recent years. The bibliography of these
investigations (Whyburn, Ayres, R. L. Moore, Kuratowski, Menger) can be
found in the works of Kuratowski and Whyburn [1] and Menger [4, Chap.
XI]. The separate treatment of purely combinatorial properties (and so
of properties of graphs) is doubtless also of use in the theory of
curves.

It should be emphasized once more that in the representation of the
theory of cut points and blocks we chose, finite and infinite graphs did
not have to be treated separately, and all our results were also proved
for infinite graphs.

§2. Separating Sets of Vertices,

particularly for bipartite graphs

The preceding material, which we shall not lean on in what follows,
allows far-reaching generalizations.

We saw (Theorem 1) that for every cut point A two disjoint subsets,
π_1 and π_2, of the set of vertices can be named such that every $\pi_1\pi_2$-path
of the graph contains A. We also express this by saying that π_1 and
π_2 are <u>separated</u> by A. As a generalization of this we introduce the
following designation. Let π be the set of vertices of a graph G. If
π_1 and π_2 are two disjoint sets of vertices, we say that

318

π_1 and π_2 are <u>separated</u> in G by the subset π' of π (or also: by the vertices of π'), if every $\pi_1\pi_2$-path of G contains at least one vertex of π'. It does not have to be assumed here that π_1 and π_2 are[8] subsets of π. π' can also be the empty set: the statement that π_1 and π_2 are separated by the empty set means that G does not contain any $\pi_1\pi_2$-path. In the definition given, furthermore, π' does not have to be disjoint from either π_1 or π_2. For example, π_1 is separated by π_1 itself from every set of vertices disjoint from π_1. The separating set π' will certainly contain only vertices of $\pi_1 + \pi_2$ if $\pi_1 + \pi_2$ is the entire set π of vertices of the graph. This case will concern us now if it is assumed that every edge of the graph joins a π_1-vertex with a π_2-vertex. (In this case the statement that π_1 and π_2 are separated by π' means that at least one endpoint of each edge of the graph belongs to π' , and that therefore π' "exhausts"[9] the totality of edges.)

> **Theorem 13** Let the set π of vertices of a graph G be split
> into two disjoint parts: $\pi = \pi_1 + \pi_2$ and let every edge of G
> join[10] a π_1-vertex with a π_2-vertex. If π_1 and π_2 cannot be
> separated by fewer than n vertices (where n is an arbitrary
> finite number), then there are n edges in G which pairwise do
> not have a common vertex.

<u>Proof</u> We assume that the maximal number k of edges of G which pairwise have no common vertex is < n; it then suffices to show that π_1 and π_2 can be separated by fewer than n vertices. Let
$K = (P_1Q_1, P_2Q_2, \ldots, P_kQ_k)$ be a set of k edges of G, where the 2k vertices P_i and Q_i are all distinct from one another. Here the P_i signify vertices of π_1, and the Q_i signify vertices of π_2; we set
$$\pi_1' = (P_1, P_2, \ldots, P_k), \quad \pi_2' = (Q_1, Q_2, \ldots, Q_k).$$
We shall call a path $A_1A_2 \ldots A_{2r}$ of G $(r \geq 1)$ a K-path, provided its second (A_2A_3), fourth (A_4A_5), ..., 2ν-th $(A_{2\nu} A_{2\nu+1})$, ..., and next-to-

319

last $(A_{2r-2}A_{2r-1})$ edge belong^{11} to K. (The remaining edges of this path then do not belong to K.) We now prove

> **Lemma** A vertex of $\pi_1 - \pi_1'$ cannot be joined with a vertex of $\pi_2 - \pi_2'$ by any K-path.

Proof of Lemma If W were such a path, then those edges could be removed from K which are contained in W, and those edges of W not contained in K (their number is greater by 1) could be added, and in this way k+1 edges of K could be obtained which pairwise have no common vertex; this stands in contradiction to the maximal property of k. (End of proof of lemma.)

Now we define a subset $\pi' = (R_1, R_2, \ldots, R_k)$ of $\pi_1' + \pi_2'$ in the following way. For $\alpha = 1, 2, 3, \ldots k$ let $R_\alpha = Q_\alpha$ if a K-path joins a vertex of $\pi_1 - \pi_1'$ with Q_α; if there is no such K-path, we set $R_\alpha = P_\alpha$. So the set π' contains one endpoint each of the edges of K. We prove that the set π' separates the sets π_1 and π_2 in G from each other, and that therefore for every edge PQ of G (where P belongs to π_1 and Q belongs to π_2) either P or Q belongs to π'. We distinguish four cases.

Case 1. P belongs to $\pi_1 - \pi_1'$ and Q belongs to $\pi_2 - \pi_2'$.

If PQ is added to the set K, then k+1 edges of G are obtained which pairwise have no common vertex (in contradiction to the maximal property of k). So this case is impossible.

Case 2. P belongs to $\pi_1 - \pi_1'$ and Q belongs to π_2'.

Then $Q = Q_\alpha$ for $\alpha = 1, 2, \ldots,$ or k and the edge PQ forms in itself a K-path which joins the vertex P of $\pi_1 - \pi_1'$ with $Q = Q_\alpha$. So $Q = Q_\alpha$ belongs to π'.

Case 3. P belongs to π_1' and Q belongs to $\pi_2 - \pi_2'$.

Then $P = P_\alpha$ for $\alpha = 1, 2, \ldots,$ or k. If there were a K-path joining a vertex P_0 of $\pi_1 - \pi_1'$ with Q_α, then a K-path joining P_0 with Q could be obtained by adding the edges $Q_\alpha P_\alpha$ and $P_\alpha Q$ to this path; but by the lemma this is impossible. So there is no K-path which joins a vertex of $\pi_1 - \pi_1'$ with Q_α. Hence $P = P_\alpha$ belongs to π'.

Case 4. P belongs to π_1' and Q belongs to π_2'.

Let $P = P_\alpha$ and $Q = Q_\beta$. If $\alpha = \beta$, then either $P = P_\alpha$ or $Q = Q_\alpha$ belongs to π'. So we may assume that $\alpha \neq \beta$. Either $P = P_\alpha$ belongs to π', or there is a K-path going from a vertex P_0 of $\pi_1 - \pi_1'$ to Q_α; in the latter case, if the edges $Q_\alpha P_\alpha$ and $P_\alpha Q_\beta$ are added to this K-path, then a K-path which goes from P_0 to Q_β is obtained, so that in this case $Q = Q_\beta$ belongs to π'.

Thus it is proved that π_1 and π_2 are separated from each other by a set π' containing k (and so < n) vertices. Hence Theorem 13 is proved.///

In a somewhat more precise way Theorem 13 can be formulated as follows:

Theorem 14 For a graph G let m be the (finite) minimal number of vertices P_1, P_2, ..., P_m with the property that every edge of G ends in one of these vertices, and let n be the maximal number of edges k_1, k_2, ..., k_n which pairwise have no common endpoint. Then if G is a bipartite graph, m = n.

Proof First of all, $m \leq n$, since by Theorem 13 there would otherwise be n + 1 edges which pairwise have no common endpoint. It is also true that $m \geq n$, since one of the vertices P_1, P_2, ..., P_m belongs to each of the edges k_1, k_2, ..., k_n as endpoint and different vertices correspond to different edges in this way.///

Theorem 13 is an important theorem which can[12] be applied to very diverse problems. In order to illustrate its range, we would first like to show that our Theorem XI.13 can be deduced directly from it.

Let G be a finite regular graph of degree g in which every edge

joins one of the vertices P_1, P_2, ..., P_n with one of the vertices
Q_1, Q_2, ..., Q_n. The sets $(P_1, P_2, ..., P_n)$ and $(Q_1, Q_2, ..., Q_n)$
cannot be separated by fewer than n vertices: if they could be
separated by $\nu < n$ vertices, so that each edge of G ends in one of
these ν vertices, then G would have at most νg edges, while it actually
has $ng > \nu g$ edges. Theorem 13, then, implies the existence of a factor
of degree 1. With this we have produced a fourth proof for Theorem
XI.13.

Theorem 13 can also be applied to great advantage in combinatorics
and set theory. As an example we now prove the following very general
theorem of P. Hall[13].

> Let $M = (T_1, T_2, ..., T_n)$ be a finite set of finite or infinite
> sets T_i, which have the property that every k of these sets T_i
> $(k = 1, 2, ..., n)$ contain[14] in all at least k distinct
> elements. Then an element a_i can be chosen from each set T_i in
> such a way that the elements a_1, a_2, ..., a_n are distinct from
> one another.

Proof A graph G can be assigned to the set M in the following way. To
every element contained in a set T_i we let correspond a point P_γ. These
points P_γ form the (finite or infinite) set π_1. To each set T_i we let
one point Q_i $(i = 1, 2, ..., n)$ correspond. The points Q_i form the
finite set π_2. We join a point (vertex) P_γ with a point (vertex) Q_i by
an edge if and only if the element corresponding to the point P_γ is
contained in T_i. No other edges are introduced. In the (bipartite)
graph G so constructed π_1 and π_2 are not separated by fewer than n
vertices. If they could be separated by m $(< n)$ vertices, of which α
vertices belong to π_1 and β vertices belong to
π_2 $(m = \alpha + \beta, \alpha \geq 0, \beta \geq 0)$, then those n - β sets T_i, to which none of
these β vertices corresponds as Q-vertices, would contain in all at

most α elements: only those elements to which the α vertices of

π_1 correspond as P-vertices. This contradicts our assumption because

$\alpha = m - \beta < n - \beta$, and so the conditions of Theorem 13 are fulfilled

for the graph G. Thus there are n edges in G without a common endpoint.

The endpoints of these edges belonging to π_2 are all the vertices of

π_2. The elements a_1, a_2, ..., a_n corresponding to the endpoints of

these n edges belonging to π_1 then have the property expressed in Hall's

Theorem.///

Also "Theorem 3" of Hall's work already mentioned can be deduced

directly from Theorem 13:

> Let the finite or infinite set M be divided into n
>
> (finitely many) pairwise disjoint sets in two ways:
>
> $$M = A_1 + A_2 + \ldots + A_n = B_1 + B_2 + \ldots + B_n$$
>
> For k = 1, 2, ..., n let any k sets B_i contain elements from at
>
> least k sets A_i. There is then a system of n elements $(x_1, x_2,$
>
> ..., $x_n)$ such that every A-set and every B-set are represented
>
> among the x_i's by an element.

Proof We let correspond a point P_i to every set A_i and a point Q_i to

every set B_i (i = 1, 2, ..., n), and join P_i with Q_j by as many edges as

A_i and B_j have common elements, so that every edge corresponds to an

element of M. In the bipartite graph G so constructed $(P_1, P_2, \ldots, P_n)$

cannot be separated from $(Q_1, Q_2, \ldots, Q_n)$ by fewer than n vertices. If

they could be separated by m (< n) vertices, of which α vertices were

P-vertices and β vertices were Q-vertices (m = α + β, $\alpha \geq 0$, $\beta \geq 0$) then

those (n - β) B_i-sets to which none of these β vertices corresponds as

Q-vertices, would contain elements from at most α A_i-sets, only those

A_i-sets to which the α P-vertices mentioned above correspond. This

contradicts our assumption because $\alpha = m - \beta < n - \beta$. So the hypotheses

of Theorem 13 are fulfilled for the graph G. Thus there are n edges in

G, which pairwise have no common endpoint. The n elements of M corresponding to these n edges yield the desired common representative system.///

Hall's Theorem just proved is a generalization of the van der Waerden formulation of Theorem XI.13 mentioned in Chap. XI, §5.

As a further application of Theorem 13 we would like to prove the following theorem, which is of interest particularly for its application to determinants -- which is the subject of the next section.

> Theorem 15 Let the set of vertices of a finite (bipartite) graph G be split into two disjoint parts
>
> $\pi_1 = (P_1, P_2, \ldots, P_n)$ and $\pi_2 = (Q_1, Q_2, \ldots, Q_n)$,
>
> and let every edge of G join a π_1 - vertex with a π_2 - vertex. Let those edges of G which belong to one of the factors of degree 1 of G form a disconnected proper (nonempty)[15] graph G*. Then there is among the numbers 1, 2, ..., (n - 1) a number r with the property that r vertices can be chosen from π_1 and n - r vertices from π_2 such that no edge of G joins two chosen vertices.

Proof First let every edge of G be contained in a factor of degree 1 of G; then G* = G. So by assumption G = $G_1 + G_2$, where neither G_1 nor G_2 is the empty graph. Since we assume the existence of a factor of degree 1 for G, we can -- since this can be obtained by the notation for the vertices -- assume that this factor is formed by the set K = $(P_1 Q_1, P_2 Q_2, \ldots, P_n Q_n)$ of edges. Along with the edge $P_i Q_j$, $P_i Q_i$ is also contained in G_1 (in G_2, respectively), and so G_1 as well as G_2 contains an edge of K. If $P_1 Q_1, P_2 Q_2, \ldots, P_r Q_r$ are contained in G_1 and $P_{r+1} Q_{r+1}, P_{r+2} Q_{r+2}, \ldots, P_n Q_n$ are contained in G_2, then there is no edge of G which would join one of the r vertices $P_1, P_2, \ldots, P_r$ with one of the n - r vertices $Q_{r+1}, Q_{r+2}, \ldots, Q_n$, so that the theorem holds in this case.

Now let an edge of G, say $P_1 Q_1$, be contained in no factor of degree 1 of G. (In this case, as we shall see, it does not need to be assumed that G* is disconnected.) Let G' be the graph which results by removing all edges of G which go to P_1 or to Q_1. In G' there cannot be n - 1 edges which pairwise have no common endpoint, since otherwise a factor of degree 1 of G would be obtained by adding $P_1 Q_1$, whereas such a factor cannot contain $P_1 Q_1$. So by Theorem 13 (where n - 1 is now to be put for n), $\pi_1' = (P_2, P_3, \ldots, P_n)$ and $\pi_2' = (Q_2, Q_3, \ldots, Q_n)$ can be separated in G' by fewer than n - 1 vertices, certainly by n - 2 vertices. If α vertices of these n - 2 vertices belong to π_1' and (n - 2 - α) belong to π_2' then there is in G', and therefore also in G, no edge which would join one of the remaining $((n - 1) - \alpha) = r$ vertices of π_1' with one of the remaining (n - 1) - (n - 2 - α) = n - r vertices of π_2'. Since $0 \leq \alpha \leq n - 2$, the inequality $1 \leq r \leq n - 1$ holds here for r = n - 1 - α. Hence Theorem 15 is also proved in the second case.///

Theorem 13 can be carried over to the case where an infinite cardinality is put for n. This can be done with the help of the well ordering theorem and transfinite induction without any particular graph theoretical considerations. Likewise Theorem 14 can be carried over to the case where m and n stand for arbitrary infinite cardinalities, where the graph does not have to be assumed to be bipartite. (The situation with regard to this is similar to the one we met in the problem of factorization of regular graphs when we went over to graphs of infinite degree in §6 of Chap. XIII.)

§3 Applications to Matrices and Determinants

We have already pointed out in Chap. IX, §8 a connection between graphs and matrices. Here we shall consider other kinds of connections, and we shall restrict ourselves to finite, undirected graphs.

A graph is completely determined with respect to its structure if for every pair of vertices (P_i, P_k) which can be formed from its vertices $P_1, P_2, ..., P_n$ the number a_{ik} (≥ 0) of its edges which join P_i with P_k is given. In this way every graph determines a square symmetric matrix $((a_{ik}))$ (i, k = 1, 2, ... n), and conversely a graph is determined with respect to its structure by such a matrix, where each a_{ik} is a nonnegative integer. Here the matrix can be modified in such a way that if i > k, then a_{ik} is replaced by $-a_{ik}$ and all a_{ii} are set equal to 0. There results a skew symmetric matrix. Cayley's Theorem, which states that if n is even this matrix is the square of a so-called Pfaff aggregate, can be connected with the factors of degree 1 of the corresponding graph. We will not go into further detail with these matters, which stem from Brunel [2], but instead we shall investigate another connection between[16] bipartite graphs and matrices and determinants.

In order to make use of our Theorem XI.13 in the theory of determinants, we let a bipartite graph G correspond to a determinant $D = |a_{ik}|$ of order n, where each a_{ik} is a nonnegative integer, in the following way. Let the vertices of G be $P_1, P_2, ..., P_n, Q_1, Q_2, ..., Q_n$; we join each vertex P_i with each vertex Q_k by a_{ik} (≥ 0) edges, without introducing any other edges. The number of edges of G which meet at P_i (at Q_k, respectively) is equal to the sum of the ith row (of the kth column, respectively) of D. So G is regular if and only if every row and every column of D give the same (positive) sum g. On the other hand every factor of degree 1 of G has the form

$(P_1 \, Q_{i_1}, \, P_2 \, Q_{i_2}, \, \ldots, \, P_n \, Q_{i_n})$, where $(i_1, \, i_2, \, \ldots, \, i_n)$ is a permutation of $(1, \, 2, \, \ldots, \, n)$. To this factor corresponds a nonzero term $a_{1i_1} \, a_{2i_2} \, \cdots \, a_{ni_n}$ of the determinant D. (The sign of the term does not matter here.) So Theorem XI.13 can be expressed in the following form[17]:

> A. If in a determinant with nonnegative coefficients every row
>
> and every column have the same positive sum, then there is at
>
> least one term of the determinant which is different from 0.

This theorem is only for whole number elements according to our above considerations. But it can be immediately extended to rational numbers and even to real numbers. Negative numbers, however, are not admitted.

As has already been done for Theorem XI.13, Theorem XI.14 can also be carried over to the corresponding square matrix. In a purely combinatorial formulation we get the following theorem:

> B. If kn figures are entered in a square table with n^2 cells in
>
> such a way that every row and every column contain exactly k
>
> figures (several figures may be entered in the same cell), this
>
> configuration is always generated by superposition of k such
>
> configurations in which every row and every column contain
>
> exactly one figure.

Applying the results of the preceding paragraph to an arbitrary (not necessarily square) matrix

$$M = ((a_{ik})) \quad (i = 1, \, 2, \, \ldots, \, p; \, k = 1, \, 2, \, \ldots, \, q)$$

a bipartite graph G is assigned[18] in the following way. We let one each of the p vertices $P_1, \, P_2, \, \ldots, \, P_p$ correspond to the rows of M and one each of the q vertices $Q_1, \, Q_2, \, \ldots, \, Q_q$ to the columns, and we introduce one edge $P_i \, Q_k$ if and only if $a_{ik} \neq 0$. No other edges are introduced. We would like to see what Theorem 13 states about the matrix M if it is applied to this graph G and if $\Pi_1 = (P_1, P_2, \ldots, P_p)$ and

$\Pi_2 = (Q_1, Q_2, \ldots, Q_q)$. If there are n edges of G which pairwise have

no common vertex, then there correspond to them n nonzero elements a_{ik}

which pairwise belong to different rows and columns of M, and whose

product, therefore, (without regard to sign) yields a term of a

subdeterminant of order n of M. What does it mean to say for a number

n, which satisfies the conditions $n \leq p$, $n \leq q$, that Π_1 and Π_2 can

be separated by fewer than n vertices in G? This means that n - 1 rows

or columns of M can be named so that all nonzero elements of M are

contained in one of these n - 1 rows and columns. If among these n - 1

rows and columns are contained p - r (≥ 0) rows and q - s (≥ 0) columns,

then all elements which have the remaining r rows in common with the

remaining s columns are zero. Because $p - r \leq n - 1 < p$, certainly

r > 0 and s > 0. Because (p - r) + (q - s) = n - 1 we have that

s = (p + q - n + 1) - r, and r assumes one of the values 1, 2, ..., p.

So translated into the language of matrices, our theorem yields the

following result:

> C. If all terms of all subdeterminants of order n of a matrix
> with p rows and q columns (where $n \leq p$, $n \leq q$) are zero, then
> there exists a positive integer r such that all elements which
> have one of a certain set of r rows in common with one of a
> certain set of (p + q - n + 1) - r columns are zero, where
> $r \leq p$.

Carrying over from Theorem 14 also gives:

> D. The minimal number of rows and columns which contain in
> their totality every nonzero element of a matrix is equal to
> the maximal number of nonzero elements which pairwise belong[19]
> to different rows and columns.

For example, for the matrix

$$\begin{vmatrix} 0 & 1 & 0 & 0 & 1 \\ 0 & 1 & 0 & 0 & 1 \\ 1 & 1 & 1 & 1 & 0 \\ 0 & 1 & 0 & 0 & 1 \end{vmatrix}$$

both of these numbers equal 3.

Here, of course, "nonzero" can be replaced by an arbitrary property of the elements, so that this theorem expresses a purely combinatorial property of matrices (of two-dimensional tables), where the elements can be arbitrary objects.

For $p = q = n$ we get from Theorem C the following theorem on determinants of Frobenius (loc. cit.), which expresses the converse of an obvious theorem.

E. If all terms of a determinant of order n are zero, then there is a positive integer r such that all elements which have one of a certain set of r rows and one of a certain set of n - r + 1 columns in common are zero, where $r \leq n$.

In a similar way, using graph theory, we would like now to prove the following theorem on determinants also stemming from Frobenius[20] and reduce it to Theorem 15 (this theorem also results from taking the converse of an immediately obvious theorem).

F. In a determinant D of order n let the nonzero elements be variables independent of one another. If D is a reducible function of its (nonzero) elements, then there exists a positive integer r such that all elements of D which have one of a certain set of r rows in common with one of a certain set of n -r columns are zero, where $r < n$.

Proof To the determinant

$$D = |a_{ik}| \quad i, k = 1, 2, \ldots, n$$

we assign a bipartite graph G with vertices P_1, P_2, ..., P_n; Q_1, Q_2,

329

..., Q_n again by the rule that G contains an edge $P_i Q_k$ if and only if $a_{ik} \neq 0$, while two P-vertices as well as two Q-vertices are never joined by an edge. To each nonzero term $a_{1i_1} a_{2i_2} \cdots a_{ni_n}$ of D (where $(i_1, i_2, \ldots, i_n)$ is a permutation of $(1, 2, \ldots, n)$) corresponds a factor of degree 1 of G (the one, which contains the edges $P_1 Q_{i_1}$, $P_2 Q_{i_2}, \ldots, P_n Q_{i_n}$) and vice versa.

Let G* be that subgraph of G which contains those edges of G contained in at least one factor of degree 1 of G. As an illustration Fig. 106 shows the graph G assigned to the determinant

$$D = \begin{vmatrix} a_{11} & a_{12} & 0 & 0 & 0 & 0 \\ 0 & a_{22} & a_{23} & 0 & 0 & 0 \\ a_{31} & 0 & a_{33} & 0 & 0 & 0 \\ 0 & a_{42} & 0 & a_{44} & a_{45} & 0 \\ 0 & 0 & a_{53} & a_{54} & a_{55} & a_{56} \\ a_{61} & 0 & 0 & 0 & 0 & a_{66} \end{vmatrix}$$

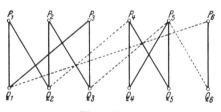

Fig. 106.

The vertical edges correspond to the elements of the main diagonal. The solid lines belong to the subgraph G*.

Now let D be a reducible polynomial. We shall prove that G* is disconnected. D depends on an (nonzero) element a_{ik} if and only if it is contained in a nonzero term of D, and therefore if and only if $P_i Q_k$

belongs to G*. If, therefore, all elements in D on which D does not

depend (in our example: a_{42}, a_{53}, a_{56}, a_{61}) are set equal to 0, then

the resulting determinant D* corresponds to G*. D*, since it represents

the same function as D, is likewise reducible: $D^* = D = D_1 D_2$, where D_1

and D_2 are both polynomials, which contain at least one variable a_{ik}.

Since the determinant is linear with respect to each variable, no

element can be contained simultaneously in both factors D_1 and D_2. The

edges $P_i Q_k$ which correspond to elements a_{ik} contained in D_1 (in D_2,

respectively) form the subgraph G_1 (G_2, respectively) of G*; every edge

of G* is contained in one and only one of the graphs G_1 and G_2. G_1 and

G_2 also have no common vertex; if P_i were such a vertex, then an edge

$P_i Q_k$ would have to belong to G_1 and an edge $P_i Q_\ell$ to G_2, and so a_{ik}

would have to be contained in D_1 and $a_{i\ell}$ in D_2. But since a_{ik} and $a_{i\ell}$

belong to the same row of D and the determinant is linear with respect

to the elements of a row, this is impossible. So G_1 and G_2 are mutually

disjoint and $G^* = G_1 + G_2$, so that G* is in fact disconnected (in our

example, Fig. 106, G* splits into three components.)

So the conditions of Theorem 15 are fulfilled for G. But this

theorem, translated into the language of determinants, states that the

common elements of r rows and n - r columns are all zeros for a certain

r. Thus Theorem F is proved.///

We wish to emphasize once more that G* is disconnected, if D is

reducible. Conversely it can be seen that G* is connected if D is

irreducible. So we have a very good practical method for deciding

whether a determinant which contains as elements only zeros and

independent variables is reducible or irreducible: it can be seen

directly by looking at the graph, whether the graph is connected or

not. We consider as a second example the determinant of order 10

$$\begin{vmatrix} a_{11} & a_{12} & a_{13} & 0 & 0 & 0 & 0 & 0 & 0 & 0 \\ a_{21} & a_{22} & 0 & 0 & 0 & 0 & a_{27} & 0 & 0 & 0 \\ a_{31} & a_{32} & a_{33} & 0 & 0 & 0 & 0 & 0 & 0 & 0 \\ 0 & 0 & a_{43} & a_{44} & 0 & 0 & 0 & 0 & 0 & a_{4,10} \\ 0 & 0 & 0 & a_{54} & a_{55} & a_{56} & 0 & 0 & 0 & 0 \\ 0 & 0 & 0 & a_{64} & a_{65} & a_{66} & 0 & 0 & 0 & 0 \\ 0 & 0 & 0 & 0 & a_{75} & a_{76} & a_{77} & 0 & 0 & 0 \\ 0 & 0 & 0 & 0 & 0 & 0 & a_{87} & a_{88} & a_{89} & 0 \\ 0 & 0 & 0 & 0 & 0 & 0 & 0 & a_{98} & a_{99} & a_{9,10} \\ 0 & 0 & 0 & 0 & 0 & 0 & 0 & a_{10,8} & a_{10,9} & a_{10,10} \end{vmatrix}$$

Even with the help of Theorem F it would be tedious to decide the question of reducibility of this determinant. Fig. 107 shows the corresponding graph G^{21}. Since every row and every column have exactly three nonzero elements, this is a regular graph, and by Theorem XI.14 every edge is contained in a factor of degree 1. So here $G^* = G$. It is easily seen that $G^* = G$ is connected. Thus our determinant of order 10 is irreducible.

In the proof given for Theorem F the characteristic properties of determinants were only partially used. If the signs ε_i in the definition of the determinant

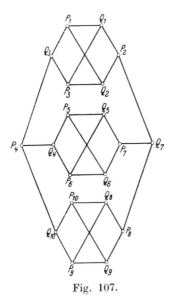

Fig. 107.

$$|a_{ik}|_1^n = \sum \varepsilon_i \, a_{1i_1} \, a_{2i_2} \cdots a_{ni_n}$$

were modified in an arbitrary way, and indeed if arbitrary numbers
distinct from 0 (independent of the a_{ik}'s) were substituted for the
ε_i's and if to this sum an arbitrary constant were added, then the
correctness of the proof given would not be impaired. This proof gives
the following more general result:

> G. In a determinant D of order n let the nonzero elements be
> independent variables. A linear integral function of the terms
> of D (as a polynomial of the elements of D) formed with nonzero
> coefficients[22] is reducible if and only if there is a positive
> integer r such that all elements of D which have one of a
> certain set of r rows in common with one of a certain set of
> n - r columns are zero, where r < n.

§4 Menger's Theorem

We shall now prove a quite general theorem, which contains Theorems
12 and 13 if these are formulated only for finite graphs. It was
discovered by Menger [1, Theorem δ].[23]

> Theorem 16 Let π_1 and π_2 be two disjoint subsets of the
> set π of vertices of a finite graph G. If π_1 and π_2 can be
> separated in G by not fewer than n vertices, then there are n
> pairwise disjoint $\pi_1\pi_2$-paths in G.

Proof If certain edges are removed from G in order, because of the
finite number of edges, a graph G* must eventually be obtained with the
property that π_1 also cannot[24] be separated from π_2 in G* by fewer than
n vertices, but that as soon as one more edge is removed, π_1 can be
separated from π_2 in the resulting graph by fewer than n vertices. It
suffices to prove the theorem for G* instead of for G, so that we may

assume right from the beginning that in every proper subgraph of G the sets π_1 and π_2 can be separated by fewer than n vertices. This property of G is designated in what follows as Property α.

First let $\pi = \pi_1 + \pi_2$. Then every $\pi_1\pi_2$-path of G contains a $\pi_1\pi_2$-edge. So if G' is obtained from G by removing an edge which joins two π_1-vertices (or two π_2-vertices), then π_1 is separated[25] from π_2 in G by a set of vertices which separates π_1 from π_2 in G'. Because of Property α there can be no edge in G which joins two π_1-vertices (or two π_2-vertices) with each other, so that the conditions of Theorem 13 are fulfilled. The n edges corresponding to this theorem are at the same time n paths which have the desired property.

So we may restrict ourselves to the second case, namely that a vertex R_0 of G belongs neither to π_1 nor to π_2. If G consists of a single edge, then the theorem to be proved is true (n = 1); so we can apply induction with respect to the number of edges.

If all edges of G going to R_0 are removed, then because of Property α, π_1 can be separated from π_2 in the resulting graph by r < n vertices $R_1, R_2, \ldots, R_r$. π_1 is separated from π_2 in G by the set $M = (R_0, R_1, R_2, \ldots, R_r)$, so that $r + 1 \geq n$; since r < n, r = n - 1. Consequently π_1 and π_2 are separated in G by a set $M = (R_0, R_1, R_2, \ldots, R_{n-1})$ of n vertices, where one of these vertices, R_0, does not belong to $\pi_1 + \pi_2$.

A vertex R_i of M is assigned to the subset M_1 or M_2 or M_0 according to whether it belongs to π_1 or π_2 or neither of these sets, respectively. So we have the partition of $M = M_0 + M_1 + M_2$ into three pairwise disjoint parts, where R_0 belongs to M_0.

We now consider all paths "of type W_1", i.e. those paths of G which join a vertex of $\pi_1 - M_1$ with a vertex of $M_0 + M_2$ and contain no vertex of M except this vertex of $M_0 + M_2$. Those edges contained in a path of type W_1 form the subgraph G_1 of G. Likewise those edges of G contained in a path which joins a vertex of $\pi_2 - M_2$ with a vertex of $M_0 + M_1$ and which contains no point of M except for this vertex of $M_0 + M_1$ (paths "of type W_2") form the subgraph G_2 of G.

We now prove that every common vertex of G_1 and G_2 belongs to M. If a path P ... A ... R_i of type W_1 and a path Q ... A ... R_j of type W_2 (P always stands for a π_1-vertex, and Q for a π_2-vertex) had a common vertex A which did not belong to M, then the subpaths P ... A and Q ... A would yield (in the sense of Theorem I.5) a $\pi_1\pi_2$-path P ... Q which contains no M-vertex; but this is impossible.

G_1 and G_2 also cannot have a common edge, for then both endpoints of this edge would have to belong to M, while according to the definition of G_1 and G_2 an edge which joins two M-vertices, can belong neither to G_1 nor to G_2.

To each vertex R_i of M belongs a $\pi_1\pi_2$-path of G which contains the vertex R_i and only the vertex R_i from M; in the opposite case the n - 1 vertices of M distinct from R_i would also already separate π_1 from π_2. If, in particular, $R_i = R_0$, then R_0 is an interior vertex of the $\pi_1\pi_2$-path assigned to it in this way, so that this path is split by R_0 into two paths, of which one is of type W_1 and the other of type W_2. So, of the two edges of this $\pi_1\pi_2$-path going to R_0, one belongs to G_1, and the other to G_2. Consequently neither G_1 nor G_2 is the empty graph, and G_2 as well as G_1 contains fewer edges than G.

In G_1 the sets $\pi_1 - M_1$ and $M_0 + M_2$ cannot be separated by a set M' containing fewer than $m_0 + m_2$ vertices (here m_0, m_1, and m_2 denote the number of elements of the sets M_0, M_1, and M_2, respectively), for if this were the case for M', then, by adding the set M_1, a set $M' + M_1$ would be obtained containing fewer than $m_0 + m_2 + m_1 = n$ vertices by which π_1 and π_2 would be separated in G. If a $\pi_1\pi_2$-path W of G contains no vertex of M_1 and so begins at a vertex P of $\pi_1 - M_1$, and if R_i is the first vertex of W counted out from P which belongs to M, and so to $M_0 + M_2$, then the subpath P ... R_i of W is a path of type W_1, and thus a path of G_1 which joins a vertex of $\pi_1 - M_1$ with a vertex of $M_0 + M_2$; this subpath must consequently contain a vertex of M' as must W.

Since G_1 has fewer edges than G, we may -- by our induction assumption -- apply the theorem to be proved, putting $m_0 + m_2$ for n, to G_1 and to the sets $\pi_1 - M_1$ and $M_0 + M_2$. This application yields the following: there are $m_0 + m_2$ pairwise disjoint paths in G_1 which each join a vertex of $\pi_1 - M_1$ with a vertex of $M_0 + M_2$. Since $m_0 + m_2$ is the number of elements of $M_0 + M_2$, each of the vertices of $M_0 + M_2$ belongs to one (and only one) of these paths as endpoint, while a vertex of M_1, since it does not belong to G_1, belongs to none of these paths. Among them let $T_1, T_2, \ldots, T_{m_2}$ be the paths going to the vertices of M_2 and let $U_1', U_2', \ldots, U_{m_0}'$ be the paths going to the vertices of M_0. In like manner $m_0 + m_1$ paths of G_2: $V_1, V_2, \ldots, V_{m_1}$; $U_1'', U_2'', \ldots, U_{m_0}''$, with corresponding properties, are defined by interchanging the roles of G_1 and G_2. If U_i' and U_i'' go to the same M_0 - vertex, then the edges of both of these paths form a $\pi_1\pi_2$-path U_i of G (i = 1, 2, ..., m_0). This follows from the fact that a common vertex of G_1 and G_2 must belong to M. For the same reason the $n = m_0 + m_1 + m_2$ $\pi_1\pi_2$-paths $U_1, U_2, \ldots, U_{m_0}$; $V_1, V_2, \ldots, V_{m_1}$; $T_1, T_2, \ldots, T_{m_2}$ of G are pairwise disjoint, by which Menger's Theorem is proved.///

A formulation of Menger's Theorem can be given which is related to Theorem 16 just as Theorem 11 is to Theorem 12.

In a somewhat more precise form Menger's Theorem can be formulated as follows corresponding to the formulation of Theorem 14:

Theorem 17 Let π_1 and π_2 be two disjoint subsets of the set of vertices of a finite graph G. The minimal number of vertices by which π_1 and π_2 can be separated in G from each other is equal to the maximal number of pairwise disjoint $\pi_1\pi_2$-paths of G.

If we denote this minimal and maximal number by m and n, respectively, there are n pairwise disjoint $\pi_1\pi_2$-paths in G. Since each of these paths must contain at least one vertex from every separating set of vertices, $m \geq n$. On the other hand, Menger's Theorem implies that $n \geq m$. So $m = n$.

Let it be emphasized that in the proof of Menger's Theorem given here, in contrast to the proofs which we gave above for Theorems 12 and 13, the assumption that the graph is finite was essential because of the application of induction. So we are led to the question of whether -- for finite n -- Menger's Theorem remains true for infinite graphs. This problem was solved[26] by P. Erdös affirmatively. In conclusion we present this proof by Erdös.

Let π_1 and π_2 be two disjoint subsets of the set π of vertices of the infinite graph G which cannot be separated in G by fewer than n vertices, where n stands for an arbitrary finite number. We assume that in contradiction to the theorem to be proved the maximal number r of the pairwise disjoint $\pi_1\pi_2$-paths of G is smaller than n. Let W_1, W_2, ..., W_r be r pairwise disjoint $\pi_1\pi_2$-paths of G. We consider the finite set M of sets of vertices $(A_1, A_2, ..., A_r)$ which have the property that A_i belongs to W_i (for i = 1, 2, ..., r). Since π_1 cannot be separated from π_2 in G by $r < n$ vertices, there is a $\pi_1\pi_2$-path U_α of G for every element $S_\alpha = (A_1, A_2, ..., A_r)$ of M which contains none of the vertices $A_1, A_2, ..., A_r$. We let such a path U_α correspond to each S_α and add the edges of these paths U_α to the edges of the paths W_1, W_2, ..., W_r. There results a finite subgraph G* of G. In G*, π_1 cannot be separated

337

from π_2 by r vertices. Such a separating set, since it must contain a vertex from each of the paths W_1, W_2, ..., W_r, would have to be an element $S_\alpha = (A_1, A_2, ..., A_r)$ of M, and the $\pi_1\pi_2$-path U_α of G* corresponding to this element S_α would not contain any of the vertices A_i. We may now apply Menger's Theorem to the finite graph G* (where we set n = r + 1), and this implies that r + 1 pairwise disjoint $\pi_1\pi_2$-paths exist in G* and so also in G. This contradicts the assumed maximal property of r. Consequently we have

<u>Theorem 18</u> Menger's Theorem (Theorem 16) also holds for

infinite graphs for an arbitrary finite number n.

Then, of course, Theorem 17 also holds for infinite graphs; but if π_1 or π_2 is an infinite set, the maximal (or minimal) number mentioned there must be assumed to be a finite number.

Notes on Chapter XIV

[1] For cut point the designations Zerlegungspunkt,
Artikulation, and cut vertex were introduced.

[2] A similar concept (membre) was defined by Sainte-Laguë [4,
p. 5]: a membre is a block, which contains only one cut point of the
original graph. The concept component introduced by Whitney [3] is
identical to the concept of block introduced here. Part I of this paper
of Whitney's overlaps repeatedly with the content of this paragraph.
The concept maximal cyclic curve (of a continuous curve) introduced by
Whyburn [1] into the new theory of (plane) continuous curves agrees --
essentially -- with our concept of block also.

[3] So if in our material that follows here we speak about
whether the edge PQ belongs to the one or the other block, we may speak
simply of the edge PQ, even if several edges join P with Q.

[4] Whitney [3] calls graphs without cut points non-separable.
The cyclicly connected continuous curves of Whyburn [1] and the
zyklischen Kontinua of Menger [4, Chapter XI] agree essentially with
this concept (although not in definition); cf. Theorem 5, which follows
here.

[5] For finite graphs see Whitney [3, Theorem 7].

[6] This "three point theorem" stems, for "connected im kleinen"
continua, from W. L. Ayres; its proof, that follows here, is by G.T.
Whyburn; see Menger [4, pp. 327-328].

[7] In the continuous interpretation of graphs, where edges also
have "interior points" this condition can be omitted; namely, if for P,
Q this condition is not satisfied, then every interior point of this
edge PQ is a common interior point of all the paths from P to Q
(Case 2).

[8] If, for example, G is a subgraph of the graph H and π_1 and π_2 are two disjoint subsets of the set of vertices of H, then, according to the definition given here, it has meaning to say that "π_1 and π_2 are separated in G by a subset π' of the set of vertices of G", even if not all the vertices of $\pi_1 + \pi_2$ belong to G. This separating occurs, for example, if no $\pi_1 \pi_2$-path of H belongs completely to G, and so, for example, if no vertex of π_1 (or of π_2) belongs to G.

[9] It can be recognized that this definition is connected with the problem of minimal vertex bases of the second kind, which we mentioned in §1 of Chapter VII.

[10] G is a bipartite graph by Theorem XI.12.

[11] This concept, on which the proof that follows here is based, was introduced (for similar reasons) in König's paper [4].

[12] I stated and proved this theorem in 1931, see König [9 and 11]. In 1932 there appeared the first complete proof of Menger's graph theorem, which we discuss in §4 and which can be viewed as a generalization of this theorem 13 (as long as it is formulated only for finite graphs).

[13] P. Hall, on representatives of subsets, Journal of the London Mathematical Society, 10 (part 1), 1935, pp. 26-30 (Theorem 1).

[14] It is clear that this condition is necessary.

[15] It can be easily seen, also with the help of Theorem 13, that the theorem also holds, if G* is the empty graph (in this case n - r can be replaced by n - r + 1); we shall not go into the details of this.

[16] We mention here also an application of graphs to the theory of determinants stemming from Polya [1].

[17] Theorem A and Theorem B, which follows here, are in König's

paper [5, §2] together with the graph theoretical proofs given here.

Frobenius then gave another proof for Theorem A, Über zerlegbare

Determinanten, Sitzungsberichte der preußischen Akademie, 1917, I, pp.

274-277. E. Egerváry gave a generalization of these theorems, Matrixok

kombinatorius tulajdonságairól (Concerning combinatorial properties of

matrices), Matematikai és Fizikai Lapok, 38, 1931, pp. 16-28 (Hungarian

with a German abstract),§3. -- In connection with Theorem A, van der

Waerden, Jahresbericht der deutschen Mathematiker - Vereinigung, 35,

1926, p. 117, posed the problem of determining the minimum (certainly

positive by Theorem A) of the function

$$\sum_{i_1, \ i_2, \ \ldots, \ i_n} a_{1i_1} \ a_{2i_2} \ \cdots \ a_{ni_n},$$

where the summation extends over all permutations $(i_1, i_2, \ldots, i_n)$ of

$(1, 2, \ldots, n)$, under the secondary conditions that $a_{ik} \geq 0$,

$$\sum_{i=1}^{n} a_{ik} = n \ (k = 1, 2, \ldots, n), \text{ and } \sum_{k=1}^{n} a_{ik} = n$$

$(i = 1, 2, \ldots, n)$. The problem remains unsolved.

[18] König [4].

[19] The author presented the proofs given here of Theorems C

and D in 1931 to the Budapest Mathematical and Physical Society, see

König [9 and 11]. Following this E. Egerváry in his above mentioned

paper (§§1 and 2) gave another proof for these theorems (using

induction) and a noteworthy generalization. Mr. L. Kalmár had also

communicated a proof to me earlier.

[20] Über Matrizen aus nicht negativen Elementen,

Sitzungsberichte der preußischen Akademie, 1912, I, pp. 456-477. This

theorem is proved in this reference "from properties borrowed from

determinants with nonnegative elements" in an extraordinarily

complicated way. I then gave in 1915 in my paper [4] an elementary
graph theoretical proof (which is replaced here by a still simpler
proof). In 1917 Frobenius also published an elementary proof (in his
paper mentioned above: Über zerlegbare Determinanten), after I had sent
him my proof (translated into German). Frobenius failed to mention this
fact or my paper [4]. He cited my paper [5], however, with the
following remark:

"The theory of graphs, by means of which Mr. König proved the above
theorem [this is our Theorem A], is in my opinion a method of little
value for the development of the theory of determinants. In this case
it leads to a quite specific theorem of little worth. What it has of
value is expressed in Theorem II [This is Frobenius' Theorem E]."

It is quite natural that the author of a book on graphs will not
subscribe to this opinion. The reasons that can be brought forth for or
against the value or lack of value of a theorem or a method always have,
more or less, a subjective character, so that it would be of little
scientific value, if we tried to attack Frobenius' viewpoint. But if
Frobenius wished to base his repudiating criticism on the applicability
of graphs to the theory of determinants on the statement that his "more
valuable" Theorem E cannot be proved graph theoretically, then the basis
of his criticism -- as we have seen -- is certainly not valid. The
graph theoretical proof which we gave for Theorem E seems to us to be a
simpler and more intuitive proof, which suits the combinatorial
character of the theorem in a natural way and also leads to a noteworthy
generalization (Theorem C).

Let it be mentioned here that we have used an idea of Frobenius' in
§2 in the proof of Theorem 15, which he used in his reduction of Theorem
F to Theorem E.

[21] This graph was given by Sainte-Laguë [2, p. 39, Fig. 9] as

an example of a bridgeless regular <u>bipartite</u> graph of degree 3, which

has no Hamiltonian path.

[22] The constant term may, of course, be zero.

[23] The proof by Menger given in this reference has holes in

it, since (P. 102, ℓℓ. 3-4) it is assumed that "K' contains a set of

vertices s, which is not contained in the set P + Q," while it is quite

possible that -- expressed in the notation used here -- every vertex of

G belongs to $\pi_1 + \pi_2$. This -- in no way simple -- case was taken care

of in our proof of Theorem 13. The further considerations, which follow

here and which will lead us to Menger's Theorem, agree essentially with

Menger's very brief reflections; see König [11, §4]. The first complete

proof of his fundamental theorem was given by Menger himself in 1932 [4,

pp. 221-228; "n - Kettensatz"]; the working out of this proof stems from

G. Nöbeling. -- Recently Hajós [1] gave a very elegant proof of Menger's

Graph Theorem. He bases his proof on the definition which specifies

when a set of vertices with respect to a vertex can be designated

"separiert." This definition of Hajós can also be applied to advantage

elsewhere in graph theory. -- For "plane continuous curves" Menger's

Theorem is stated in Rutt [1], where this "general theorem" is ascribed

to J. R. Kline. -- One of the main results of an article by Whitney [3],

namely his Theorem 7, follows immediately from Menger's Theorem, and

yet, as it seems, not conversely.

[24] As was already emphasized, this statement also has meaning

if certain vertices of $\pi_1 + \pi_2$ do not belong to the subgraph G*.

[25] In my paper [11] in lines 20 and 21 of p. 174 the symbols G

and G' were erroneously interchanged.

[26] See König [11, §4].

Bibliography

Of the literature used, only those works are included in this list, in which the concept of graph appears directly. Other works are cited in the text and in the notes. In the works for which several appearances or editions are cited, statements concerning page numbers, etc. refer to the place last named in this list.

Ahrens (W.),   [1]Über das Gleichungssystem einer Kirchhoffschen galvanischen Stromverzweigung, Mathematische Annalen, 49, 1897, pp. 311-324.

_____   [2] Mathematische Unterhaltungen und Spiele, Leipzig, 1st edition, 1901; latest edition: Vol. 1, 3rd edition, 1921; Vol. II, 2nd edition, 1918

Alexander (J.W.), [1] Combinatorial Analysis Situs, Transactions of the American Mathematical Society, 28, 1926, pp. 301-329.

_____   (see Veblen).

Baker (R.P.),   [1] Cayley diagrams on the anchor ring, American Journal of Mathematics, 53, 1931, pp. 645-669.

Ball (Rouse W.), [1] Mathematical Recreations and Problems, 1st edition London 1892. Latest French edition: Récréations mathématiques et problèmes des temps anciens et modernes, traduit par J. Fitz - Patrick, Vols. I-III, Paris 1926/27.

Baltzer (R.),   [1] Eine Erinnerung an Möbius und seinen Freund Weiske, Berichte der k. sächsischen Gesellschaft der Wissenschaften, 37, 1885, pp. 1-6.

344

Birkhoff (G.D.), [1] On the number of ways of colouring a map, Proceedings of the Edinbourgh Mathematical Society, (2), part II, 1930, pp. 83-91.

Brahana (H.R.), [1] A proof of Petersen's theorem, Annals of Mathematics (2), 19, 1917, pp. 59-63.

Brunel (G.), [1] (Sur un problème de combinaisons), Extraits des Procès - Verbaux des séances de la Société des Sciences de Bordeaux, 1893/94, pp. XIV-XV.

——————— [2] Analysis Situs. Recherches sur les réseaux. Mémoires de la Société des Sciences de Bordeaux (4), 5, 1895, pp. 165-215.

——————— [3] Construction d'un réseau donné l'aide d'un nombre déterminé de traits, Procès - Verbaux des séances de la Société des Sciences de Bordeaux, 1895/96, pp. 62-65.

Cayley (A.), [1] On the theory of the analytical forms called trees, Philosophical Magazine, 13, 1857, pp. 172-176 - Mathematical Papers, Cambridge 1889-1897, Vol. III, pp. 242-246.

——————— [2] On the mathematical theory of isomers, Philosophical Magazine, 47, 1874 pp. 444-446 = Mathematical Papers, Vol. IX, pp. 202-204.

——————— [3] On the analytical forms called trees, with application to the theory of chemical combinations, Report of the British Association for the Advancement of Science, 1875, pp. 257-305 = Mathematical Papers, Vol.

IX, pp. 427-460

Cayley, (A.)

[4] Solution of problem 5208, Mathematical Questions with Solutions from the Educational Times, 27, 1877, pp. 81-83 = Mathematical Papers, Vol. X, pp. 598-600.

——————

[5] Desiderata and suggestions, Nr. 2: The theory of groups, graphical representation, American Journal of Mathematics, 1, 1878, pp. 174-176 = Mathematical Papers, Vol. X, pp. 403-405.

——————

[6] On the theory of groups, Proceedings of the London Mathematical Society, 9, 1878, pp. 126-133 = Mathematical Papers, Vol. X, pp. 323-330.

——————

[7] On the analytical forms called trees, American Journal of Mathematics, 4, 1881, pp. 266-268 = Mathematical Papers, Vol. XI, pp. 365-367.

——————

[8] On the theory of groups, American Journal of Mathematics, 11, 1889, pp. 139-157 = Mathematical Papers, Vol. XII, pp. 639-656.

——————

[9] A theorem on trees, Quarterly Journal of Mathematics, 23, 1889, pp. 376-378 = Mathematical Papers, Vol. XIII, pp. 26-28.

Chuard (J.),

[1] Questions d'Analysis Situs (Thèse, Lausanne 1921), Rendiconti del Circolo Matematico di Palermo, 46, 1922, pp. 185-224.

——————

(see Dumas).

Clausen (Th.),   [1] (Without title) Astronomische

                          Nachrichten, 21, 1844, p. 216

Dehn (M.) and Heegaard (P.),[1] Analysis Situs,Encyklopädie der

                          mathematischen Wissenschaften, Vol. III 1_1,

                          pp. 153-220, 1907.

Dumas (G.) and Chuard (J.),[1] Sur les homologies de Poincaré, Comptes

                          Rendus, Paris, 171, 1920, pp. 1113-1116.

Errera (A.),   [1] Du coloriage des cartes et de quelques

                          questions d'Analysis Situs, Thèse, Bruxelles,

                          1921.

_____   [2] Une démonstration du théorème de

                          Petersen, Mathesis, 36, 1922, pp. 56-61.

_____   [3] Un théorème sur les liaisons, Comptes

                          Rendus, Paris, 177, 1923, pp. 489-491.

_____   [4] Exposé historique du problème des quatres

                          couleurs, Periodico di Matematiche (4), 7,

                          1927, pp. 20-41.

Euler (L.),   [1] Solutio problematis ad geometriam situs

                          pertinentis, Commentarii Academiae

                          Petropolitanae, 8, 1736 (1741), pp. 128-140 =

                          Opera omnia, Ser. I, Vol. 7, pp. 1-10.

                          French translation by Coupy in Nouvelles

                          Annales de Mathématiques, 10, 1851, pp. 106-

                          118. German translation in Speiser:

                          Klassische Stücke der Mathematik, Zürich

                          1927, pp. 127-138.

Frink (Orrin, Jr.),   [1], A proof of Petersen's theorem, Annals of

                          Mathematics (2), 27, 1926, pp. 491-493.

Hajós (G.),   [1] Zum Mengerschen Graphensatz, Acta
                     Litterarum ac Scientiarum (Sectio Scientiarum
                     Mathematicarum), Szeged, 7, 1934, pp. 44-47.

Heawood (P.J.),   [1] Map-colour theorem, Quarterly Journal of
                     Mathematics, 24, 1890, pp. 332-338 and 29,
                     1898, pp. 270-285.

Heegaard (P.),   (see Dehn).

Heffter (L.),   [1]Über das Problem der Nachbargebiete,
                     Mathematische Annalen, 38, 1891, pp. 477-508.

Hertz (P.),   [1]Über Axiomensysteme für beliebige
                     Satzsysteme, Mathematische Annalen, 87, 1922,
                     pp. 246-269.

Hierholzer (C.),   [1] Ueber die Möglichkeit, einen Linienzug
                     ohne Wiederholung und ohne Unterbrechung zu
                     umfahren, Mathematische Annalen, 6, 1873, pp.
                     30-32.

Hilton (H.),   [1] An introduction to the theory of groups
                     of finite order, Oxford 1908.

Jordan (C.),   [1] Sur les assemblages de lignes, Journal
                     für die reine und angewandte Mathematik, 70,
                     1869, pp. 185-190.

Kempe (A.B.),   [1] On the geographical problem of the four
                     colours, American Journal of Mathematics, 2,
                     1879, pp. 193-200.

Kirchhoff (G.),   [1] Über die Auflösung der Gleichungen, auf
                     welche man bei der Untersuchung der linearen
                     Verteilung galvanischer Ströme geführt wird,
                     Annalen der Physik und Chemie, 72, 1847, pp.
                     497-508 = Gesammelte Abhandlungen, Leipzig

1882, pp. 22-33.

König (D.),

[1] Vonalrendszerek Kétoldalú felületeken
(Graphs on two-sided surfaces), Mathematikai
és Természettudományi Értesítö, 29, 1911, pp.
112-117 (Hungarian).

_____

[2] A vonalrendszerek nemszámáról (Concerning
the genus number of graphs), Mathematikai és
Természettudományi Értesítö, 29, 1911, pp.
345-350 (Hungarian).

_____

[3] Sur un problème de la théorie générale
des ensembles et la théorie des graphes
(Communication made Apr. 7, 1914 in Paris to
the Congrès de Philosophie Mathématique),
Revue de Métaphysique et de Morale, 30, 1923,
pp. 443-449.

_____

[4] Vonalrendszerek és determinánsok (Graphs
and determinants), Mathematikai és
Természettudományi Értesítö, 33, 1915, pp.
221-229 (Hungarian)

_____

[5] Über Graphen und ihre Anwendung auf
Determinanten- theorie und Mengenlehre,
Mathematische Annalen, 77, 1916, pp. 453-465.

_____

[6] Sur les rapports topologiques d'un
problème d'analyse combinatoire, Acta
Litterarum ac Scientiarum (Sectio Scientiarum
Mathematicarum), Szeged, 2, 1924, pp. 32-38.

_____

[7] Sur les correspondances multivoques des
ensembles, Fundamenta Mathematicae, 8, 1926,
pp. 114-134.

König (D.),

[8] Über eine Schlußweise aus dem Endlichen ins Unendliche (Punktmengen. - Kartenfärben. - Verwandtschaftsbeziehungen. - Schachspiel), Acta Litterarum ac Scientiarum (Sectio Scientiarum Mathematicarum), Szeged, 3, 1927, pp. 121-130.

[9] Graphok és matrixok (Graphs and matrices), Matematikai és Fizikai Lapok, 38, 1931, pp. 116-119 (Hungarian with a German abstract)

[10] Egy végességi tétel és alkalmazásai (A finiteness theorem with two applications), Matematikai és Fizikai Lapok, 39, 1932, pp. 27-29 (Hungarian with a German abstract).

[11] Über trennende Knotenpunkte in Graphen (nebst Anwendungen auf Determinanten und Matrizen), Acta Litterarum ac Scientiarum (Sectio Scientiarum Mathematicarum), Szeged, 6, 1933, pp. 155-179.

und Valkó (St.), [1] Über mehrdeutige Abbildungen von Mengen, Mathematische Annalen, 95, 1926, pp. 135-138.

Kowalewski (A.),

[1] W.R. Hamiltons Dodekaederaufgabe als Buntordnungsproblem, Sitzungsberichte der Akademie in Wien, 126, 1917, pp. 67-90.

[2] Topologische Deutung von Buntordnungsproblemen, Sitzungsberichte der Akademie in Wien, 126, 1917, pp. 963-1007.

Kowalewski (G.),   [1] Mathematica delectans, I: Boss-Puzzle und verwandte Spiele, Leipzig 1921.

_____   [2] Alte und neue mathematische Spiele, Leipzig 1930.

Kürschák (J.),   [1] Lóugrás a végtelen sakktáblán (The knight's moves on an infinite chess board), Mathematikai és Physikai Lapok, 33, 1926, pp. 117-119 (Hungarian)

_____   [2] Rösselsprung auf dem unendlichen Schachbrette, Acta Litterarum ac Scientiarum (Sectio Scientiarum Mathematicarum), Szeged, 4, 1928, pp. 12-13.

Kuratowski (C.),   [1] Sur le problème des courbes gauches en Topologie, Fundamenta Mathematicae, 15, 1930, pp. 271-283.

_____   et Whyburn (G.T.), [1] Sur les éléments cycliques et leurs applications, Fundamenta Mathematicae, 16, 1930, pp. 305-331.

Listing (J.B.),   [1] Vorstudien zur Topologie, Göttingen Studien, 1847; auch separat erschienen: Göttingen, 1848.

_____   [2] Der Census räumlicher Complexe oder Verallgemeinerung des Eulerschen Satzes von den Polyedern, Göttinger Abhandlungen, 10, 1862.

Lucas (E.),   [1] Récréations Mathématiques, I-IV, Paris 1882-1894.

_____   [2] Théorie des nombres, I, Paris 1891.

Maschke (H.),   [1] The representation of finite groups ...

by Cayley's color diagrams, American Journal

of Mathematics, 18, 1896, pp. 156-194.

Menger (K.),   [1] Zur allgemeinen Kurventheorie, Fundamenta

Mathematicae, 10, 1927, pp. 96-115.

_____ [2] Über reguläre Baumkurven, Mathematische

Annalen, 96, 1927, pp. 572-582.

_____ [3] Über plättbare Dreiergraphen und Potenzen

nichtplättbarer Graphen, Anzeiger der

Akademie der Wissenschaften in Wien, 67,

1930, pp. 85-86 und Ergebnisse eines

Mathematischen Kolloquiums, Heft 2, 1930, pp.

30-31.

_____ [4] Kurventheorie (unter Mitarbeit von G.

Nöbeling), Leipzig und Berlin 1932.

Petersen (J.),   [1] Die Theorie der regulären Graphen, Acta

Mathematica, 15, 1891, pp. 193-220.

_____ [2] Sur le théorème de Tait, L'Intermédiaire

des Mathématiciens, 5, 1898, pp. 225-227.

_____ [3] Les 36 officiers, Annuaire des

mathématiciens 1901-02, Paris 1902, pp. 413-

427.

Poincaré (H.),   [1] (Premier) complémemt à l'Analysis Situs,

Rendiconti del Circolo Matematico di Palermo,

13, 1899, pp. 285-343.

_____ [2] Second complément à l'Analysis Situs,

Proceedings of the London Mathematical

Society, 32, 1901, pp. 277-308.

de Polignac (C.),   [1] Formules et considérations diverses se
                           rapportant á la théorie des ramifications,
                           Bulletin de la Société Mathématique de
                           France, 8, 1880, pp. 120-124 and 9, 1881, pp.
                           30-42.

———————————   [2] Sur le théorème de Tait, Bulletin de la
                           Société Mathématique de France, 27, 1899, pp.
                           142-145.

Pólya (G.),   [1] Lösung der Aufgabe 386 (von J. Schur),
                           Archiv der Mathematik und Physik (3), 24,
                           1916, pp. 369-375.

Prüfer (H.),   [1] Neuer Beweis eines Satzes über
                           Permutationen, Archiv der Mathematik und
                           Physik (3), 27, 1918, pp. 142-144.

Rademacher (H.)   (see Steinitz).

———————————   und Toeplitz (O.), [1] Von Zahlen und
                           Figuren, Berlin 1930.

Rédei (L.),   [1] Ein Kombinatorischer Satz, Acta
                           Litterarum ac Scientiarum (Sectio Scientiarum
                           Mathematicarum), Szeged, 7, 1934, pp. 39-43.

Reidemeister (K.),   [1] Einführung in die kombinatorische
                           Topologie (Die Wissenschaft, Vol. 86),
                           Braunschweig 1932.

Reynolds (C.N.),   [1] On the problem of coloring maps in four
                           colors, Annals of Mathematics (2), 28, 1927,
                           pp. 1-15 and 477-492.

Rutt (N.E.),   [1] Concerning the cut points of a continuous
                           curve when the arc curve AB contains exactly
                           N independent arcs, American Journal of

Mathematics, 51, 1929, pp. 217-246.

Sainte-Laguë (A.),   [1] Les réseaux, Comptes Rendus, Paris, 176, 1923, pp. 1202-1205.

[2] Les réseaux, Toulouse 1924

[3] Les réseaux unicursaux et bicursaux, Comptes Rendus, Paris, 182, 1926, pp. 747-750.

[4] Les réseaux (ou graphes), Mémorial des Sciences Mathématiques, Fasc. 18, Paris 1926.

[5] Géométrie de situation et jeux, Mémorial des Sciences Mathématiques, Fasc. 41, Paris 1929.

Schönberger (T.),   [1] Ein Beweis des Petersenschen Graphensatzes, Acta Litterarum ac Scientiarum (Sectio Scientiarum Mathematicarum), Szeged, 7, 1934, pp. 51-57.

Skolem (Th.),   [1] Gruppierungen, kombinatorische Reziprozitäten, Paarsysteme. Nachtrag (Kap. 15) zu Netto: Lehrbuch der Kombinatorik, 2nd ed., Leipzig and Berlin 1927.

v. Staudt (G.K. Chr.),   [1] Geometrie der Lage, Nürmberg 1847.

Steinitz (E.),   [1] Vorlesungen über die Theorie der Polyeder (unter Einschluß der Elemente der Topologie), aus dem Nachlaß herausgegeben und ergänzt von H. Rademacher (Grundlehren der math. Wiss., Vol. 41), Berlin 1934.

Sylvester (J.J.),   [1] On recent discoveries in mechanical conversion of motion, Proceedings of the Royal Institution of Great Britain, 7, 1873-

75, pp. 179-198 = Mathematical Papers,

Cambridge 1904-1912, vol. III, pp. 7-25.

Sylvester (J.J.),   [2] Problem 5208, Mathematical Questions with

their Solutions from the Educational Times,

27, 1877; printed in Cayley's Mathematical

Papers, Vol. X, p. 598.

_____   [3] On an application of the new atomic

theory to the graphical representation of the

invariants and covariants of binary quantics,

American Journal of Mathematics, 1, 1878, pp.

64-125 = Mathematical Papers, vol. III, pp.

148-206.

_____   [4] On the geometrical forms called trees,

Johns Hopkins University Circulars, I, 1882,

pp. 202-203 = Mathematical Papers, vol. III,

pp. 640-641.

Tait (P.G.),   [1] Note on a theorem in geometry of

position, Transactions of the Royal Society

of Edinbourgh, 29, 1880, pp. 657-660 =

Scientific Papers, Cambridge 1898-1900, vol.

I, pp. 408-411.

_____   [2] Listing's Topologie, Philosophical

Magazine (5), 17, 1884, pp. 30-46 =

Scientific Papers, Vol. II, pp. 85-98.

Tarry (G.),   [1] Le problème des labyrinthes, Nouvelles

Annales de Mathématiques (3), 14, 1895, pp.

187-190.

Terquem (O.),   [1] Sur les polygones et les polyèdres

étoilés, polygones funiculaires (d'après M.

Poinsot), Nouvelles Annales de Mathématiques, 8, 1849, pp. 68-74

Toeplitz (O.) (see Rademacher).

Ulam (S.), Remark on the generalised Bernstein's theorem, Fundamenta Mathematicae, 13, 1929, pp. 281-283.

Valkó (St.), (see König).

Veblen (O.), [1] An application of modular equations in Analysis Situs, Annals of Mathematics (2), 14, 1912, pp. 86-94.

———————— [2] Analysis Situs, The Cambridge Colloquium 1916, part II, New York, 1st edition 1922, 2nd edition 1931.

———————— and Alexander (J.W.), [1] Manifolds of N dimensions, Annals of Mathematics (2), 14, 1913, pp. 163-178.

van der Waerden (B.L.), [1] Ein Satz über Klasseneinteilungen von endlichen Mengen, Abhandlungen aus dem mathematischen Seminar der Hamburgischen Universität, 5, 1927, pp. 185-188.

Weyl (H.), [1] Repartición de corriente en una red conductora (Introducción al Análisis situs combinatorio), Revista Matemática Hispano-Americana, 5, 1923, pp. 153-164.

Whitney (H.), [1] A theorem on graphs, Annals of Mathematics (2), 32, 1931, pp. 378-390.

———————— [2] Congruent graphs and connectivity of graphs, American Journal of Mathematics, 54, 1932, pp. 150-168.

Whitney (H.),

[3] Non-separable and planar graphs, Transactions of the American Mathematical Society, 34, 1932, pp. 339-362 (an abstract appeared in Proceedings of the National Academy of Sciences of the U.S.A., 17, 1931, pp. 125-127).

_____

[4] Planar Graphs, Fundamenta Mathematicae, 21, 1933, pp. 73-84.

Whyburn (G.T.),

[1] Cyclicly connected continuous curves, Proceedings of the National Academy of Sciences of the U.S.A., 13, 1927, pp. 31-38.

_____

(see Kuratowski).

Wiener (Chr.),

[1] Ueber eine Aufgabe aus der Geometria situs, Mathematische Annalen, 6, 1873, pp. 29-30.

List of technical terms

The numbers give the pages of the definitions.

Index of Names

The letter B refers to the Bibliography.

Dénes König: A Biographical Sketch

T. Gallai

Dénes König was born on September 21, 1884 in Budapest. His father was Gyula König, the well-known professor of mathematics at the Technische Hochschule in Budapest. The early development of his mathematical talent was stimulated not only by his father but also by his outstanding teachers, Manó Beke and Miklós Szijjártó.

König published his first work in 1899, when he was still a high school student, in the journal *Matematikai és Fizikai Lapok*. It dealt with two extreme value problems in an elementary way. After graduation in 1902 he won first prize in the mathematical student competition "Eötvös Loránd". The first volume of his little book *Matematikai Mulatságok* (Mathematical Entertainments) appeared the same year. This book and its second volume, which appeared in 1905, were very successful collections.

König spent the first four semesters of his university studies in Budapest and the last five at Gottingen. His notebook from 1904-05 contains notes of Minkowski's lectures on analysis situs. According to these notes, Minkowski wished to present a proof of the Four Color Theorem by Wernicke, but during the discussion it turned out that the proof was defective, and that only a proof of the Five Color Theorem could be presented. The lectures of this outstanding mathematician certainly exerted a great influence on König's choice of problems to work on. They were not his first encounter with combinatorial topology, however, because he had already met such problems in collecting material for *Mathematical Entertainments*.

König received his doctorate in 1907 based on a dissertation in geometry. In the same year he began to work at the Technische Hochschule in Budapest, and remained a member of the faculty until his death. He started as an assistant in problem sessions, was promoted to "oberassistent" in 1910, and became a Privatdocent in 1911, teaching nomography, analysis situs, set theory, real numbers and functions, and graph theory. Although the name "graph theory" appears in the catalogue for the first time in 1927, this subject had been represented in his lectures since 1911, falling within the realm of analysis situs. He also gave guest mathematics lectures for architecture and chemistry students; these lectures appeared in book form in 1920. In 1935 he was named full professor at the Technische Hochschule.

Like his father, Dénes König played an active role in the mathematical life of Hungary. The Mathematical and Physical Society named for Loránd Eötvös, which he joined in 1907, was the most important vehicle for his activities. He was a member of the judging committee for mathematics school contests, in charge of collecting the contest problems and organizing the contests, from 1915 to 1942 and in 1942 became the chairman of the committee. In 1933 he was elected to the position of secretary of the society, and he remained in that post until his death. In the same year he began to edit the society journal. At that time the editors had not only to edit but also to procure the necessary money. The very small state subsidy was not enough to publish the journal, and the difference had to be solicited from commercial and private sources.

In memory of their father who died in 1913, König and his brother György endowed the Gyula König Prize for young mathematicians in 1918. The endowment later became devaluated, but the medal remained a sign of high scientific recognition. König regarded the support of young mathematicians as part of his editorial duties and gave every possible help towards the publication of their doctoral dissertations and other works.

His greatest support to the younger generation was his teaching. The lectures he gave as university instructor had only a small audience, but the few enthusiastic listeners became acquainted with several modern branches of mathematics and were introduced to a new mathematical discipline, graph theory, as it was being developed. König had the ability to present his thoughts well: he knew how to emphasize the essential points, how to awaken interest, and —as the leading expert on graph theory—how to ask exciting questions. Under the influence of his lectures, many mathematicians turned to this field. The fact that so many Hungarian researchers are well known in the field testifies to this influence. König's activity played an important role in the graph theoretical work of László Edyed, Pál Erdös, Tibor Gallai, György Hájos, Jószef Kraus, Tibor

Szele, Pál Turán, Endre Vázsonyi, and many others.

König was a cheerful, sociable man with sparkling humor, who enjoyed telling anecdotes. He liked his colleagues and was an indispensable participant in the Coffee House Club of mathematicians. In 1944—after the occupation of Hungary by the Nazis—he worked to help persecuted mathematicians. The events of October 15, 1944* brought an end to his plans. On October 19 he chose death over persecution.

Dénes König's most significant results were connected with graph theory. Several basic theorems of this branch of mathematics are named after him. No less significant are his accomplishments in the popularization and recognition of graph theory. It can be said with good reason that graph theory was developed by his activity from an arbitrary branch of recreational mathematics to a recognized new discipline of the mathematical sciences. Belittling opinions could not discourage him: his belief in the "future" of graph theory was unshakable. Even now the words with which he opened his lectures on graph theory in the fall of 1930 sound in my ears: "Graph theory is one of the most interesting of mathematical disciplines."

Landmarks of his graph theoretical activity

1905 His first graph theoretic work appeared in which he proved a map coloring theorem.

1911 Two papers on the imbedding of graphs in surfaces. Here he introduced the concept of the genus of a graph.

1914 In Paris he presented his well-known theorem on the factors of a finite regular bipartite graph (the text of this address did not appear in print until 1923).

1916 His results, supplemented by applications, published, including the first proofs of theorems concerning infinite graphs.

1926 He and István Valkó succeeded in extending the validity of his factor theorems to infinite graphs. From the proof he separated out the well-known Infinity Lemma and sketched several interesting applications.

---

* This was the date on which the Nazi-affiliated Hungarian National Socialist Party took over the country. König, as the son of a Jew, expected to share the fate of other European Jews at that time.

1931 Published theorem on the separation of vertices of bipartite graphs. This may be his most frequently cited theorem.

1936 *Theory of finite and infinite graphs* published in Leipzig.

The book was the first work of truly scientific level concerned exclusively with graph theory. Until its publication only isolated journal articles, articles in encyclopedias, and popular books on recreational mathematics had dealt with graph theoretical problems. Their methods of handling the subject often lacked mathematical precision.

The book grew out of the lectures König gave at the Technische Hochschule in Budapest. He worked on it for many years with the greatest care. It is not known exactly when he began writing, but certainly a significant portion of the book was ready in 1930. He investigated all discoverable sources of the widely scattered literature and collected nearly all the essential results of combinatorial graph theory. The carefully thought out arrangement of the material, the mathematically accurate treatment, and the virtually complete index to the literature assured this. The style, free of formalism and easily understandable, and the interesting problems awakened the interest of many young mathematicians in graph theory. The impact of the book was really felt only in the postwar years. In 1950 it was reprinted in the USA. Through two decades König's work was the only monograph on the subject (until 1958, when *Theory of Graphs and their Applications* by C. Berge was published).

The sharp, critical spirit and the mathematical exactness are characteristic not only of the book but also of all of König's works. He had a flair for finding the kind of omissions that are usually detected only with difficulty and perseverance in the proofs in graph theory. He also had a flair for discovering the graph theoretical essence of results from the most diverse fields and was continually on the hunt for possible applications and for connections with other branches of mathematics. This is reflected in the choice of material for his book. He devoted great care to tracking down literature; his quotations are exact and detailed. The mathematical investigation is augmented by numerous footnotes that acquaint the reader with the historical background of the problem.